Lecture Notes in Electrical Engineering

Volume 240

For further volumes:
http://www.springer.com/series/7818

James J. (Jong Hyuk) Park
Joseph Kee-Yin Ng · Hwa Young Jeong
Borgy Waluyo
Editors

Multimedia and Ubiquitous Engineering

MUE 2013

Volume I

Editors
James J. (Jong Hyuk) Park
Department of Computer Science
Seoul University of Science
   and Technology (SeoulTech)
Seoul
Republic of South Korea

Joseph Kee-Yin Ng
Department of Computer Science
Hong Kong Baptist University
Kowloon Tong
Hong Kong SAR

Hwa Young Jeong
Humanitas College
Kyung Hee University
Seoul
Republic of South Korea

Borgy Waluyo
School of Computer Science
Monash University
Clayton, VIC
Australia

ISSN 1876-1100   ISSN 1876-1119 (electronic)
ISBN 978-94-007-6737-9   ISBN 978-94-007-6738-6 (eBook)
DOI 10.1007/978-94-007-6738-6
Springer Dordrecht Heidelberg New York London

Library of Congress Control Number: 2013936524

© Springer Science+Business Media Dordrecht(Outside the USA) 2013
This work is subject to copyright. All rights are reserved by the Publisher, whether the whole or part of
the material is concerned, specifically the rights of translation, reprinting, reuse of illustrations,
recitation, broadcasting, reproduction on microfilms or in any other physical way, and transmission or
information storage and retrieval, electronic adaptation, computer software, or by similar or dissimilar
methodology now known or hereafter developed. Exempted from this legal reservation are brief
excerpts in connection with reviews or scholarly analysis or material supplied specifically for the
purpose of being entered and executed on a computer system, for exclusive use by the purchaser of the
work. Duplication of this publication or parts thereof is permitted only under the provisions of
the Copyright Law of the Publisher's location, in its current version, and permission for use must
always be obtained from Springer. Permissions for use may be obtained through RightsLink at the
Copyright Clearance Center. Violations are liable to prosecution under the respective Copyright Law.
The use of general descriptive names, registered names, trademarks, service marks, etc. in this
publication does not imply, even in the absence of a specific statement, that such names are exempt
from the relevant protective laws and regulations and therefore free for general use.
While the advice and information in this book are believed to be true and accurate at the date of
publication, neither the authors nor the editors nor the publisher can accept any legal responsibility for
any errors or omissions that may be made. The publisher makes no warranty, express or implied, with
respect to the material contained herein.

Printed on acid-free paper

Springer is part of Springer Science+Business Media (www.springer.com)

Message and Organization Committee

Message from the MUE 2013 General Chairs

MUE 2013 is the FTRA 7th event of the series of international scientific conferences. This conference will take place in May 9–11, 2013, in Seoul Korea. The aim of the MUE 2013 is to provide an international forum for scientific research in the technologies and application of Multimedia and Ubiquitous Engineering. It is organized by the Korea Information Technology Convergence Society in cooperation with Korea Information Processing Society. MUE2013 is the next event in a series of highly successful international conferences on Multimedia and Ubiquitous Engineering, MUE-12 (Madrid, Spain, July 2012), MUE-11 (Loutraki, Greece, June 2011), MUE-10 (Cebu, Philippines, August 2010), MUE-09 (Qingdao, China, June 2009), MUE-08 (Busan, Korea, April 2008), and MUE-07 (Seoul, Korea, April 2007).

The papers included in the proceedings cover the following topics: *Multimedia Modeling and Processing, Ubiquitous and Pervasive Computing, Ubiquitous Networks and Mobile Communications, Intelligent Computing, Multimedia and Ubiquitous Computing Security, Multimedia and Ubiquitous Services, Multimedia Entertainment, IT and Multimedia Applications.* Accepted and presented papers highlight new trends and challenges of Multimedia and Ubiquitous Engineering. The presenters showed how new research could lead to novel and innovative applications. We hope you will find these results useful and inspiring for your future research.

We would like to express our sincere thanks to Steering Chairs: James J. (Jong Hyuk) Park (SeoulTech, Korea), Martin Sang-Soo Yeo (Mokwon University, Korea). Our special thanks go to the Program Chairs: Eunyoung Lee (Dongduk Women's University, Korea), Cho-Li Wang (University of Hong Kong, Hong Kong), Borgy Waluyo (Monash University, Australia), Al-Sakib Khan Pathan (IIUM, Malaysia), SangHyun Seo (University of Lyon 1, France), all Program Committee members and all the additional reviewers for their valuable efforts in the review process, which helped us to guarantee the highest quality of the selected papers for the conference.

We cordially thank all the authors for their valuable contributions and the other participants of this conference. The conference would not have been possible without their support. Thanks are also due to the many experts who contributed to making the event a success.

May 2013

Young-Sik Jeong
Leonard Barolli
Joseph Kee-Yin Ng
C. S. Raghavendra
MUE 2013 General Chairs

Message from the MUE 2013 Program Chairs

Welcome to the FTRA 7th International Conference on Multimedia and Ubiquitous Engineering (MUE 2013), to be held in Seoul, Korea on May 9–11, 2013. MUE 2013 will be the most comprehensive conference focused on the various aspects of multimedia and ubiquitous engineering. MUE 2013 will provide an opportunity for academic and industry professionals to discuss recent progress in the area of multimedia and ubiquitous environment. In addition, the conference will publish high quality papers which are closely related to the various theories and practical applications in multimedia and ubiquitous engineering. Furthermore, we expect that the conference and its publications will be a trigger for further related research and technology improvements in these important subjects.

For MUE 2013, we received many paper submissions; after a rigorous peer review process, we accepted 62 articles with high quality for the MUE 2013 proceedings, published by Springer. All submitted papers have undergone blind reviews by at least two reviewers from the technical program committee, which consists of leading researchers around the globe. Without their hard work, achieving such a high-quality proceeding would not have been possible. We take this opportunity to thank them for their great support and cooperation. We would like to sincerely thank the following invited speaker who kindly accepted our invitations, and, in this way, helped to meet the objectives of the conference: Prof. Hui-Huang Hsu, Tamkang University, Taiwan. Finally, we would like to thank all of you for your participation in our conference, and also thank all the authors, reviewers, and organizing committee members. Thank you and enjoy the conference!

Eunyoung Lee, Korea
Cho-Li Wang, Hong Kong
Borgy Waluyo, Australia
Al-Sakib Khan Pathan, Malaysia
SangHyun Seo, France
MUE 2013 Program Chairs

Organization

Honorary Chair:	Makoto Takizawa, Seikei University, Japan
Steering Chairs:	James J. Park, SeoulTech, Korea
	Martin Sang-Soo Yeo, Mokwon University, Korea
General Chairs:	Young-Sik Jeong, Wonkwang University, Korea
	Leonard Barolli, Fukuoka Institute of Technology, Japan
	Joseph Kee-Yin Ng, Hong Kong Baptist University, Hong Kong
	C. S. Raghavendra, University of Southern California, USA
Program Chairs:	Eunyoung Lee, Dongduk Women's University, Korea
	Cho-Li Wang, University of Hong Kong, Hong Kong
	Borgy Waluyo, Monash University, Australia
	Al-Sakib Khan Pathan, IIUM, Malaysia
	SangHyun Seo, University of Lyon 1, France
Workshop Chairs:	Young-Gab Kim, Korea University, Korea
	Lei Ye, University of Wollongong, Australia
	Hiroaki Nishino, Oita University, Japan
	Neil Y. Yen, The University of Aizu, Japan
Publication Chair:	Hwa Young Jeong, Kyung Hee University, Korea
International Advisiory Committee:	Seok Cheon Park, Gachon University, Korea
	Borko Furht, Florida Atlantic University, USA
	Thomas Plagemann, University of Oslo, Norway
	Roger Zimmermann, National University of Singapore, Singapore
	Han-Chieh Chao, National Ilan University, Taiwan
	Hai Jin, HUST, China
	Weijia Jia, City University of Hong Kong, Hong Kong
	Jianhua Ma, Hosei University, Japan
	Shu-Ching Chen, Florida International University, USA

	Hamid R. Arabnia, The University of Georgia, USA
	Stephan Olariu, Old Dominion University, USA
	Albert Zomaya, University of Sydney, Australia
	Bin Hu, Lanzhou University, China
	Yi Pan, Georgia State University USA
	Doo-soon Park, SoonChunHyang University, Korea
	Richard P. Brent, Australian National University, Australia
	Koji Nakano, University of Hiroshima, Japan
	J. Daniel Garcia, University Carlos III of Madrid, Spain
	Qun Jin, Waseda University, Japan
	Kyung-Hyune Rhee, Pukyong National University, Korea
Publicity Chairs:	Chengcui Zhang, The University of Alabama at Birmingham, USA
	Michele Ruta, Politecnico di Bari, Italy
	Bessam Abdulrazak, Sherbrooke University, Canada
	Junaid Chaudhry, Universiti Teknologi Malaysia, Malaysia
	Bong-Hwa Hong, Kyung Hee Cyber University, Korea
	Won-Joo Hwang, Inje University, Korea
Local Arrangement Chairs:	HyunsungKim, Kyungil University, Korea
	Eun-Jun Yoon, Kyungil University, Korea
Invited Speaker:	Hui-Huang Hsu, Tamkang University, Taiwan
Program Committee:	A Ra Khil, Soongsil University, Korea
	Afrand Agah, West Chester University of Pennsylvania, USA
	Akihiro Sugimoto, National Institute of Informatics, Japan
	Akimitsu Kanzaki, Osaka University, Japan
	Angel D. Sappa, Universitat Autonoma de Barcelona, Spain
	Bartosz Ziolko, AGH University of Science and Technology, Poland
	Bin Lu, West Chester University, USA

Brent Lagesse, BBN Technologies, USA
Ch. Z. Patrikakis, Technological Education Institute of Piraeus, Greece
Chang-Sun Shin, Sunchon National University, Korea
Chantana Chantrapornchai, Silpakorn University, Thailand
Chih Cheng Hung, Southern Polytechnic State University, USA
Chulung Lee, Korea University, Korea
Dakshina Ranjan Kisku, Asansol Engineering College, India
Dalton Lin, National Taipei University, Taiwan
Dariusz Frejlichowski, West Pomeranian University of Technology, Poland
Debzani Deb, Winston-Salem State University, USA
Deqing Zou, Huazhong University of Science and Technology, China
Ezendu Ariwa, London Metropolitan University, United Kingdom
Farid Meziane, University of Salford, UK
Florian Stegmaier, University of Passau, Germany
Francisco Jose Monaco, University of Sao Paulo, Brazil
Guillermo Camara Chavez, Universidade Federal de Minas Gerais, Brazil
Hae-Young Lee, ETRI, Korea
Hai Jin, Huazhong University of Science and Techn, China
Hangzai Luo, East China Normal University, China
Harald Kosch, University of Passau, Germany
Hari Om, Indian School of Mines University, India
Helen Huang, The University of Queensland, Australia
Hermann Hellwagner, Klagenfurt University, Austria
Hong Lu, Fudan University, China

Jeong-Joon Lee, Korea Polytechnic University, Korea
Jin Kwak, Soonchunhyang University, Korea
Jinye Peng, Northwest University, China
Joel Rodrigue, University of Beira Interior, Portugal
Jong-Kook Kim, Korea University, Korea
Joyce El Haddad, Universite Paris-Dauphine, France
Jungong Han, Civolution Technology, the Netherlands
Jun-Won Ho, Seoul Women's University, Korea
Kilhung Lee, Seoul National University of Science and Technology, Korea
Klaus Schoffmann, Klagenfurt University, Austria
Ko Eung Nam, Baekseok University, Korea
Lidan Shou, Zhejiang University, China
Lukas Ruf, CEO Consecom AG, Switzerland
Marco Cremonini, University of Milan, Italy
Maria Vargas-Vera, Universidad Adolfo Ibanez, Chile
Mario Doeller, University of applied science, Germany
Maytham Safar, Kuwait University, Kuwait
Mehran Asadi, Lincoln University of Penssylvania, USA
Min Choi, Chungbuk National University, Korea
Ming Li, California State University, USA
Muhammad Younas, Oxford Brookes University, UK
Namje Park, Jeju National University, Korea
Neungsoo Park, Konkuk University, Korea
Ning Zhou, University of North Carolina, USA
Oliver Amft, TU Eindhoven, Netherlands
Paisarn Muneesawang, Naresuan University, Thailand
Pascal Lorenz, University of Haute Alsace
Quanqing Xu, Quanqing Xu, Data Storage Institute, A*STAR, Singapore
Rachid Anane, Coventry University, UK

Rainer Unland, University of Duisburg-Essen, Germany

Rajkumar Kannan, Affiliation Bishop Heber College, India

Ralf Klamma, RWTH Aachen University, Germany

Ramanathan Subramanian, Advanced Digital Sciences Center, Singapore

Reinhard Klette, The University of Auckland, New Zealand

Rene Hansen, Aalborg University, Denmark

Sae-Hak Chun, Seoul National University of Science and Technology, Korea

Sagarmay Deb, University of Southern Queensland, Australia

Savvas Chatzichristofis, Democritus University of Thrace, Greece

Seung-Ho Lim, Hankuk University of Foreign Studies, Korea

Shingo Ichii, University of Tokyo, Japan

Sokratis Katsikas, University of Piraeus, Greece

SoonSeok Kim, Halla Universty, Korea

Teng Li, Baidu Inc., China

Thomas Grill, University of Salzburg, Austria

Tingxin Yan, University of arkansas, USA

Toshihiro Yamauchi, Okayama University, Japan

Waleed Farag, Indiana University of Pennsylvania, USA

Wee Siong Ng, Institute for Infocomm Research, Singapore

Weifeng Chen, California University of Pennsylvania, USA

Weifeng Zhang, Nanjing University of Posts and Telecommunication, China

Wesley De Neve, Ghent University iMinds and KAIST,

Won Woo Ro, Yonsei University, Korea

Wookho Son, ETRI, Korea

Wei Wei, Xi'an University of Technology, China

Xubo Song, Oregon Health and Science University, USA
Yan Liu, The Hong Kong Polytechnic University, Hong Kong
Yijuan Lu, Texas State University, USA
Yingchun Yang, Zhejiang University, China
Yong-Yoon Cho, Sunchon University, Korea
Yo-Sung Ho, GIST, Korea
Young-Hee Kim, Korea Copyright Commission, Korea
Young-Ho Park, Sookmyung Women's University, Korea
Zheng-Jun Zha, National University of Singapore, Singapore
Zhu Li, Samsung Telecom America, USA

Message from ATACS-2013 Workshop Chair

Welcome to the Advanced Technologies and Applications for Cloud Computing and Sensor Networks (ATACS-2013), which will be held from May 9 to 11, 2013 in Seoul, Korea.

The main objective of this workshop is to share information on new and innovative research related to advanced technologies and applications in the areas of cloud computing and sensor networks. Many advanced techniques and applications in these two areas have been developed in the past few years. Sensor networks are becoming increasingly large and produce vast amounts of raw sensing data, which cannot be easily processed, analyzed, or stored using conventional computing systems. Cloud computing is one promising technique of efficiently processing these sensing data to create useful services and applications. The convergence of cloud computing and sensor networks requires new and innovative infrastructure, middleware, designs, protocols, services, and applications. ATACS-2013 will bring together researchers and practitioners interested in both the technical and applied aspects of Advanced Techniques and Application for Cloud computing and Sensor networks. Furthermore, we expect that the ATACS-2013 and its publications will be a trigger for further related research and technology improvements in this important subject.

ATACS-2013 contains high quality research papers submitted by researchers from all over the world. Each submitted paper was peer-reviewed by reviewers who are experts in the subject area of the paper. Based on the review results, the Program Committee accepted eight papers.

I hope that you will enjoy the technical programs as well as the social activities during ATACS-2013. I would like to send our sincere appreciation to all of the Organizing and Program Committees who contributed directly to ATACS-2013. Finally, we special thank all the authors and participants for their contributions to make this workshop a grand success.

Joon-Min Gil
Catholic University of Daegu
ATACS-2013 Workshop Chair

ATACS-2013 Organization

General Chair: Joon-Min Gil, Catholic University of Daegu, Korea
Program Chair: Jaehwa Chung, Korea National Open University, Korea
Publicity Chair: Dae Won Lee, Seokyeong University, Korea
Program Committee: Byeongchang Kim, Catholic University of Daegu, Korea
Hansung Lee, Electronics and Telecommunications Research Institute (ETRI), Korea
HeonChang Yu, Korea University, Korea
Jeong-Hyon Hwang, State University of New York at Albany, USA
JongHyuk Lee, Samsung Electronics, Korea
Ki-Sik Kong, Namseoul University, Korea
KwangHee Choi, LG Uplus, Korea
Kwang Sik Chung, Korea National Open University, Korea
Mi-Hye Kim, Catholic University of Daegu, Korea
Shanmugasundaram Hariharan, TRP Engineering College (SRM Group), India
Sung-Hwa Hong, Mokpo National Maritime University, Korea
Sung Suk Kim, Seokyeong University, Korea
Tae-Gyu Lee, Korea Institute of Industrial Technology (KITECH), Korea
Tae-Young Byun, Catholic University of Daegu, Korea
Ui-Sung Song, Busan National University of Education, Korea
Yong-Hee Jeon, Catholic University of Daegu, Korea
Yunhee Kang, Baekseok University, Korea
Zhefu Shi, University of Missouri, USA

Message from PSSI-2013 Workshop Chair

The organizing committee of *the FTRA International Workshop on Pervasive Services, Systems and Intelligence* (*PSSI 2013*) would like to welcome all of you to join the workshop as well as the FTRA MUE 2013. Advances in information and communications technology (ICT) have presented a dramatic growth in merging the boundaries between physical space and cyberspace, and go further to improve mankind's daily life. One typical instance is the use of smartphones. The modern smartphone is equipped with a variety of sensors that are used to collect activities, locations, and situations of its user continuously and provide immediate help accordingly. Some commercial products (e.g., smart house, etc.) also demonstrate the feasibility of comprehensive supports by deploying a rapidly growing number of sensors (or intelligent objects) into our living environments. These developments are collectively best characterized as ubiquitous service that promises to enhance awareness of the cyber, physical, and social contexts. As such, researchers (and companies as well) tend to provide tailored and precise solutions (e.g., services, supports, etc.) wherever and whenever human beings are active according to individuals' contexts. Making technology usable by and useful to, via the ubiquitous services and correlated techniques, humans in ways that were previously unimaginable has become a challenging issue to explore the picture of technology in the next era.

This workshop aims at providing a forum to discuss problems, studies, practices, and issues regarding the emerging trend of pervasive computing. Researchers are encouraged to share achievements, experiments, and ideas with international participants, and furthermore, look forward to map out the research directions and collaboration in the future.

With an amount of submissions (13 in exact), the organizing chairs decided to accept six of them based on the paper quality and the relevancy (acceptance rate at 46 %). These papers are from Canada, China, and Taiwan. Each paper was reviewed by at least three program committee members and discussed by the organizing chairs before acceptance.

We would like to thank three FTRA Workshop Chair, Young-Gab Kim from Korea University, Korea for the support and coordination. We thank all authors for submitting their works to the workshop. We also appreciate the program committee members for their efforts in reviewing the papers. Finally, we sincerely welcome all participants to join the discussion during the workshop.

James J. Park
Neil Y. Yen
Workshop Co-Chairs

FTRA International Workshop on Pervasive Services, Systems and Intelligence (PSSI-13)

Workshop Organization

Workshop Chairs::	James J. Park (Seoul National University of Science and Technology, Korea)
	Neil Y. Yen (The University of Aizu, Japan)
Program Committee::	Christopher Watson, Durham University, United Kingdom
	Chengjiu Yin, Kyushu University, Japan
	David Taniar, Monash University, Australia
	Jui-Hong Chen, Tamkang University, Taiwan
	Junbo Wang, the University of Aizu, Japan
	Lei Jing, the University of Aizu, Japan
	Marc Spaniol, Max-Planck-Institute for Informatic, Germany
	Martin M. Weng, Tamkang University, Taiwan
	Nigel Lin, Microsoft Research, United States
	Ralf Klamma, RWTH Aachen University, Germany
	Vitaly Klyuev, the University of Aizu, Japan
	Xaver Y. R. Chen, National Central University, Taiwan
	Wallapak Tavanapong, Iowa State University, United States
	Renato Ishii, Federal University of Mato Grosso do Sul, Brazil
	Nicoletta Sala, U. of Lugano, Switzerland and Università dell'Insubria Varese, Italy
	Yuanchun Shi, Tsinghua University, China
	Robert Simon, George Mason University, USA

Contents

Part I Multimedia Modeling and Processing

Multiwedgelets in Image Denoising 3
Agnieszka Lisowska

**A Novel Video Compression Method Based on Underdetermined
Blind Source Separation** 13
Jing Liu, Fei Qiao, Qi Wei and Huazhong Yang

Grid Service Matching Process Based on Ontology Semantic 21
Ganglei Zhang and Man Li

Enhancements on the Loss of Beacon Frames in LR-WPANs 27
Ji-Hoon Park and Byung-Seo Kim

**Case Studies on Distribution Environmental Monitoring
and Quality Measurement of Exporting Agricultural Products** 35
Yoonsik Kwak, Jeongsam Lee, Sangmun Byun, Jeongbin Lem,
Miae Choi, Jeongyong Lee and Seokil Song

**Vision Based Approach for Driver Drowsiness Detection
Based on 3DHead Orientation** 43
Belhassen Akrout and Walid Mahdi

Potentiality for Executing Hadoop Map Tasks on GPGPU via JNI ... 51
Bongen Gu, Dojin Choi and Yoonsik Kwak

**An Adaptive Intelligent Recommendation Scheme for Smart
Learning Contents Management Systems** 57
Do-Eun Cho, Sang-Soo Yeo and Si Jung Kim

Part II Ubiquitous and Pervasive Computing

An Evolutionary Path-Based Analysis of Social Experience Design ... 69
Toshihiko Yamakami

Block IO Request Handlingfor DRAM-SSD in Linux Systems 77
Kyungkoo Jun

**Implementation of the Closed Plant Factory System Based
on Crop Growth Model** 83
Myeong-Bae Lee, Taehyung Kim, HongGeun Kim, Nam-Jin Bae,
Miran Baek, Chang-Woo Park, Yong-Yun Cho and Chang-Sun Shin

Part III Ubiquitous Networks and Mobile Communications

**An Energy Efficient Layer for Event-Based Communications
in Web-of-Things Frameworks.** 93
Gérôme Bovet and Jean Hennebert

A Secure Registration Scheme for Femtocell Embedded Networks ... 103
Ikram Syed and Hoon Kim

Part IV Intelligent Computing

**Unsupervised Keyphrase Extraction Based Ranking Algorithm
for Opinion Articles** 113
Heungmo Ryang and Unil Yun

**A Frequent Pattern Mining Technique for Ranking Webpages
Based on Topics** 121
Gwangbum Pyun and Unil Yun

**Trimming Prototypes of Handwritten Digit Images with Subset
Infinite Relational Model** 129
Tomonari Masada and Atsuhiro Takasu

Ranking Book Reviews Based on User Influence. 135
Unil Yun and Heungmo Ryang

**Speaker Verification System Using LLR-Based Multiple
Kernel Learning** 143
Yi-Hsiang Chao

Edit Distance Comparison Confidence Measure for Speech Recognition 151
Dawid Skurzok and Bartosz Ziólko

Weighted Pooling of Image Code with Saliency Map for Object Recognition 157
Dong-Hyun Kim, Kwanyong Lee and Hyeyoung Park

Calibration of Urine Biomarkers for Ovarian Cancer Diagnosis 163
Yu-Seop Kim, Eun-Suk Yang, Kyoung-Min Nam, Chan-Young Park, Hye-Jung Song and Jong-Dae Kim

An Iterative Algorithm for Selecting the Parameters in Kernel Methods 169
Tan Zhiying, She Kun and Song Xiaobo

A Fast Self-Organizing Map Algorithm for Handwritten Digit Recognition 177
Yimu Wang, Alexander Peyls, Yun Pan, Luc Claesen and Xiaolang Yan

Frequent Graph Pattern Mining with Length-Decreasing Support Constraints 185
Gangin Lee and Unil Yun

An Improved Ranking Aggregation Method for Meta-Search Engine 193
Junliang Feng, Junzhong Gu and Zili Zhou

Part V   Multimedia and Ubiquitous Computing Security

Identity-Based Privacy Preservation Framework over u-Healthcare System 203
Kambombo Mtonga, Haomiao Yang, Eun-Jun Yoon and Hyunsung Kim

A Webmail Reconstructing Method from Windows XP Memory Dumps 211
Fei Kong, Ming Xu, Yizhi Ren, Jian Xu, Haiping Zhang and Ning Zheng

On Privacy Preserving Encrypted Data Stores 219
Tracey Raybourn, Jong Kwan Lee and Ray Kresman

Mobile User Authentication Scheme Based on Minesweeper Game . . . 227
Taejin Kim, Siwan Kim, Hyunyi Yi, Gunil Ma and Jeong Hyun Yi

Design and Evaluation of a Diffusion Tracing Function for Classified Information Among Multiple Computers 235
Nobuto Otsubo, Shinichiro Uemura, Toshihiro Yamauchi and Hideo Taniguchi

DroidTrack: Tracking Information Diffusion and Preventing Information Leakage on Android 243
Syunya Sakamoto, Kenji Okuda, Ryo Nakatsuka and Toshihiro Yamauchi

Three Factor Authentication Protocol Based on Bilinear Pairing 253
Thokozani Felix Vallent and Hyunsung Kim

A LBP-Based Method for DetectingCopy-Move Forgery with Rotation 261
Ning Zheng, Yixing Wang and Ming Xu

Attack on Recent Homomorphic Encryption Scheme over Integers ... 269
Haomiao Yang, Hyunsung Kim and Dianhua Tang

A New Sensitive Data Aggregation Scheme for Protecting Data Integrity in Wireless Sensor Network 277
Min Yoon, Miyoung Jang, Hyoung-il Kim and Jae-woo Chang

Reversible Image Watermarking Based on Neural Network and Parity Property 285
Rongrong Ni, H. D. Cheng, Yao Zhao, Zhitong Zhang and Rui Liu

A Based on Single Image Authentication System in Aviation Security 293
Deok Gyu Lee and Jong Wook Han

Part VI Multimedia and Ubiquitous Services

A Development of Android Based Debate-Learning System for Cultivating Divergent Thinking 305
SungWan Kim, EunGil Kim and JongHoon Kim

Development of a Lever Learning Webapp for an HTML5-BasedCross-Platform 313
TaeHun Kim, ByeongSu Kim and JongHoon Kim

Contents

Looking for Better Combination of Biomarker Selection and Classification Algorithm for Early Screening of Ovarian Cancer 321
Yu-Seop Kim, Jong-Dae Kim, Min-Ki Jang,
Chan-Young Park and Hye-Jeong Song

A Remote Control and Media Sharing System Based on DLNA/UPnP Technology for Smart Home. 329
Ti-Hsin Yu and Shou-Chih Lo

A New Distributed Grid Structure for k-NN Query Processing Algorithm Based on Incremental Cell Expansion in LBSs 337
Seungtae Hong, Hyunjo Lee and Jaewoo Chang

A New Grid-Based Cloaking Scheme for Continuous Queries in Centralized LBS Systems 345
Hyeong-Il Kim, Mi-Young Jang, Min Yoon and Jae-Woo Chang

New Database Mapping Schema for XML Document in Electronic Commerce 353
Eun-Young Kim and Se-Hak Chun

A Study on the Location-Based Reservation Management Service Model Using a Smart Phone 359
Nam-Jin Bae, Seong Ryoung Park, Tae Hyung Kim, Myeong Bae Lee,
Hong Gean Kim, Mi Ran Baek, Jang Woo Park,
Chang-Sun Shin and Yong-Yun Cho

A Real-time Object Detection System Using Selected Principal Components 367
Jong-Ho Kim, Byoung-Doo Kang, Sang-Ho Ahn,
Heung-Shik Kim and Sang-Kyoon Kim

Trajectory Calculation Based on Position and Speed for Effective Air Traffic Flow Management 377
Yong-Kyun Kim, Deok Gyu Lee and Jong Wook Han

Part VII   Multimedia Entertainment

Design and Implementation of a Geometric Origami Edutainment Application. 387
ByeongSu Kim, TaeHun Kim and JongHoon Kim

Part VIII IT and Multimedia Applications

Gamification Literacy: Emerging Needs for Identifying Bad Gamification 395
Toshihiko Yamakami

Automatic Fixing of Foot Skating of Human Motions from Depth Sensor 405
Mankyu Sung

Part VIII IT and Multimedia Applications

A Study on the Development and Application of Programming Language Education for Creativity Enhancement: Based on LOGO and Scratch 415
YoungHoon Yang, DongLim Hyun, EunGil Kim, JongJin Kim and JongHoon Kim

Design and Implementation of Learning Content Authoring Framework for Android-Based Three-Dimensional Shape 423
EunGil Kim, DongLim Hyun and JongHoon Kim

A Study on GUI Development of Memo Function for the E-Book: A Comparative Study Using iBooks 431
Jeong Ah Kim and Jun Kyo Kim

Relaxed Stability Technology Approach in Organization Management: Implications from Configured-Control Vehicle Technology 439
Toshihiko Yamakami

Mapping and Optimizing 2-D Scientific Applications on a Stream Processor 449
Ying Zhang, Gen Li, Hongwei Zhou, Pingjing Lu, Caixia Sun and Qiang Dou

Development of an Android Field Trip Support Application Using Augmented Reality and Google Maps 459
DongLim Hyun, EunGil Kim and JongHoon Kim

Implementation of Automotive Media Streaming Service Adapted to Vehicular Environment 467
Sang Yub Lee, Sang Hyun Park and Hyo Sub Choi

Contents

xxiii

The Evaluation of the Transmission Power Consumption Laxity-Based (TPCLB) Algorithm 477
Tomoya Enokido, Ailixier Aikebaier and Makoto Takizawa

The Methodology for Hardening SCADA Security Using Countermeasure Ordering 485
Sung-Hwan Kim, Min-Woo Park, Jung-Ho Eom and Tai-Myoung Chung

Development and Application of STEAM Based Education Program Using Scratch: Focus on 6th Graders' Science in Elementary School 493
JungCheol Oh, JiHwon Lee and JongHoon Kim

Part IX Advanced Technologies and Applications for Cloud Computing and Sensor Networks

Performance Evaluation of Zigbee Sensor Network for Smart Grid AMI 505
Yong-Hee Jeon

P2P-Based Home Monitoring System Architecture Using a Vacuum Robot with an IP Camera 511
KwangHee Choi, Ki-Sik Kong and Joon-Min Gil

Design and Simulation of Access Router Discovery Process in Mobile Environments 521
DaeWon Lee, James J. Park and Joon-Min Gil

Integrated SDN and Non-SDN Network Management Approaches for Future Internet Environment 529
Dongkyun Kim, Joon-Min Gil, Gicheol Wang and Seung-Hae Kim

Analysis and Design of a Half Hypercube Interconnection Network 537
Jong-Seok Kim, Mi-Hye Kim and Hyeong-Ok Lee

Aperiodic Event Communication Process for Wearable P2P Computing 545
Tae-Gyu Lee and Gi-Soo Chung

Broadcasting and Embedding Algorithms for a Half Hypercube Interconnection Network 553
Mi-Hye Kim, Jong-Seok Kim and Hyeong-Ok Lee

Obstacle Searching Method Using a Simultaneous Ultrasound Emission for Autonomous Wheelchairs 561
Byung-Seop Song and Chang-Geol Kim

Part X Future Technology and its Application

A Study on Smart Traffic Analysis and Smart Device Speed Measurement Platform 569
Haejong Joo, Bonghwa Hong and Sangsoo Kim

Analysis and Study on RFID Tag Failure Phenomenon............ 575
Seongsoo Cho, Son Kwang Chul, Jong-Hyun Park and Bonghwa Hong

Administration Management System Design for Smart Phone Applications in use of QR Code............................... 585
So-Min Won, Mi-Hye Kim and Jin-Mook Kim

Use of Genetic Algorithm for Robot-Posture 593
Dong W. Kim, Sung-Wook Park and Jong-Wook Park

Use of Flexible Network Framework for Various Service Components of Network Based Robot 597
Dong W. Kim, Ho-Dong Lee, Sung-Wook Park and Jong-Wook Park

China's Shift in Culture Policy and Cultural Awareness........... 601
KyooSeob Lim

China's Cultural Industry Policy............................. 611
WonBong Lee and KyooSeob Lim

Development of Mobile Games for Rehabilitation Training for the Hearing Impaired 621
Seongsoo Cho, Son Kwang Chul, Chung Hyeok Kim and Yunho Lee

A Study to Prediction Modeling of the Number of Traffic Accidents.. 627
Young-Suk Chung, Jin-Mook Kim, Dong-Hyun Kim and Koo-Rock Park

Part XI Pervasive Services, Systems and Intelligence

A Wiki-Based Assessment System Towards Social-Empowered Collaborative Learning Environment 633
Bruce C. Kao and Yung Hui Chen

Universal User Pattern Discovery for Social Games: An Instance on Facebook 641
Martin M. Weng and Bruce C. Kao

Ubiquitous Geography Learning Smartphone System for 1st Year Junior High Students in Taiwan 649
Wen-Chih Chang, Hsuan-Che Yang, Ming-Ren Jheng and Shih-Wei Wu

Housing Learning Game Using Web-Based Map Service 657
Te-Hua Wang

Digital Publication Converter: From SCORM to EPUB 665
Hsuan-pu Chang

An Intelligent Recommender System for Real-Time Information Navigation 673
Victoria Hsu

Part XII Advanced Mechanical and Industrial Engineering, and Control I

Modal Characteristics Analysis on Rotating Flexible Beam Considering the Effect from Rotation 683
Haibin Yin, Wei Xu, Jinli Xu and Fengyun Huang

The Simulation Study on Harvested Power in Synchronized Switch Harvesting on Inductor 691
Jang Woo Park, Honggeun Kim, Chang-Sun Shin, Kyungryong Cho, Yong-Yun Cho and Kisuk Kim

An Approach for a Self-Growing Agricultural Knowledge Cloud in Smart Agriculture 699
TaeHyung Kim, Nam-Jin Bae, Chang-Sun Shin, Jang Woo Park, DongGook Park and Yong-Yun Cho

Determination of Water-Miscible Fluids Properties 707
Zajac Jozef, Cuma Matus and Hatala Michal

Influence of Technological Factors of Die Casting on Mechanical Properties of Castings from Silumin 713
Stefan Gaspar and Jan Pasko

Active Ranging Sensors Based on Structured Light Image for Mobile Robot 723
Jin Shin and Soo-Yeong Yi

Improved Composite Order Bilinear Pairing on Graphics Hardware 731
Hao Xiong, Xiaoqi Yu, Yi-Jun He and Siu Ming Yiu

Deployment and Management of Multimedia Contents Distribution Networks Using an Autonomous Agent Service 739
Kilhung Lee

Part XIII Advanced Mechanical and Industrial Engineering, and Control II

Design Optimization of the Assembly Process Structure Based on Complexity Criterion 747
Vladimir Modrak, Slavomir Bednar and David Marton

Kinematics Modelling for Omnidirectional Rolling Robot 755
Soo-Yeong Yi

Design of Device Sociality Database for Zero-Configured Device Interaction 763
Jinyoung Moon, Dong-oh Kang and Changseok Bae

Image Processing Based a Wireless Charging System with Two Mobile Robots 769
Jae-O Kim, Chan-Woo Moon and Hyun-Sik Ahn

Design of a Reliable In-Vehicle Network Using ZigBee Communication 777
Sunny Ro, Kyung-Jung Lee and Hyun-Sik Ahn

Wireless Positioning Techniques and Location-Based Services: A Literature Review 785
Pantea Keikhosrokiani, Norlia Mustaffa, Nasriah Zakaria and Muhammad Imran Sarwar

Part XIV Green and Human Information Technology

Performance Analysis of Digital Retrodirective Array Antenna System in Presence of Frequency Offset 801
Junyeong Bok and Heung-Gyoon Ryu

A Novel Low Profile Multi-Band Antenna for LTE Handset 809
Bao Ngoc Nguyen, Dinh Uyen Nguyen, Tran Van Su,
Binh Duong Nguyen and Mai Linh

Digital Signature Schemes from Two Hard Problems 817
Binh V. Do, Minh H. Nguyen and Nikolay A. Moldovyan

Performance Improvements Using Upgrading Precedences in MIL-STD-188-220 Standard 827
Sewon Han and Byung-Seo Kim

Blind Beamforming Using the MCMA and SAG-MCMA Algorithm with MUSIC Algorithm. 835
Yongguk Kim and Heung-Gyoon Ryu

Performance Evaluation of EPON-Based Communication Network Architectures for Large-Scale Offshore Wind Power Farms. 841
Mohamed A. Ahmed, Won-Hyuk Yang and Young-Chon Kim

A User-Data Division Multiple Access Scheme 849
P. Niroopan, K. Bandara and Yeon-ho Chung

On Channel Capacity of Two-Way Multiple-hop MIMO Relay System with Specific Access Control. 857
Pham Thanh Hiep, Nguyen Huy Hoang and Ryuji Kohno

Single-Feed Wideband Circularly Polarized Antenna for UHF RFID Reader 863
Pham HuuTo, B. D. Nguyen, Van-Su Tran, Tram Van
and Kien T. Pham

Experimental Evaluation of WBAN Antenna Performance for FCC Common Frequency Band with Human Body. 871
Musleemin Noitubtim, Chairak Deepunya and Sathaporn Promwong

Performance Evaluation of UWB-BAN with Friis's Formula and CLEAN Algorithm 879
Krisada Koonchiang, Dissakan Arpasilp and Sathaporn Promwong

A Study of Algorithm Comparison Simulator for Energy Consumption Prediction in Indoor Space 887
Do-Hyeun Kim and Nan Chen

Energy Efficient Wireless Sensor Network Design and Simulation for Water Environment Monitoring 895
Nguyen Thi Hong Doanh and Nguyen Tuan Duc

An Energy Efficient Reliability Scheme for Event Driven Service in Wireless Sensor Actuator Networks 903
Seungcheon Kim

Efficient and Reliable GPS-Based Wireless Ad Hoc for Marine Search Rescue System 911
Ta Duc-Tuyen, Tran Duc-Tan and Do Duc Dung

Improved Relay Selection for MIMO-SDM Cooperative Communications 919
Duc Hiep Vu, Quoc Trinh Do, Xuan Nam Tran and Vo Nguyen Quoc Bao

Freshness Preserving Hierarchical Key Agreement Protocol Over Hierarchical MANETs 927
Hyunsung Kim

A Deployment of RFID for Manufacturing and Logistic 935
Patcharaporn Choeysuwan and Somsak Choomchuay

Real Time Video Implementation on FPGA 943
Pham Minh Luan Nguyen and Sang Bock Cho

Recovery Algorithm for Compressive Image Sensing with Adaptive Hard Thresholding 949
Viet Anh Nguyen and Byeungwoo Jeon

Estimation Value for Three Dimension Reconstruction 957
Tae-Eun Kim

Gesture Recognition Algorithm using Morphological Analysis 967
Tae-Eun Kim

Omnidirectional Object Recognition Based Mobile Robot Localization ... 975

Sungho Kim and In So Kweon

Gender Classification Using Faces and Gaits ... 983

Hong Quan Dang, Intaek Kim and YoungSung Soh

Implementation of Improved Census Transform Stereo Matching on a Multicore Processor ... 989

Jae Chang Kwak, Tae Ryong Park, Yong Seo Koo and Kwang Yeob Lee

A Filter Selection Method in Hard Thresholding Recovery for Compressed Image Sensing ... 997

Phuong Minh Pham, Khanh Quoc Dinh and Byeungwoo Jeon

Facial Expression Recognition Using Extended Local Binary Patterns of 3D Curvature ... 1005

Soon-Yong Chun, Chan-Su Lee and Sang-Heon Lee

Overview of Three and Four-Dimensional GIS Data Models ... 1013

Tuan Anh Nguyen Gia, Phuoc Vinh Tran and Duy Huynh Khac

Modeling and Simulation of an Intelligent Traffic Light System Using Multiagent Technology ... 1021

Tuyen T. T. Truong and Cuong H. Phan

A Numerical Approach to Solve Point Kinetic Equations Using Taylor-Lie Series and the Adomian Decomposition Method ... 1031

Hag-Tae Kim, Ganduulga, Dong Pyo Hong and Kil To Chong

Regional CRL Distribution Based on the LBS for Vehicular Networks ... 1039

HyunGon Kim, MinSoo Kim, SeokWon Jung and JaeHyun Seo

Study of Reinforcement Learning Based Dynamic Traffic Control Mechanism ... 1047

Zheng Zhang, Seung Jun Baek, Duck Jin Lee and Kil To Chong

Understanding and Extending AUTOSAR BSW for Custom Functionality Implementation ... 1057

Taeho Kim, Ji Chan Maeng, Hyunmin Yoon and Minsoo Ryu

A Hybrid Intelligent Control Method in Application of Battery Management System 1065
T. T. Ngoc Nguyen and Franklin Bien

Interpretation and Modeling of Change Patterns of Concentration Based on EEG Signals 1073
JungEun Lim, Soon-Yong Chun and BoHyeok Seo

Design of Autonomic Nerve Measuring System Using Pulse Signal 1081
Un-Ho Ji and Soon-Yong Chun

Semiconductor Monitoring System for Etching Process 1091
Sang-Chul Kim

Enhancing the Robustness of Fault Isolation Estimator for Fault Diagnosis in Robotic Systems 1099
Ngoc-Bach Hoang and Hee-Jun Kang

Software-Based Fault Detection and Recovery for Cyber-Physical Systems 1107
Jooyi Lee, Ji Chan Maeng, Byeonghun Song, Hyunmin Yoon, Taeho Kim, Won-Tae Kim and Minsoo Ryu

Sample Adaptive Offset Parallelism in HEVC 1113
Eun-kyung Ryu, Jung-hak Nam, Seon-oh Lee, Hyun-ho Jo and Dong-gyu Sim

Comparison Between SVM and Back Propagation Neural Network in Building IDS 1121
Nguyen Dai Hai and Nguyen Linh Giang

Anomaly Detection with Multinomial Logistic Regression and Naïve Bayesian 1129
Nguyen Dai Hai and Nguyen Linh Giang

Implementation of Miniaturized Automotive Media Platform with Vehicle Data Processing 1137
Sang Yub Lee, Sang Hyun Park, Duck Keun Park, Jae Kyu Lee and Hyo Sub Choi

Design of Software-Based Receiver and Analyzer System for DVB-T2 Broadcast System 1147
M. G. Kang, Y. J. Woo, K. T. Lee, I. K. Kim, J. S. Lee and J. S. Lee

Contents

**Age-Group Classification for Family Members Using
Multi-Layered Bayesian Classifier with Gaussian Mixture Model** . . . 1153
Chuho Yi, Seungdo Jeong, Kyeong-Soo Han and Hankyu Lee

**Enhancing Utilization of Integer Functional Units
for High-Throughput Floating Point Operations
on Coarse-Grained Reconfigurable Architecture** 1161
Manhwee Jo, Kyuseung Han and Kiyoung Choi

**An Improved Double Delta Correlator for BOC Signal
Tracking in GNSS Receivers** 1169
Pham-Viet Hung, Dao-Ngoc Chien and Nguyen-Van Khang

**Implementation of Automatic Failure Diagnosis
for Wind Turbine Monitoring System Based on Neural Network** . . . 1181
Ming-Shou An, Sang-June Park, Jin-Sup Shin,
Hye-Youn Lim and Dae-Seong Kang

**Development of Compact Microphone Array
for Direction-of-Arrival Estimation** 1189
Trình Quốc Võ and Udo Klein

**Design and Implementation of a SoPC System
for Speech Recognition** 1197
Tran Van Hoang, Nguyen Ly Thien Truong,
Hoang Trang and Xuan-Tu Tran

Index 1205

Part I
Multimedia Modeling and Processing

Multiwedgelets in Image Denoising

Agnieszka Lisowska

Abstract In this paper the definition of a multiwedgelet is introduced. The multiwedgelet is defined as a vector of wedgelets. In order to use a multiwedgelet in image approximation its visualization and computation methods are also proposed. The application of multiwedgelets in image denoising is presented, as well. As follows from the experiments performed multiwedgelets assure better denoising results than the other known state-of-the-art methods.

Keywords Multiwedgelets · Wedgelets · Multiresolution · Denoising

1 Introduction

Geometrical multiresolution methods of image approximation are widely used in these days. It follows from the multiscale nature of the world, especially of digital images. Such methods can better adapt to image singularities than the well known wavelets theory [1]. Many new, geometrical, representations have been proposed recently. They can be divided into two groups. The one is based on nonadaptive methods of computing, with the use of frames, like brushlets [2], ridgelets [3], curvelets [4], contourlets [5], shearlets [6]. In the second group the approximations are computed in an adaptive way. The majority of the representations are based on dictionaries, examples include wedgelets [7], beamlets [8], second order wedgelets [9, 10], platelets [11], surflets [12], smoothlets [13]. However, recently also the adaptive schemes based on basis have been proposed, like bandelets [14], grouplets [15], tetrolets [16]. More and more "X-lets" have been still defined.

A. Lisowska (✉)
Institute of Computer Science, University of Silesia,
ul. Bedzinska 39 41-200 Sosnowiec, Poland
e-mail: alisow@ux2.math.us.edu.pl
URL: http://www.math.us.edu.pl/al/eng_index.html

J. J. (Jong Hyuk) Park et al. (eds.), *Multimedia and Ubiquitous Engineering*,
Lecture Notes in Electrical Engineering 240, DOI: 10.1007/978-94-007-6738-6_1,
© Springer Science+Business Media Dordrecht(Outside the USA) 2013

Many theories, which are based on functions, are further extended on vectors of functions. In general, from the mathematical point of view, it is rather a simple task. However, the practical application of such theories is not easy, especially in the area of image processing. It follows mainly from the fact that an image is a two dimensional object and can be represented by a set of functions in a natural way. A set of vectors seems to be used rather in the representation of a set of images. So, the main question is not how to extend a theory on vectors but how to apply such an extended theory to image processing?

In this paper the answer to the above question is presented. In more details, firstly, the definition of multiwedgelet is proposed as a vector of wedgelets. Because the visualization of a multiwedgelet is not straightforward, the method of visualization is also proposed. It is used further in the image approximation. In order to justify the usefulness of the proposed approach the application of multiwedgelets to image denoising is presented. As follows from the experiments, the proposed method outperforms the known methods like wedgelets, second order wedgelets, curvelets and wavelets [17, 18].

2 Multiwedgelets

Let us define an image domain $D = [0, 1] \times [0, 1]$. Next, let us denote function $h(x)$ defined within D as the "horizon", that is any smooth function defined on the interval $[0, 1]$. In practical applications it is sufficiently to assume that the function h is of C^2 class.

Further, consider the characteristic function

$$H(x, y) = \mathbf{1}\{y \leq h(x)\}, \quad 0 \leq x, y \leq 1. \tag{1}$$

Then function H is called a "horizon function" if h is a "horizon". Function H models a black and white image with a horizon where the image is white above the horizon and black below.

Having an image domain $D = [0, 1] \times [0, 1]$ one can, in some sense, discretize it on different levels of multiresolution. Consider the dyadic square $D(j_1, j_2, i)$ as the two dimensional interval

$$D(j_1, j_2, i) = [j_1/2^i, (j_1 + 1)/2^i] \times [j_2/2^i, (j_2 + 1)/2^i], \tag{2}$$

where $j_1, j_2 \in \{0, \ldots, 2^i - 1\}$, $i \in \mathbb{N}$. Note that $D(0, 0, 0)$ denotes the whole image domain D, that is the square $[0, 1] \times [0, 1]$. On the other hand $D(j_1, j_2, I)$ for $j_1, j_2 \in \{0, \ldots, N\}$ denote appropriate pixels from $N \times N$ grid, where N is dyadic (it means that $N = 2^I$). From this moment on let us consider a domain of an image as such $N \times N$ grid of pixels.

2.1 Basic Definitions

Having assumed that an image domain is the square $[0,1] \times [0,1]$ and that it consists of $N \times N$ pixels (or, more precisely, squares of size $1/N$) one can note that on each border of any square $D(j_1,j_2,i), j_1,j_2 \in \{0,\ldots,2^i-1\}, i \in \{0,\ldots,\log_2 N\}$ the vertices with distance equal to $1/N$ can be denoted. Every two such vertices in any fixed square may be connected to form a straight line b—an edge (also called a *beamlet* after the work [8]).

Let us denote then $B_{D(j_1,j_2,i)}$ as the set of all nondegenerated beamlets (that is no lying on the same side of a square border) within $D(j_1,j_2,i)$ for any $j_1,j_2 \in \{0,\ldots,2^i-1\}, i \in \mathbb{N}$. Consider then a vector of beamlets $\mathbf{b}^M_{j_1,j_2,i} = [b^1_{j_1,j_2,i},\ldots,b^M_{j_1,j_2,i}]$, $M \in \mathbb{N}$. We call vector $\mathbf{b}^M_{j_1,j_2,i}$ a *multibeamlet* if for all $k \in \{1,\ldots,M\}$ $b^k_{j_1,j_2,i} \in B_{D(j_1,j_2,i)}$ for fixed $j_1,j_2 \in \{0,\ldots,2^i-1\}, i \in \mathbb{N}$. In Fig. 1 some examples of multibeamlets are presented.

Let us consider a beamlet b. It splits any square D (we skip the subscripts denoting the location and the scale for a moment for better clarity) into two pieces. Let us consider one of the two pieces which is bounded by lines connecting in turn in clockwise direction, from the lower left corner, the first of the two edge vertices and the second one. Let us define then the indicator function of that piece

$$W(x,y) = \mathbf{1}\{y \leq b(x)\}, \quad (x,y) \in D. \tag{3}$$

Such a function we call a *wedgelet* defined by beamlet b [7].

Let us denote then $W_{D(j_1,j_2,i)}$ as the set of all nondegenerated wedgelets within $D(j_1,j_2,i)$ for any $j_1,j_2 \in \{0,\ldots,2^i-1\}, i \in \mathbb{N}$. Consider then a vector of wedgelets $\mathbf{W}^M_{j_1,j_2,i} = [W^1_{j_1,j_2,i},\ldots,W^M_{j_1,j_2,i}]$, $M \in \mathbb{N}$. We call vector $\mathbf{W}^M_{j_1,j_2,i}$ a *multi-wedgelet* if for all $k \in \{1,\ldots,M\}$ $W^k_{j_1,j_2,i} \in W_{D(j_1,j_2,i)}$ for fixed $j_1,j_2 \in \{0,\ldots,2^i-1\}, i \in \mathbb{N}$.

Consider the complete quadtree image partition. Each segment can be represented by two numbers i and j where $j \in \{0,\ldots,4^i-1\}$ and $i \in \{0,\ldots,\log_2 N\}$. In other words, the pair of subscripts (j_1,j_2) from the above considerations can be replaced by the subscript j. Having defined a multiwedgelet and renumerating subscripts for better clarity one can define the dictionary of multiwedgelets as the following set

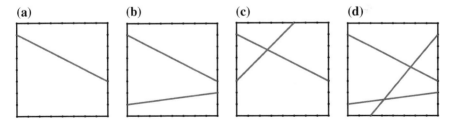

Fig. 1 Examples of multibeamlets: **a** $M = 1$, **b–c** $M = 2$, **d** $M = 3$

$$W_M = \{\mathbf{W}_{i,j}^M : i \in \{0, \ldots, \log_2 N\}, j \in \{0, \ldots, 4^i - 1\}\}, \tag{4}$$

where $\mathbf{W}_{i,j}^M = [W_{i,j}^1, \ldots, W_{i,j}^M]$.

Image approximation is performed in two steps. In the first step, for each quadtree partition segment, the best multiwedgelet has to be found (in the mean of the smallest Mean Square Error approximation), forming the complete quadtree with the optimal multiwedgelet parameters in each node. In the second step the bottom-up tree pruning algorithm has to be applied [7] in order to obtain the optimal image representation.

There is one drawback related to the image approximation by multiwedgelets. It is related to finding a reasonable method of a multiwedgelet visualization. Since the visualization of one wedgelet is quite simple, the vector of wedgelets has to be visualized in a tricky way. Below such a method is presented.

2.2 Multiwedgelet Visualization

Let us consider the example presented in Fig. 2 for $M = 3$. The multiwedgelet W is defined as $\mathbf{W} = [W^1, W^2, W^3]$ where wedgelets W^i are based on appropriate beamlets b^i for $i \in \{1, 2, 3\}$. If the wedgelets are defined as

$$W^i = \begin{cases} h_1^i, y \leq b^i, \\ h_2^i, y > b^i, \end{cases} \text{ for } i \in \{1, \ldots, M\},$$

the appropriate colors are defined as

$$c^a = \frac{1}{M} \sum_{k=1}^{M} h_u^k \quad \text{for } a \in \{1, \ldots, \text{Number of Areas}\}, u \in \{1, 2\}. \tag{5}$$

In other words, the colors are the means of all wedgelets colors.

Let us note that such correlation between multiwedgelets coefficients and image colors causes that the image is defined as a mean of all wedgelets of the multiwedgelet. Indeed, for image segment F one obtains $F = \frac{1}{M} \sum_{k=1}^{M} W^k$.

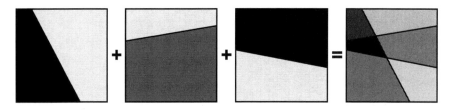

Fig. 2 The method of a multiwedgelet visualization ($M = 3$)

2.3 Multiwedgelet Computation

It can be defined plenty of methods of a multiwedgelet computation. In the paper the one is proposed. It is based on the fast wedgelet computation proposed in [19]. In order to compute multiwedgelet parameters one needs to proceed in the following way. Firstly, compute the wedgelet transform for the first wedgelet of a given multiwedgelet, then compute the wedgelet transform for the second wedgelet of the multiwedgelet for a slightly translated support (i.e. one pixel up and left), then do the same for all the rest wedgelets of multiwedgelet, each time changing slightly the support (by translating it in different directions). The support manipulation causes that the optimal wedgelets of multiwedgelet are different.

Let us note that the computational complexity of the proposed method is $O(N^2\log_2 N)$ for an image of size $N \times N$ pixels. It follows from the fact that the method is based on the fast wedgelet transform [19] and it is performed M times. Since, usually, $M = 3$ in practice it can be treated as a constant. The measured computation time of the multiwedgelet transform is as follows: for $M = 1$ it equals 1.8 s, for $M = 2$ it equals 3.2 s and for $M = 3$ it equals 4.7 s for an image of size 256×256 pixels. The computations were performed on Intel Core2 Duo 2 GHz processor.

3 Experimental Results

In order to perform numerical computations the standard set of test images, presented in Fig. 3, was used. The images were additionally contaminated by Gaussian noise with zero mean and different values of variance with the help of Matlab Image Processing Toolbox. All computations were made with the help of the software written in C++ Builder 6 Environment.

In Table 1 the numerical results of image denoising for different methods and different values of noise variance are presented. The same set of images was used in the paper [18] from which it follows that the curvelets-based method of image denoising assures better results than denoising by wavelets. From Table 1 it follows that the method of image denoising by multiwedgelets (for $M = 3$) outperforms the curvelets and wedgelets-based methods, it even outperforms the second order wedgelets-based method (wedgeletsII) [18]. Only in the case of images with strongly curvature geometry, like "Circles" and "Blobs", second order wedgelets-based method assures better results of denoising than multiwedgelets. It is very natural since second order wedgelets were designed to best approximate images with curvature geometry.

In Fig. 4 two plots, arbitrarily chosen, of image denoising by multiwedgelets are presented. As one can see the use of multiwedgelets outperforms the use of wedgelets ($M = 1$). The larger the value of M the better the result of image denoising. However, from the performed experiments follows that the choice of

Fig. 3 The benchmark images, respectively: "Balloons", "Bird", "Monarch", "Peppers", "Circles", "Blobs" [18]

$M = 3$ is the most flexible one. It gives satisfactory results and is not computationally expansive.

Sample results of image denoising are presented in Fig. 5 for images "Bird" and "Monarch" contaminated by zero-mean Gaussian noise with variance $V = 0.022$. The images were denoised by curvelets, second order wedgelets and multiwedgelets, respectively. As one can see the method based on multiwedgelets assured the best denoising results, both visually and in the mean of PSNR values (Table 1).

4 Conclusions

In this paper the new theory of multiwedgelets was presented. Instead of considering wedgelets, the vectors of these functions were used. In order to use them in image approximation the computation and visualization methods were also proposed.

In this paper also the application of multiwedgelets to image denoising was presented. In comparison to the other known state-of-the-art methods (like curvelets, wedgelets, second order wedgelets—directly, and wavelets—indirectly via comparison to the paper [18]) the method based on multiwedgelets assures the best denoising results. Probably, it is possible to apply multiwedgelets in other image

Multiwedgelets in Image Denoising

Table 1 Numerical results of image denoising for methods based on curvelets, wedgelets, second order wedgelets and multiwedgelets for different values of noise variance (PSNR)

Image	Method	\multicolumn{8}{c}{Noise variance}							
		0.001	0.005	0.010	0.015	0.022	0.030	0.050	0.070
Balloons	Curvelets	24.97	24.22	23.87	23.28	21.54	20.04	17.46	15.89
	Wedgelets	30.50	26.10	24.03	23.17	22.29	21.72	20.60	19.94
	Wedg.II	30.40	25.92	24.00	23.12	22.26	21.71	20.67	19.97
	Multiwed.	29.23	26.14	24.59	23.72	22.91	22.33	21.24	20.45
Bird	Curvelets	20.34	20.95	26.94	25.87	23.40	21.02	18.09	16.19
	Wedgelets	34.24	30.24	28.76	28.05	27.35	26.82	25.71	25.21
	Wedg.II	34.07	30.24	28.76	28.02	27.29	26.79	25.66	25.09
	Multiwed.	34.55	31.05	29.70	28.95	27.99	27.50	26.47	25.72
Monarch	Curvelets	24.14	24.99	24.15	23.37	21.77	20.38	17.53	15.95
	Wedgelets	30.47	26.20	24.32	23.27	22.33	21.63	20.50	19.70
	Wedg.II	30.38	26.21	24.39	23.40	22.37	21.71	20.56	19.71
	Multiwed.	28.32	25.82	24.54	23.60	22.83	21.91	21.01	20.41
Peppers	Curvelets	22.52	25.04	23.91	24.41	22.89	20.85	17.65	15.95
	Wedgelets	31.71	27.44	25.82	24.89	24.10	23.41	22.43	21.75
	Wedg.II	31.56	27.31	25.81	24.79	24.04	23.37	22.36	21.68
	Multiwed.	31.63	27.81	26.58	25.64	24.83	24.21	22.99	22.32
Circles	Curvelets	25.56	22.35	23.90	21.49	20.41	18.71	16.61	15.32
	Wedgelets	41.97	35.32	31.84	29.95	28.21	26.84	24.56	23.18
	Wedg.II	43.19	36.60	32.60	30.51	28.59	26.93	24.58	23.17
	Multiwed.	33.16	32.22	30.67	29.58	28.31	26.96	24.66	23.06
Blobs	Curvelets	14.83	28.43	30.71	25.65	23.64	21.55	18.33	16.41
	Wedgelets	44.23	36.31	33.74	32.85	31.77	30.97	29.18	27.51
	Wedg.II	45.12	37.49	34.52	33.51	31.95	31.43	29.20	27.55
	Multiwed.	38.53	35.53	34.22	33.14	32.33	31.01	29.84	27.67

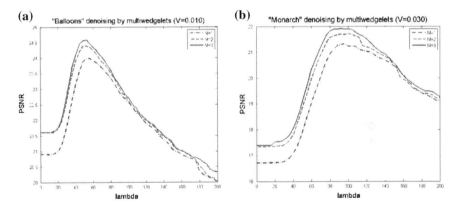

Fig. 4 Denoising by multiwedgelets for different values of M for images contaminated by zero-mean Gaussian noise: **a** "Balloons", $V = 0.010$, **b** "Monarch", $V = 0.030$. Let us note that $M = 1$ denotes wedgelets

Fig. 5 Examples of image denoising: (*upper row*) "Bird", (*lower row*) "Monarch"; by **a**, **d** curvelets, **b**, **e** second order wedgelets, **c**, **f** multiwedgelets. The images were contaminated by zero-mean Gaussian noise with variance $V = 0.022$

processing tasks, like image compression or edge detection. It is the open problem for future research.

As follows from the performed experiments, the potential of multiwedgelets can be quite large. The methods proposed in this paper and the parameters fixed are both not optimal but multiwedgelets still outperform the state-of-the-art methods in such an image processing task like image denoising. There is still much to do in improving the multiwedgelets theory and in finding new applications.

References

1. Mallat S (1999) A wavelet tour of signal processing. Academic Press, San Diego
2. Meyer FG, Coifman RR (1997) Brushlets: a tool for directional image analysis and image compression. Appl Comput Harm Anal 4:147–187
3. Candés E (1998) Ridgelets: theory and applications, Ph.D. thesis, Department of Statistics, Stanford University, Stanford
4. Candés E, Donoho D (1999) Curvelets—a surprisingly effective nonadaptive representation for objects with edges. In: Cohen A, Rabut C, Schumaker LL (ed) Curves and surface fitting, Vanderbilt University Press, Saint-Malo, pp 105–120
5. Do MN, Vetterli M (2003) Contourlets. In: Stoeckler J, Welland GV (ed) Beyond wavelets, Academic Press, San Diego, pp 83–105
6. Labate D, Lim W, Kutyniok G, Weiss G (2005) Sparse multidimensional representation using shearlets. Proc SPIE 5914:254–262

7. Donoho DL (1999) Wedgelets: nearly-minimax estimation of edges. Ann Stat 27:859–897
8. Donoho DL, Huo X (2000) Beamlet pyramids: a new form of multiresolution analysis, suited for extracting lines, curves and objects from very noisy image data. In: Proceedings of SPIE, vol 4119
9. Lisowska A (2000) Effective coding of images with the use of geometrical wavelets. In: Proceedings of the decision support systems conference, Zakopane, Poland
10. Lisowska A (2005) Geometrical wavelets and their generalizations in digital image coding and processing. Ph.D. thesis, University of Silesia, Poland
11. Willet RM, Nowak RD (2003) Platelets: a multiscale approach for recovering edges and surfaces in photon limited medical imaging. IEEE Trans Med Imaging 22:332–350
12. Chandrasekaran V, Wakin MB, Baron D, Baraniuk R (2004) Surflets: a sparse representation for multidimensional functions containing smooth discontinuities. In: IEEE international symposium on information theory, Chicago
13. Lisowska A (2011) Smoothlets—multiscale functions for adaptive representations of images. IEEE Trans Image Process 20(7):1777–1787
14. Pennec E, Mallat S (2005) Sparse geometric image representations with bandelets. IEEE Trans Image Process 14(4):423–438
15. Mallat S (2009) Geometrical grouplets. Appl Comput Harm Anal 26(2):161–180
16. Krommweh J (2009) Image approximation by adaptive tetrolet transform and International conference on sampling theory and applications,
17. Demare L, Friedrich F, Führ H, Szygowski T (2005) multiscale wedgelet denoising algorithms, Proceedings of SPIE, San Diego, Wavelets XI, Vol. 5914, X1-12
18. Lisowska A (2008) Image denoising with second order wedgelets. Intern J Signal Imaging Syst Eng 1(2):90–98
19. Lisowska A (2011) Moments-based fast wedgelet transform. J Math Imaging Vis 39(2):180–192. Springer

A Novel Video Compression Method Based on Underdetermined Blind Source Separation

Jing Liu, Fei Qiao, Qi Wei and Huazhong Yang

Abstract If a piece of picture could contain a sequence of video frames, it is amazing. This paper develops a new video compression approach based on underdetermined blind source separation. Underdetermined blind source separation, which can be used to efficiently enhance the video compression ratio, is combined with various off-the-shelf codecs in this paper. Combining with MPEG-2, video compression ratio could be improved slightly more than 33 %. As for combing with H.264, twice compression ratio could be achieved with acceptable PSNR, according to different kinds of video sequences.

Keywords Underdetermined blind source separation · Sparse component analysis · Video surveillance system · Video compression

1 Introduction

Digital video is famous for its abundant information and its rigid demand for the bandwidth and process power as well. It leads to the emergence of multiple video coding standards, such as MPEG-2, H.264. However, new thoughts can be applied to compress video as well.

Blind Source Separation (BSS) provides a solution to recover original signals from mixed signals. It can be used in multiple fields, such as wireless communication to separate mixed radio signals, biomedicine to separate fetal electrocardiogram signals recorded by sensors, and typical "cocktail party" problem to separate mixed speech signals.

J. Liu · F. Qiao (✉) · Q. Wei · H. Yang
Department of Electronic Engineering, Tsinghua University, Beijing 100084, China
e-mail: qiaofei@tsinghua.edu.cn

J. J. (Jong Hyuk) Park et al. (eds.), *Multimedia and Ubiquitous Engineering*,
Lecture Notes in Electrical Engineering 240, DOI: 10.1007/978-94-007-6738-6_2,
© Springer Science+Business Media Dordrecht(Outside the USA) 2013

Independent Component Analysis (ICA) was widely accepted as a powerful solution of BSS since the past 20 years [1]. In 1999, A. Hyvarinen presented an improved ICA algorithm, called FastICA [2]. A detailed overview of many algorithms on BSS is made and their usages on image processing are presented as well [3]. However, few researchers focused on utilizing BSS into video processing.

In this paper, we apply Underdetermined BSS (UBSS, meaning the number of original signals is more than that of mixed signals) to compress video sequences. A new codec defined as Underdetermined Blind Source Separation based Video Compression (UBSSVC) is developed. As we explained later in detail, UBSSVC has good performance on video compression.

This paper is organized as follows. The next section briefly reviews BSS problem. In Sect. 3, detailed structure of UBSSVC is stated. And Sect. 4 shows simulation results. Finally, Sect. 5 summarizes the superior and deficiency of this video compression method. Also, future work is proposed in this section.

2 Blind Source Separation and Solution to UBSS

BSS was first established by J. Herault and C. Jutten in 1985 [1]. It can be described as following: multiple signals from separate sources s are somehow mixed into several other signals, defined as mixed signals x. Here n represents the number of source signals and m represents the number of mixed signals. The objective of BSS is to design an inverse system to get the estimation of source signals. The reason for the "Blind" here is neither the source signals nor the mixed process is known to the observer. The mixing model can be expressed as,

$$x = As; \quad A \in R^{m \times n}, s \in R^{n \times T} \tag{1}$$

where A is an $m \times n$ mixing matrix. Both A and s are unknown, while x is known to observer.

Independent Component Analysis (ICA) is the main solution for overdetermined $(n < m)$ and standard $(n = m)$ case. However, it is not suitable for UBSS $(n > m)$. Other methods like Spare Component Analysis (SCA) [4–7] and overcomplete ICA [8, 9], are investigated for UBSS recent years.

In this work, SCA is adopted to solve the UBSS. SCA uses the sparsity of source signals to compensate information loss in the mixing process. So specific assumptions of mixing matrix A and source matrix s should be considered as follows [4]. Assumption 1: any $m \times m$ square sub-matrix of mixing matrix $A \in R^{m \times n}$ is nonsingular; assumption 2: there are at most $m - 1$ nonzero elements of any column of matrix s. If the above assumptions are satisfied, the source matrix s can be recovered by SCA.

Let x_i, $i = 1, 2, \ldots, m$ and s_i, $i = 1, 2, \ldots, n$ represent mixed signals and source signals respectively; and a_j, $j = 1, \ldots, n$ is the jth column of mixing matrix A. Therefore, the mixing process can also be described as following.

A Novel Video Compression Method

$$x(t) = (x_1(t) \quad x_2(t) \quad \cdots \quad x_m(t))^T = a_1 s_1(t) + a_2 s_2(t) + \cdots + a_n s_n(t) \quad (2)$$

Given the mixing matrix A satisfies the assumption 1, any $m-1$ columns of A span a m-dimensional linear hyperplane $\mathcal{H}_q$, which can be denoted as $\mathcal{H}_q = \{h | h \in R^m, \lambda_{ik} \in R, h = \lambda_{i_1} a_{i_1} + \cdots + \lambda_{i_{m-1}} a_{i_{m-1}}\}$, where $q = 1, \ldots, C_n^{m-1}$. If source matrix s satisfies assumption 2, it is reasonable to suppose that at the t moment, all source signals except for $s_{i_1}, s_{i_2}, \ldots, s_{i_{m-1}}$ are zero, where $\{i_1, i_2, \ldots, i_{m-1}\} \subset \{1, 2, \ldots, n\}$. Consequently, at t moment, Eq. (2) can be rewritten as

$$(x_1(t) \quad x_2(t) \quad \cdots \quad x_m(t))^T = a_{i_1} s_{i_1}(t) + a_{i_2} s_{i_2}(t) + \cdots + a_{i_{m-1}} s_{i_{m-1}}(t) \quad (3)$$

From (3), it can be concluded that the tth column vector of observed signals matrix x is in one of C_n^{m-1} hyperplanes $\mathcal{H}$. Therefore, mixed frames can be recovered by the following algorithm.

(a) Get the set $\mathcal{H}$ of C_n^{m-1} m-dimensional hyperplanes which are spanned by any $m-1$ columns of A;
(b) j repeat from 1 to m,

    (i) If x_j, which stands for the jth column of mixed signals matrix x, is in a hyperplane $\mathcal{H}_q$, then the following equation can be gotten

$$x_j = \sum_{v=1}^{m-1} \lambda_{i_v, j} a_{i_v} \quad (4)$$

    (ii) Comparing Eqs. (3) and (4), s_i, the ith column of source signals matrix s, can be recovered: its components are $\lambda_{i_v, j}$ in the place $i_v, v = 1, \ldots, m-1$, and other components equal to zero.

3 Proposed UBSSVC Method

As explained above, for UBSS the number of mixed signals is less than that of source signals. Therefore, the mixing process of UBSS could be used to compress video sequences, and the separating process is used to decode the compressed video sequences.

3.1 Mapping UBSS to Video Compression

Consider a video sequence with L frames, $s_1, s_2, \ldots, s_L$, where $s_i \in R^T$ is a T-pixel frame; we firstly divide the L video frames into b groups and in each group there are n frames. The encoder first chooses a matrix $A \in R^{m \times n} (m < n)$ to mix n frames in each group. Thus, the compression ratio is n/m.

At the encoder side, unlike the traditional scenario of the UBSS issue, the mixing process is factitious in this proposed method. Thus, a specific mixing matrix A, known by both encoder and decoder, is chosen to mix raw video frames.

For standard BSS, there is only one restriction of mixing matrix A, that the columns of A should be mutually independent. However, in the proposed method, matrix A not only needs to satisfy the assumption 1, but also has to decrease the information loss in mixing process. Thus, in different mixed frames, the weight of different original frames should be varied. As each component of a row of A can be treated as the weight of every original frame in a mixed frame, the components of a row of A should be varied largely from each other. Experiments will be done to show A's influence on the separation results in Sect. 4.

At the decoder side, the matrix A is known exactly, so the frames' order of recovered video sequence is not disturbed by mixing process and separating process, which is different from traditional BSS.

To ensure that the frames could satisfy the assumption 2, mixed frames are first transformed by a 2-D discrete Haar wavelet transform. And then SCA is used to recover the sparse high frequency components, while the recovered low frequency components are equal to multiply generalized inverse of mixing matrix A by mixed low frequency components.

3.2 Proposed UBSSVC Structure

The compression ratio for UBSS is only n/m. Therefore, to enhance the compression ratio more, we proposed UBSSVC framework that combines UBSS and conventional codec together, shown in Fig. 1.

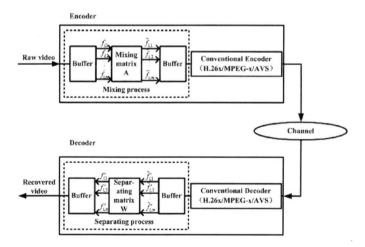

Fig. 1 Framework of UBSSVC

At the encoder side, n frames, $f_{i,1}, f_{i,2},\ldots, f_{i,n}$, are mixed into m frames, $\tilde{f}_{i,1}, \tilde{f}_{i,2},\ldots, \tilde{f}_{i,m}$. And then these mixed frames are encoded by traditional encoder such as MPEG-2, H.264. The buffer before mixing is used to buffer enough frames for being mixed. And the buffer after mixing is for storing mixed frames temporarily so that they can be encoded by conventional encoder one by one.

In the proposed decoder structure, received data is firstly decoded by traditional decoder; then the source recovery algorithm of underdetermined BSS is applied to recover original video sequence. In the separating process, m frames, $\tilde{f}'_{i,1}, \tilde{f}'_{i,2},\ldots, \tilde{f}'_{i,m}$, are separated to n frames, $f'_{i,1}, f'_{i,2},\ldots, f'_{i,n}$. The function of two buffers in decoder is similar to that of those two buffers in encoder.

4 Experiment Results

In order to validate this approach, multiple simulations are performed on four standard test video sequences: hall, container, foreman and football. The football sequence has the largest temporal variations, followed by foreman, and container ranks the third, while the hall sequence contains the most slowly scene variations. The first 40 frames of each sequence are used for test. Peak-Signal-to-Noise Ratio (PSNR) is used to evaluate the performance of recovery algorithm.

In the experiments, we just show an example of mixing 4 video frames into 3 frames, so the compression ratio of UBSS in the experiments is just 4/3. The mixing matrix $A \in R^{3\times 4}$, shown in (5), is chosen to mix raw video sequence, where $k \in Z, k \neq 0$. The mixing process is performed as follows: continuous 4 frames are taken as source signals s, then A multiplies by s to calculate the mixed frames x. 30 mixed frames are generated after the mixing process. And then the above algorithm is applied to separate these mixed frames. Figure 2 shows the recovery PSNR on different video test sequences when $k = 0.5$–5. These plots show that the value of k has little influence on the separation PSNR. Although for some sequence, such as football, PSNR is a little low, it is still enough for monitor applications, which don't have very strict demands on high resolution.

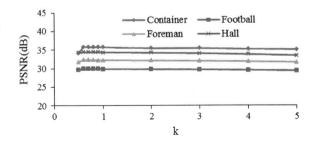

Fig. 2 Separation PSNR related to different values of k on four videos

$$A = \frac{1}{k} \begin{pmatrix} 0.30 & 0.45 & 0.15 & 0.10 \\ 0.35 & 0.15 & 0.05 & 0.45 \\ 0.30 & 0.05 & 0.45 & 0.20 \end{pmatrix} \quad (5)$$

Experiments are done as well to show the k's effect to the compression ratio and separation PSNR of UBSSVC+MPEG-2 which means that the conventional codec in Fig. 1 is MPEG-2, UBSSVC+H.264 which means that the conventional codec in Fig. 1 is H.264. Results are shown in Figs. 3, 4, 5 and 6. From the results, the k values indeed affect the UBSSVC+MPEG-2 and UBSSVC+H.264 compression ratio. That's because with the increment of k, most pixels values of the mixed frames approach to zero. Therefore, the compression ratios of MPEG-2 and H.264 for these mixed frames are much larger than that for the original frames. Meanwhile, it leads to a higher distortion. So the decoding PSNR decreases with the k increment when $k > 0.7$. For these four different test sequences, the largest PSNR and lowest compression ratio are almost gotten at the point $k = 0.7$. However, even the lowest compression ratio is larger than the corresponding

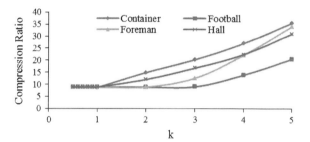

Fig. 3 Compression ratio of UBSSVC+MPEG-2 related to different values of k on four video

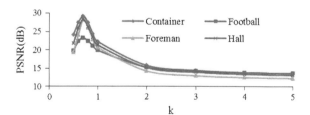

Fig. 4 Decoding PSNR of UBSSVC+MPEG-2 related to different values of k on four videos

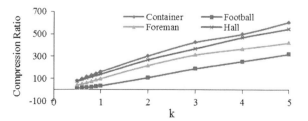

Fig. 5 Compression ratio of UBSSVC+H.264 related to different values of k on four videos

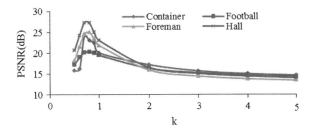

Fig. 6 Decoding PSNR of UBSSVC+H.264 related to different values of k on four videos

Table 1 Compression results of test videos

	Hall (dB)	Container (dB)	Foreman (dB)	Football (dB)
MPEG-2	6.48	6.48	6.48	6.48
UBSSVC+MPEG-2 (k = 0.7)	8.84	8.84	8.84	8.84
H.264	76.74	92.28	62.04	12.15
UBSSVC+H.264 (k = 0.7)	102.41	161.06	74.25	25.38

Table 2 PSNR results of test videos

	Hall (dB)	Container (dB)	Foreman (dB)	Football (dB)
MPEG-2	37.87	30.86	27.04	27.11
UBSSVC+MPEG-2 (k = 0.7)	28.23	29.18	28.15	23.23
H.264	36.19	35.43	35.13	31.7
UBSSVC+H.264 (k = 0.7)	27.26	24.16	25.01	20.29

compression ratio of MPEG-2 and H.264. Tables 1 and 2 show the comparison results. The PSNRs of UBSSVC+H.264 (k = 0.7), and UBSSVC+MPEG-2 (k = 0.7) is lower than those of H.264 and MPEG-2 respectively. Although the PSNR value is a little low, it is enough for some applications which don't have strict demands on high resolution, such as video surveillance system.

5 Conclusion

This paper initially develops the novel video compression approach UBSSVC. Furthermore, experiments are conducted to validate the efficiency of recovery algorithm, the influence of *k* values on separation PSNR and to measure the video compression ratio improvements of UBSSVC. The proposed method is suitable for video surveillance system perfectly. Firstly, it can achieve higher video compression ratio to decrease the bandwidth resource utilization. Secondly, the computation complexity of mixing process at encoder side is low, when improving the video compression ratio. What's more, the mixing and separating process of UBSS has great potential in low-complexity video compression.

However, the presented new method still has more issues to be improved in our future work. Like the largest compression ratio the UBSS can achieve, and how to improve the compression ratio gained by the mixing process and enhance the separating results of video quality.

References

1. Hyvarinen A, Karhunen J, Oja E (2001) Independent component analysis. Wiley, New York
2. Hyvarinen A (1999) Fast and robust fixed-point algorithms for independent component analysis. IEEE Trans Neural Netw 10:626–634
3. Cichocki A, Amari SI (2002) Adaptive blind signal and image processing: learning algorithms and applications. Wiley, Chichester
4. Georgiev P, Theis F, Cichocki A (2005) Sparse component analysis and blind source separation of underdetermined mixtures. IEEE Trans Neural Netw 16:992–996
5. Li YQ, Cichocki A, Amari SI, Shishkin S, Cao JT, Gu FJ (2004) Sparse representation and its applications in blind source separation. In: Thrun S, Saul K, Scholkopf B (eds) Advances in neural information processing systems 16, vol 16. MIT Press, Cambridge, pp 241–248
6. Ren M.-r, Wang P (2009) Underdetermined blind source separation based on sparse component in electronic computer technology, 2009 international conference on. pp 174–177
7. Zhenwei S, Huanwen T, Yiyuan T (2005) Blind source separation of more sources than mixtures using sparse mixture models. Pattern Recogn Lett 26:2491–2499
8. Lee TW, Lewicki MS, Girolami M, Sejnowski TJ (1999) Blind source separation of more sources than mixtures using overcomplete representations. IEEE Signal Process Lett 6:87–90
9. Waheed K, Salem FM (eds) (2003) Algebraic independent component analysis: An approach for separation of overcomplete speech mixtures. In: Proceedings of the IEEE, international joint conference on neural networks 2003, vols 1–4. New York, pp 775–780

Grid Service Matching Process Based on Ontology Semantic

Ganglei Zhang and Man Li

Abstract Little prior communication between supplier and demander makes recognition of grid service ability helpful to make full use of some rich resources, such as software, hardware, information and others. Description and matching process about grid service based on ontology semantic were proposed, and its relative advantages comparing with the existing grid discovery mechanisms were also proved.

Keywords Grid computing · Ontology · Service matching

1 Introduction

Grid is a non centralized control across heterogeneous platform of collaborative working. It uses a standard, open and wide application protocol and seamless service quality. In addition, Grid technology allows sharing and collaboration in dynamic virtual Organizations. A new grid standard OGSA introduced Web Service technology, which has become industry standard, and presented the concept of grid service. It packages all software, hardware, network resources in a grid environment in the form of service, to provide a relatively uniform abstract interface. Grid resources are extremely rich, but if a specific user wants to find out adaptive service, he will face many problems. Some technologies such as UDDI, MDS2 mechanism in Globus Toolkit2 based on LDAP and Index Service based on service data matching. Their common shortcoming is little communication of

---

G. Zhang (✉) · M. Li
Shandong Huayu Vocational and Technical College, DeZhou 253000, China
e-mail: zhangganglei@163.com

M. Li
e-mail: xdclm@126.com

J. J. (Jong Hyuk) Park et al. (eds.), *Multimedia and Ubiquitous Engineering*,
Lecture Notes in Electrical Engineering 240, DOI: 10.1007/978-94-007-6738-6_3,
© Springer Science+Business Media Dordrecht(Outside the USA) 2013

service providers and demand side in advance [1]. Therefore, simple keywords matching can not provide enough flexibility and reasoning ability, and it is difficult to express the true ability of grid services. What's more, the retrieval result is not satisfied. This paper discussed the issues of description and retrieval of Grid Service on the basis of these results.

2 Grid Service and Its Ontology Description

2.1 Ontology Theory and Description Language

Ontology is derived from philosophy, intended to refer to the objective existence of nature. In computer science, an ontology is a group of shared concepts and deficit formal specifications. The type of concept and its bounds are defined definitely and ontology should be understood by machine. Sharing represents the concept must be public recognition, but not for private individuals or group. In addition, Concept represents ontology is the reflection of real world. Concepts such as class, object and inheritance are included in the development of ontology [2].

OWL Web ontology language is a universal language proposed by W3C organization after the standardization of DAML and OIL which is used to express Web Ontology. Compared with the original extensible markup language (XML) and resource description framework (RDF) technology, OWL is more convenient to express the semantic, and its ability of reading information is stronger. Its three sub-languages OWL Lite, OWL DL and OWL Full are respectively provided with different classification and binding capacity, supporting different levels of modeling needs.

2.2 Grid Service and Globus

Grid service is a Web service and provides a series of definitely defined interfaces to solve problems such as establishment of dynamic service, management of life circle, and notice. It can simplify some problems about large-scale network. Grid service used SOAP and WSDL technologies, and has some extension on the following aspects [3].

Provide status service. In Web Service, service never save the status information about users, while Grid Service can provide status service for users, and can generate a service instance and a mechanism of data inquiry.

Provide instant service. In Web Service, all the services are forever, while grid service not only provides forever services, but also provides instant services, that is, service instance and resource distribution only be carried out when the users demand them. Resources and instances can be released after used by the customers to improve the availability.

2.3 Ontology Description of Grid Service

OWL-S. OWL-S is a kind of Web Service based on OWL, and it is derived from DAML-S. It provides a series of markup languages, and these languages can describe the features and abilities of Web Service exactly. According to the definition of OWL-S, a Web Service is described by presents, described by and supports, and their Range are respectively ServiceProfiles, ServiceModel and ServiceGrounding. They stated what the service can do, how to do and how to use service.

Extension of OWL-S. Grid service can be considered as an formal extension of Web Service. On one hand, OWL-S Service Profiles has its advantages on the description of ontology semantics. On the other hand, some extension is needed to describe the unique features of grid service. An extensible ontology class Grid-ServiceProfiles is introduced to provide a standard for the description of service ability of grid.

3 Semantic Matching Based on Ontology

Semantic matching is to analyze the descriptions of two groups of semantics, and judge their conformation level. In factual application, requirement semantics description proposed by demand side is matched with description files from service publisher side to find out required service information. Ontology semantic information, the basis of matching, has strict concept definition and little ambiguity to realize strict matching. In addition, inheritance and intersection among concepts provide guarantee for the flexibility of matching. These are not exist in simple keywords matching. The matching includes main concept matching and service ability matching.

3.1 Basic Concept Matching

Grid service semantic matching is based on the concept model about its each property, while concept matching mainly depends on their positions in inheritance relationship. A matching function concept_match(C1, C2) is defined, and its return results can be the following values.

Exact. It is an accurate matching, that is, C1 and C2 are the same concepts. In OWL, they point at the same URI node or the relative class through <owl:sameClassOf>.

PlugIn. C1 is the sub-concept of C2, that is C1 $\subseteq$ C2, that is, C1 and C1 are combined by <rdfs:subClassOf>.

Subsume. In contrast with PlugIn, the meaning of Subsume is that C2 is the sub-concept of C1.

Intersection. There is an intersection between Cl and C2. It is defined by to <owl:intersectionOf> in OWL.

Fail. In addition to the above four matching results, all the other results are Fail. It means the result is failed.

The five matching results are in descending order according to matching degree. It can be seen from definitions that Exact matching is the most accurate and the strictest matching, while the others provides alternatives with varying degrees when accurate matching can not be met. In addition, results of the middle three matching methods have transitivity. The more degrees of transition a matching results through, the lower the similarity degree is. In order to improve matching efficiency, the degrees can be limited by service requester.

3.2 Service Ability Matching

Overall evaluation of matching degree of service capability is seen as follows. Overall capacity matching should mainly consider category, input, output and data information. For the information of services provider, because it does not belong to semantic, it is only be matched when the demand side need.

For the the information of class service, if it is in the OWL framework, matches were nothing special; if it is in the external classification system,

There are two ways: one is mapping the existing classification system to its equivalent OWL ontology description, the other is to introduce external classification system with plugs. At present, the two programs are with high cost.

For the information of output, output description of demand side are removed one by one, and the corresponding matching results of its belonging concept are looked for in output of release side. If each item of output information can find a match, overall matching of output information is considered successful.

The process of input information matching is similar with output information matching. However, the judgment way is extracting each item of input information of release side to find matching results in input description of demand side. It is equivalent to swapping the positions of release side and demand side in output matching function. This is because whether output information is matched is relative to whether the service can be accepted. While matching degree of input information is related to whether the service can be normally executed.

Service data can be seen as a special kind of output information, so output information matching mode is adaptive to it.

Finally is the sorting of the whole service ability. It mainly refers to classification and service data whose return value is not Fail. Because of low importance of service data, only its matching results are returned as parameters. The sorting process followed by category information, input data, output data and service data.

4 Conclusions

This paper introduced the improved matching process and ontology semantic description of grid service ability. It is easy to be carried out under GT3 framework and has great flexibility and accuracy. However, it should be improved on specific matching information, the reliability of semantic information and the automation.

References

1. Foster I, Kesselman C, Nick JM et al (2003) The physiology of the grid:an open grid services architecture for distributed systems integration. http://www.globus.org
2. http://www.w3c.org/2002/ws [EB/OL] (2004)
3. UDDI: Universal Description, Discovery and Integration (2004) http://www.uddi.org

Enhancements on the Loss of Beacon Frames in LR-WPANs

Ji-Hoon Park and Byung-Seo Kim

Abstract In beacon-enabled LR-WPANs, the high reliability of beacon frame transmission is required because all transmissions are controlled by the information in the beacon frame. However, the process for the beacon-loss scenario is not carefully considered in the standard. The method proposed in this paper allows a node not receiving a beacon frame to keep transmit its pending frames only within the minimum period of CAP based on the previously received beacon frame while the standard prevents the node from sending any pending frame during a whole superframe. The method is extensively simulated and proven the enhancements on the performances.

Keywords LR-WPAN · Beacon · Sensor · IEEE802.15.4

1 Introduction

Applications using such IEEE802.15.4 standards-based Low Rate-Wireless Personal Area Networks (LR-WPANs) have been increasing in broad areas including public safety, home entertainment system, home automation systems, ubiquitous building systems, traffic information systems, and so on. IEEE802.15.4 standard [1] defines two types of LR-WPANs: beacon–enabled and nonbeacon-enabled networks. Any transmission of any node in the beacon-enabled LR-WPANs is controlled by the information in the beacon frames transmitted by the central PAN

J.-H. Park (✉)
Nongshim Data System, XXX, Korea
e-mail: topsicrit@daum.net

B.-S. Kim
Department of Computer and Information Communications Engineering, Hongik University, Hongik, Korea
e-mail: jsnbs@hongik.ac.kr

J. J. (Jong Hyuk) Park et al. (eds.), *Multimedia and Ubiquitous Engineering*, Lecture Notes in Electrical Engineering 240, DOI: 10.1007/978-94-007-6738-6_4, © Springer Science+Business Media Dordrecht(Outside the USA) 2013

coordinator. Therefore, the high reliability of beacon transmission is essential for beacon-enabled LR-WPANs.

However, there are some factors to cause the loss of beacons such as collisions between beacons and interferences from other devices. The focus of this paper is the beacon loss due to the letter. The loss of beacon due to interference occurs because many communication networks like LR-WPANs and Wireless Local Area Networks (WLANs) and even microwaves uses same frequency bands of 2.4 GHz which is called Industrial Scientific Medical (ISM) band [2]. As a consequence, LR-WPANs experience severe interferences from other devices. The performance degradations of LR-WPAN due to interferences are reported by many experiments and studies as shown in [3–9]. As the electric power grid systems recently utilize LR-WPANs and WLANs, the interference issues in the power grid system is reported as shown in [9]. Especially, as the number of deployed WLANs rapidly increases, the impacts on LR-WPANs of interferences from WLANs are actively researched in [4–8] and it is shown that LR-WPANs coexisting with WLANs experience 10–100 % degradations on the performances depending on the distances between LR-WPANs and WLANs, locations, the channels used by LR-WPANs, and the traffic loads of WLANs. There are many studies to avoid the interference. To resolve the problem, the most of methods switch the channels to non-interference channel.

While all aforementioned method proposed methods to avoid a beacon loss, no aforementioned studies mentions the process when a device fails to receive a beacon. Even though the many solutions have been proposed, the beacon can still be lost because of the channel characteristics like noise, fading, Doppler effects and so on. Based on IEEE802.15.4 standard, devices failed to receive a beacons have to hold their pending transmissions during a superframe associated with the beacon which cause the performance degradations. Therefore, we need to a better method to improve the network performances when the beacon is lost.

In this paper, a method is proposed to improve the performances of beacon-enabled LR-WPANs by allowing nodes to transmit its pending frames during a Contention Access Period (CAP).

In Sect. 2, IEEE802.15.4 standard-based LR-WPANs and the process when the beacon is lost are introduced. In Sect. 3, the proposed protocol is described. After evaluating the performances of the proposed method with extensive simulations, finally conclusions are made.

2 Preliminary Researches

2.1 IEEE802.15.4 Standard

In beacon-enabled LR-WPANs in IEEE802.15.4 standard, the time is subdivided into consecutive superframes. The structure of the superframe is shown in Fig. 1. The standard optionally allows the superframe to be divided in two parts: active

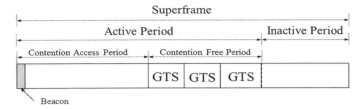

Fig. 1 Superframe structure for beacon-enabled LR-WPANs

and inactive periods. Actual data packets between devices are transmitted during active period while any packet transmission is prohibited during the inactive period for saving the power. Therefore, all devices with pending packets have to hold their transmissions until the next active period. The active period is composed of 16 slots and each slot is 960us based on [1]. The superframe is started with the transmission of beacon from a PAN coordinator. The beacon is used to synchronize with participating devices, to identify the WPAN, and to inform the participating devices the structure of the superframes. After beacon transmission, contention access period (CAP) is followed. The CAP adopts the contention-based data transmissions like carrier sense multiple access with collision avoidance (CSMA/CA). After CAP, Contention Free Period (CFP) is follows. CFP is composed of multiple Guaranteed Time Slots (GTSs). The lengths of GTSs are varied unlike time slots in conventional Time Division Multiple Access (TDMA)-based system. The maximum number of GTSs in CFP is 7 and a GTS can occupy more than one slot. GTSs in CFP are allocated by the PAN coordinator when devices requests. The information on the GTS allocation is included in the beacon. During GTS, only designated device transmits its packet without contention and collision.

2.2 Beacon Loss in IEEE802.15.4 Standard

The process when a device fails to receive a beacon is not clearly described in the standard and any literatures except for the case in which GTSs are allocated in the superframe. When a device's GTSs are allocated in a superframe and it loses a beacon, the device is not allowed to transmit its packet during its GTS. Since a beacon contains the information on superframe structure like period of CAP, the allocation of GTS, and so on, and the superframe structure can vary in every superframe, if a device fails to receive a beacon, it can be assumed that it needs better to hold its transmissions during the superframe to prevent from collisions with other scheduled transmissions. This assumption is clear for the cases that the network parameters like the number of nodes, traffic loads, etc. are frequently fluctuated.

Furthermore, based on the IEEE802.15.4 standard [1], if a device does not receive beacons an *aMaxLostBeacons* times, it declares synchronization loss and starts orphan channel scan after discarding all buffered packets in MAC layer. The orphan channel scan scans the channels in a specified set of logical channels to search a PAN coordinator to re-associate with. When starting the orphan channel scan, the device sends an orphan notification command, and waits a PAN coordinator realignment command from a PNC within a *macResponseWaitTime* symbols. This process is repeated for the channels in the set of logical channels.

Overall, the losses of beacons cause holding devices' transmissions as well as the synchronization loss, and as consequences it degrades the network performances.

3 Proposed Method

In this paper, we propose an enhancement on IEEE802.15.4 standard-based and beacon-enabled LR-WPAN for the case that the beacons are not received by participating devices. As mentioned in Sect. 2.2, the loss of beacon causes two issues: holding transmissions and re-association. This paper focuses to the first issue.

The proposed method in this paper is to allow devices failed to receive a beacon (hereinafter the device is called 'failed-device') to transmit their pending frames during the minimum period of CAP. The proposed method is focused at the scenario that the lengths of superframe size and active period are fixed which traffic is not much fluctuated like sensor networks.

As described in Sect. 2.1, the active period is composed of CAP and CFP. If the beacon is not received, any transmission in CFP by the failed-device causes a problem because each GTS in CFP is assigned to a specific device and the assigned device transmits its own data without any collision. If a failed-device transmits its data in CFP because it does not know the current structure of the superframe, it causes collisions with transmission that is supposed to be collision-free. In order to prevent this case, any device failed to receive beacon is allowed to transmit its data only in CAP while the device discard all pending frames defined IEEE802.15.4 standard. However, CAP can be varied due to varying the length of CFP. Therefore, the period that a failed-device can be allowed to transmit is defined as follows:

$$T_{BeaconLoss} = aNumSuperframeSlots - \text{MaxNumofSymbol}_{CFP} \qquad (1)$$

where *aNumSuperframeSlots* is the number of slots in active period defined in IEEE802.15.4 standard and $\text{MaxNumofSymbol}_{CFP}$ is a maximum number of symbols that can be assigned for CFP.

4 Performance Evaluations

In this section the enhancements by using the proposed method is evaluated in terms of throughput. The proposed method is compared with IEEE802.15.4-base LR-WPANs. For the simplicity in the mathematical analysis, the saturated traffic model and no CFP is considered. The synchronization loss is not considered because both of the proposed and IEEE802.15.4-based LR-WPANs have same effects on the loss of synchronizations. The throughput of the proposed method is

$$Thr_p = \frac{D(1 - PER_D)(1 - PER_B) + D'(1 - PER_D)PER_B}{T}, \tag{2}$$

where D and D' represent the amounts of data transmitted during active periods when a beacon is successfully and unsuccessfully received, respectively, and T means the duration of two superframes. The reason of two superframe durations is that one superframe with successfully receiving a beacon and with the miss of beacon. In addition, PER_D and PER_B represent packet error rates of data and beacons, respectively. The throughput of the IEEE802.15.4-base WPANs is

$$Thr_{IEEE} = \frac{D(1 - PER_D)(1 - PER_B)}{T}. \tag{3}$$

Because the amount of transmitted data is proportional to active period and $T_{BeaconLoss}$ in (3), the ratio of D and D' is the ratio of active and $T_{BeaconLoss}$ periods in a superframe. When the ratio of D' to D is γ, the enhancements obtained by using the proposed method is

$$E = \frac{Thr_p - Thr_{IEEE}}{Thr_{IEEE}} = \gamma \frac{PER_B}{1 - PER_B}. \tag{4}$$

Based on the Eq. (6), the performance enhancement in the throughput depends on the PER_B and the ratio of D' to D.

Figure 2 shows the enhancements in throughput performances as a function of data packet error rates and the lengths of beacons. γ is set to 7/15 because active period is composed of 16 slots, a beacon uses one slot, and the maximum slots for CFP is recommended 7 in [1]. As shown in Eq. (6) the performance enhancements are depending on the packet error rate of beacons. As shown in Fig. 2, the performance enhancements increase as the error rate of beacons increase. The proposed method allows devices to keep transmitting their pending frames during the minimum required times while the conventional method does not. Therefore, as the number of missed beacons increase, the conventional method loses opportunities for transmitting devices' pending frames, so that the proposed method shows the better performances. The enhancements are from 2.5 % with 5 % PER_D to 31 % with 60 % PER_D. As the measurement studies for interference issue with WLANs are shown in [3–9], the PER_D is varied from 10 % to 100 %. Therefore,

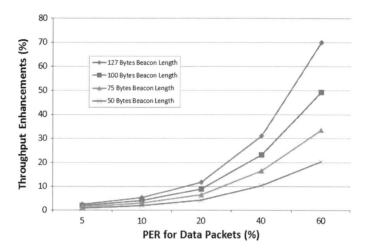

Fig. 2 Throughput enhancements as function of PER for data packets and the lengths of beacons

even though the standard [1] requires 10 % PER_D, analyzing performances over temporal high PER_D scenarios is valuable. Even at 10 % PER_D, 6 % improvements is achieved.

5 Conclusions

The reliability in the beacon transmissions is very critical on the performance of Beacon-enabled LR-WPANs because the loss of beacon causes for devices to hold their transmissions during the superframe. Unlike specification in the standard, the method proposed in the paper allows devices to transmit its pending packet only during the minimum period of CAP that is guaranteed in the superframe. Therefore, the proposed method improves the network performances.

Acknowledgments This research is supported in part by the National Research Foundation of Korea (NRF) grant funded by the Korea government (MEST) (2012-0003609) and in part by the International Science and Business Belt Program through the Ministry of Education, Science and Technology (2012K001556).

References

1. IEEE Std 802.15.4, Part 15.4: wireless medium access control (MAC) and Physical Layer (PHY) specification for Low Rate
2. Lau S-Y, Lin T-H, Huang T-Y, Ng I-H, Huang P (2009) A measurement study of zigbee-based indoor localization systems under RF interference. In: Proceedings of the 4th ACM

international workshop on experimental evaluation and characterization (WINTECH'09), pp 35–42, Beijing 20–25 Sept 2009

3. Howitt I, Gutierrez JA (2003) IEEE 802.15.4 Low rate -wireless personal area network coexistence issues. In: Proceedings of the IEEE WCNC'03, vol 3, pp 1481–1486

4. Sikora A, Groza, VF (2005) Coexistence of IEEE802.15.4 with other Systems in the 2.4 GHz-ISM-Band. In: Proceedings of the IEEE instrumentation and measurement technology conference (IMTC'05), vol 3, pp 1786–1791, Canada, 17–19 May 2005

5. Shin S et al, Packet error rate analysis of IEEE 802.15.4 under IEEE 802.11b interference. In: Proceedings of the WWIC'05. pp 279–288

6. Yoon DG, Shin SY, Kwon WH, Park HS (2006) Packet error rate analysis of IEEE 802.1 lb under IEEE 802.15.4 interference. In: Proceedings of the IEEE 63rd vehicular technology conference, pp 1186–1190, Australia, 7–10 May 2006

7. Petrova M, Gutierrez JA (2006) IEEE 802.15.4 Low rate—wireless personal area network coexistence issues. In: Proceedings of the IEEE WCNC'06, USA

8. Yuan W, Wang X, Linnartz J-PMG (2007) A coexistence model of IEEE 802.15.4 and IEEE 802.11b/g. In: 14th IEEE symposium on communications and vehicular technology in the Benelux, 15 Nov. 2007

9. Shin SY, Park HS, Kwon WH (2007) Mutual interference analysis of IEEE 802.15.4 and IEEE 802.11b. Comput Netw 51(12):3338–3353, 22 August 2007

10. Stanciulescu G, Farhangi H, Palizban A et al (2012) Communication technologies for BCIT Smart Microgrid. 2012 IEEE PES innovative smart grid technologies (ISGT), 16–20 Jan 2012

11. The network simulator NS-2, Web site http://www.isi.edu/nsnam/ns

Case Studies on Distribution Environmental Monitoring and Quality Measurement of Exporting Agricultural Products

Yoonsik Kwak, Jeongsam Lee, Sangmun Byun, Jeongbin Lem, Miae Choi, Jeongyong Lee and Seokil Song

Abstract In this paper, we present monitoring the distribution environmental factors for exporting agricultural products in real time based on sensor networks and packaging technologies, and how the distribution environmental factors would influence the quality of agricultural products has been studied by measuring the actual quality of agricultural products when this distribution process has been completed. For this, sensor nodes and communication system, optimized to a monitoring process for the distribution environmental factors of agricultural products, have been designed and implemented. With the paprika exported to overseas, information on temperature/humidity/path (distribution environmental factors) has been monitored in real time. The possibility of utilization of sensor networks based distribution environmental factors monitoring technology could be verified through such case studies.

Y. Kwak (✉) · S. Song
Department of Computer Engineering, Korea National University of Transportation, ChungJu, South Korea
e-mail: yskwak@ut.ac.kr

S. Song
e-mail: sisong@ut.ac.kr

J. Lee · S. Byun · J. Lem
Marketing Policy Division, Ministry for Food, Agriculture, Forestry and Fisheries, Gwacheon, South Korea
e-mail: gnothi@hanmail.net

M. Choi
Postharvest Research Scientist, National Institute of Horticultural & Herbal Science, RDA, Suwon, South Korea
e-mail: choma818@korea.kr

J. Lee
Agribusiness Development Team, FACT, Suwon, South Korea
e-mail: dfy0928@daum.net

J. J. (Jong Hyuk) Park et al. (eds.), *Multimedia and Ubiquitous Engineering*,
Lecture Notes in Electrical Engineering 240, DOI: 10.1007/978-94-007-6738-6_5,
© Springer Science+Business Media Dordrecht(Outside the USA) 2013

Keywords Monitoring · Distribution environmental factors · Agricultural products

1 Introduction

Due to the rapid globalization of the world economy and an aggravated competition between countries, to secure competitiveness and differentiate itself in the agricultural and fishery industry field, various efforts have been made. To secure competitiveness in the agricultural and fishery industry fields is being more and more important to the both of nations and related producers because of WTO and FTA systems. Thus, in an effort to secure competitiveness for each country and its producers, the agricultural industry has been promoted as a nation's growing potential industry. Subsequently, each country has invested heavily in agricultural technology. On the other hand, the significance is being emphasized even more as the agricultural and fishery products are rendered to resources or utilized as a scale of nation's competitiveness.

Consequently, for the past couple of years, information and communication technologies in agriculture industry have been received heavy attention as research issues. One of the research issues is sensor network based monitoring system for distribution and cultivation of agricultural products. In this paper, we develop a wireless sensor network (WSN) based monitoring system for agricultural products distribution. The monitoring system gathers physical environment during agricultural products distribution such as temperature and humidity which are main factors to affect the freshness of agricultural products. Subsequently, we analyze the monitoring results how the physical environment affect the commercial value of agricultural products with various packaging techniques [1–5].

We deploy our developed system in a container box that where paprika boxes packaged with some techniques are loaded. Among the exporting agricultural products in Korea, paprika has been the one of the biggest export quantities. In the year 2000, the amount of exported paprika was 2,207 ton, but the amount has increased by 733 % to 16,168 ton in the year 2010. However, in case of the exporting quantity of paprika, the needs of pioneering a new market is on the rise, as the domestic producers and export related industries are greatly influenced by change of the Japanese market as more than 99 % of it is biased to Japan.

Also, although various recent attempts to venture Australian or the U.S. markets have been done, the distribution period and freshness maintenance problems are still at the forefront as a prerequisite. In particular, in terms of the distribution period problem, long period transportation period is required as the transportation period by shipment takes 20 days to the U.S., 25 days to Canada and 23 days to Australian. Thus, long period transportation is impossible since the shelf life of paprika is only 2 weeks.

In order to solve such problems, efforts are continues to secure quality through freshness maintenance, during the long term distribution period, utilizing the research and developed management technology (packaging) after the harvest. Additionally, as efforts to increase export quantity and to construct a stable production supply system, export nations of diversified are being made consistently, and also, there are efforts to secure price in the overseas market (high quality, high price) [6, 7].

In this paper, determination process of the harvest time, chlorine dioxide processing and MA packaging technology have been applied as the packaging technology, and the distribution environmental factors of the distribution process from harvest to sales have been monitored in real time using the sensor network technologies. In other word, a temperature and humidity sensor was installed inside the container during the full-period distribution process of paprika, and change of the environmental factors (temperature, humidity) is measured on the whole of the distribution period. The information on temperature and humidity is measured with 30 min interval, and at each measurement, the location information is measured and stored at the same time. By doing this, what kind of temperature and humidity at which location was measured could be monitored in real-time? Additionally based on the obtained data for distribution environmental factors, through analyzing relationship of the changes for quality and temperature/humidity changes after the distribution, figure out how the changes of temperature/humidity during the distribution process had influenced the paprika quality. Quality management strategies of exporting agricultural products were proposed based on this.

2 Packaging and Sensor Networks Technology

Table 1 shows the post-harvest management and packaging technologies used in paprika transportation process. Two packaging technologies haves been applied to our experiments. The first technology is to sterilize the top and surface parts of paprika with chlorine dioxide for 30 min with 0.1 ppm. The second technique is to maintain the internal humidity of the box to an adequate level using the internal packaging materials such as MA packaging. The last one applies the both technologies together. As shown in Table 1, for the agricultural products used in the process, the harvest time of paprika are controlled using color chart.

For the sensor network technologies, temperature and humidity sensors and the communication hub that transmits the data collected by the sensors to remote server is used for real time monitoring the distribution environmental factors during the export.

400 MHz frequency bandwidth is used for the sensor nodes, and through the super low power technology, they are designed to sufficiently operate during the transportation period to Australia. Additionally, in case of the sensor nodes, they are designed and realized classified into fixed and box feeding types. The fixing

Table 1 Packaging technologies

Applied technologies	Contents	Remark
Determine harvest period (color chart)	Adjust harvest period using a color chart Yellow: For export 3, for domestic demand 4– 5 Red: For export 4, for domestic demand 5	
Chlorine dioxide processing	Sterilize paprika's top and surface parts Processing condition: 0.1 ppm, 30 min	
MA packaging	Maintain adequate humidity within the box using the internal packaging materials External packaging materials: 4 air drains Internal packaging materials: 0.03 mm PE film, Vent ratio: 1.5– 2.0 %	

Case Studies on Distribution Environmental Monitoring

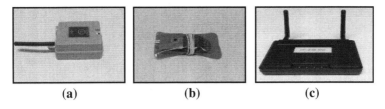

(a)　　　　　　　　(b)　　　　　　　　(c)

Fig. 1 Sensor nodes and communication hub. **a**, **b** Node. **c** Communication hub

type is designed to use by fixing into storage and the box feeding type is designed suing soft plastic bags to minimize the impact to the quality of agricultural products such as paprika. Fig. 1 shows the sensor nodes (fixing type, box feeding type) and the communication hub used in the experiment.

In order to transmit temperature and humidity data to the remote server in real time, the communication hub is equipped with a WCDMA module. However, transmission in open waters is not possible and can transmit the data to the server through automatic roaming when the ship anchors in the port. Also, it is designed to transmit environmental factors simultaneously through a GPS equipped in the communication hub to figure out the location where temperature and humidity data is collected. Through this, it is possible to analyze how the external environments affect to internal environments of the containers.

3 Case Studies

In this paper, the distribution environmental factors (temperature, humidity, location data) have been monitored in real time for the paprika, agricultural products exported to overseas, as the exported object from its domestic origin to Australia passing Busan Port. Figure 2 shows the sensor nodes and the communication hub that are deployed in the container box. Figure 2a shows the container box and Fig. 2b shows the fixing type sensor nodes and the communication hub. The fixing type sensor nodes to acquire the temperature and humidity data inside the container are shown in Fig. 2c and d, they have been installed at the middle height from the container floor. Figure 2e shows the loaded paprika boxes and Figure 2f shows the antenna installed at the outside of the container box to enable data transmission of the communication hub through the mobile telephone network.

Figure 3 illustrates the mapping of the map and the data of the GPS mounted on the communication hub during the real-time monitoring on the distribution environmental factors. The map shows the transfer route of the container box by a transportation means.

Figure 4 shows the graph of the temperature and humidity data collected by the sensor nodes when selecting pre-installed sensor nodes. The figure shows which temperature and humidity has been measured at each location by interlocking with the graph. In the figure, the red at the right side is the humidity, and the blue at the

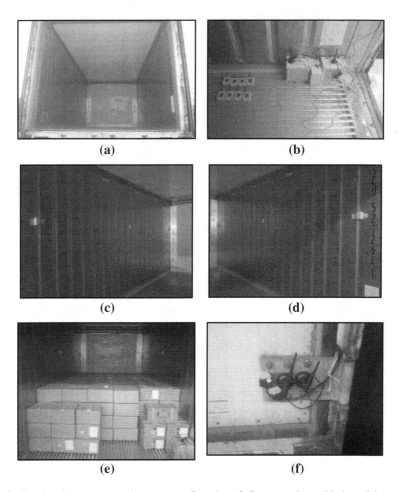

Fig. 2 Deployed sensor networks system. **a** Container. **b** Sensor nodes and hub. **c, d** Deployed sensor nodes. **e** Paprika box. **f** Installed antenna

Fig. 3 Information of GPS

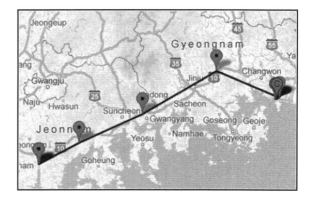

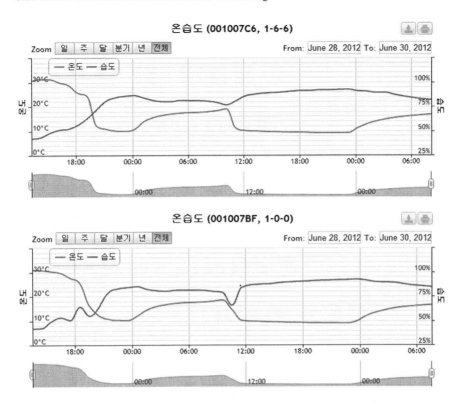

Fig. 4 Information on temperature and humidity

left side is the temperature. As can be known from the picture, it is the result obtained by real-time monitoring of the temperature change during transportation started from the initial loading of room's temperature 30°, and it can be known that it is maintained at the temperature/humidity pre-set by the user as time passed starting from the initial humidity of 30 %. Also, a drastic change of temperature and humidity at a certain period of time is observed and this implies that a random event (container power cut, container door problem etc.) occurs during transportation.

4 Conclusion

From the case studies, taking a real-time action was possible by monitoring various environmental problems likely to occur during the transportation process through applying packaging and sensor network technologies as the advancement technology for agricultural and fisheries products, and securing the commercial value was also possible through quality management.

Through real-time monitoring the distribution environmental factors for high value-added agricultural products, the conclusion that sales routes of paprika as well as agricultural products which require long time transportation could be ventured if the quality and freshness of agricultural products could be adequately maintained and estimated. Furthermore, it is determined to diversify exporting countries, enable stable production and construct of supply systems, and thus can promote profits of farmers and stabilization of market price.

The obtained information could be utilized as various reference data and such a technology is planned to be applied diversely to prosperous exporting agricultural products in the future.

Acknowledgments This research was supported by Technology Development Program for 'Bio- Industry Technology Development', Ministry for Food, Agriculture, Forestry and Fisheries, Republic of Korea.

References

1. Hwang J, Shin C, Yoe H (2010) Study on an agricultural environment monitoring server system using wireless sensor networks. Sensors 10(12):11198–11211
2. Ruiz-Garcia L, Lunadei L, Barreiro P, Robla JI (2009) A review of wireless sensor technologies and applications in agriculture and food industry: state of the art and current trends. Sensors 9(6):4728–4750
3. Akyildiz IF, Su W (2002) A survey on sensor networks. IEEE Commun Mag 40(8):102–114
4. Culler D, Estrin D, Srivastava M (2004) Overview of sensor networks. Computer 37(8):41–49
5. Yoneki E, Bacon J (2005) A survey of wireless sensor network technologies: research trends and middleware's role. Technical Report, University of Cambridge
6. Kwak YS (2010) Design and implementation of sensor node hardware platform based on sensor network environments. J Korea Navig Inst 14(2):227–232
7. Kwak YS (2011) Design and implementation of the control system of automatic spry based on sensor network environments. J Korea Navig Inst 15(1):91–96

Vision Based Approach for Driver Drowsiness Detection Based on 3D Head Orientation

Belhassen Akrout and Walid Mahdi

Abstract The increasing number of accidents is attributed to several factors, among which is the lack of concentration caused by fatigue. The driver drowsiness state can be detected with several ways. Among these methods, we can quote those which analyze the driver eyes or head by video or studying the EEG signal. We present, in this paper an approach which makes it possible to determine the orientation of the driver head to capture the drowsiness state. This approach is based on the estimation of head rotation angles in the three directions yaw, pitch and roll by exploiting only three points face features.

Keywords Driver drowsiness detection · 3D head orientation · Perspective Projection · Haar features · Harris detector

1 Introduction

In literature, many systems based on video analysis have proposed for drowsiness detecting [1]. Special attention is given to the measures related to the speed of eye closure. Indeed, the analysis of the size of the iris that changes its surface according to its state in the video allows the determination of the eye closure [2]. Other work is based on detecting the distance between the upper and the lower eyelids in order to locate eye blinks. This distance decreases if the eyes are closed

---

B. Akrout (✉) · W. Mahdi
Laboratory MIRACL, Institute of Computer Science and Multimedia of Sfax,
Sfax University, Sfax, Tunisia
e-mail: akrout_belhassen@yahoo.fr

W. Mahdi
e-mail: walid.mahdi@isimsf.rnu.tn

and increases when they are open [3]. These so-called single-variable approaches can prevent the driver in case of prolonged eye closure, of its reduced alertness. The second type of approach is called multi-variable [4, 5]. In this context, the maximum speed reached by the eyelid when the eye is closed (velocity) and the amplitude of blinking calculated from the beginning of blink until the maximum blinking are two indications that have been studied by Murray [6]. Takuhiro [7] uses an infrared camera and suggests five levels of vigilance namely non-drowsy, slightly drowsy, sleepy, rather sleepy, very sleepy and asleep. Picot [4] presents a synthesis of different sizes as the duration to 50 %, the PERCLOS 80 %, the frequency of blinking and the velocity amplitude ratio. These variables are calculated every second on a sliding window of the length of 20 s. Some multi-variable approaches require technical cooperation between the hardware and the driver. Moreover, these methods need the use of wide range of parameters, which calls for more data for learning. Other studies estimate orientation of the head driver [8] to detect drowsiness state. These researches are based on the face shape and calculate the local descriptors such as eyes, mouth and nose to estimate head angles rotation. In this paper, we present an approach called geometric, based on the nose tip and mouth corners to determine the angles (Yaw, Pitch and Roll).

2 Proposed Approach for 3D Head Pose Estimation

Our approach requires primordial stages in order to pose estimation. The first step allows detecting driver nose and mouth with Haar features method [9]. Interest points of the face are located by Harris detector [10] to define mouth corners. The center of the box which encompasses the nose is the tip. We can calculate 3D rotation angles for head driver, from these three points.

2.1 Proposed Perspective Model

We present, in this section, a perspective model to estimate the driver head pose. We suppose that the subject is installed in front of fixed and calibrated RBG camera. f is the focal point (Fig. 1). We suppose that Θ, β and φ are the head angle rotations for X, Y and Z axes respectively. We consider that the image plane is parallel to X–Y axis of our subject. Let B_1 and B_2 the two corners points of the mouth and N the tip nose in 3D space. Projections of these last points in the image plane are b_1, b_2 and n respectively.

Fig. 1 Geometric system coordinates for 3D head pose and its projection in the image plane

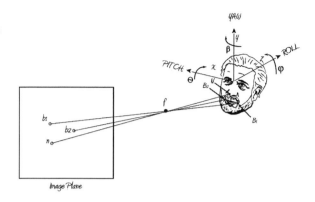

2.2 Roll (φ) Estimation

Let b_1 and b_2 the two corners points of the mouth, O with (X_0, Y_0) coordinates is the center of the segment $[b_1, b_2]$. Let A the distance between O and b_1, H the distance between O and the projection of point b_1 on the horizontal axis which passes by the point O (Fig. 2). The rotation angle φ of head Roll on Z axis is calculated as follows

$$\varphi = \arccos\left(A/H\right) \quad (1)$$

with

$$A = \sqrt{(X_{b1} - X_o)^2 + (Y_{b1} - Y_o)^2} \text{ and } H = |X_{b1} - X_o| \quad (2)$$

Fig. 2 Calculate angle φ starting from the two points of mouth corners

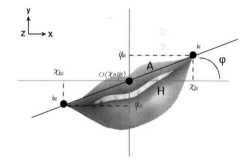

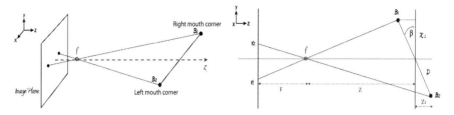

Fig. 3 Representation of mouth corners in 3D space (*top*), their projections in 2D image plane (*bottom*)

2.3 Yaw (β) Estimation

For the estimation of the angle β, we propose a new method based on the perspective projection. Let us pose D is the distance between the center of the mouth and one of two mouth corners in 3D space (Fig. 3). With F is the focal distance from RGB camera. v_1 and v_2 are the two points mouth corners projections in the image plane respectively given by Eqs. 3 and 4.

$$v_1 = \frac{FX_1}{Z - Z_1} = \frac{F \times D \times \cos(\beta)}{Z - D \times \sin(\beta)} \qquad (3)$$

and

$$v_2 = \frac{FX_1}{Z + Z_1} = \frac{F \times D \times \cos(\beta)}{Z + D \times \sin(\beta)} \qquad (4)$$

We can calculate the distance Z according to v_1 and v_2

$$Z = \frac{D \times (F \times \cos(\beta) + v_1 \times \sin(\beta))}{v_1}, \quad Z = \frac{D \times (F \times \cos(\beta) - v_2 \times \sin(\beta))}{v_2} \qquad (5)$$

Insofar Z is not known, Eq. 6 is advantageously replaced by Eq. 6.

$$\frac{D \times v_2 \times (F \times \cos(\beta) + v_1 \times \sin(\beta))}{v_1 \times v_2} + \frac{D \times v_1 \times (v_2 \times \sin(\beta) - F \times \cos(\beta))}{v_1 \times v_2} = 0 \qquad (6)$$

When D is factoring, we obtain

$$\frac{D \times (S + C)}{v_1 \times v_2} = 0 \qquad (7)$$

With S is represented as

$$S = v_2 \times (F \times \cos(\beta) + v_1 \times \sin(\beta)) \qquad (8)$$

While C represents the following equation

$$C = v_1 \times (v_2 \times \sin(\beta) - F \times \cos(\beta)) \tag{9}$$

The distance D between the center of the mouth and one of its corners is strongly different from zero. On the other hand, the denominator of Eq. 7 is also different from zero, we can conclude that

$$S + C = 0 \tag{10}$$

By development of Eq. 10 gives the value of the angle β is calculated finally according to Eq. 11

$$\beta = arctg\left(\frac{F \times (v_2 - v_1)}{-2v_2v_1}\right) \tag{11}$$

Equation 11 shows that we can calculate the yaw head driver without the influence of the distance between the driver and camera. This equation depends only on the classical camera calibration to determine the focal distance calculated only once.

2.4 Pitch (Θ) Estimation

The estimation of angle Θ, depend on the tip of the noise, presented with point N and one of the mouth corners, let us take in our case the point B_1 in 3D space. We suppose K the image of point N relative to (B_1B_2) axis. Let the distance $P = ([NK]/2)$.

The points e_1 and e_2 are the projection of the points N and K in the image plane respectively (Fig. 4). We obtain the following equation

$$e_1 = \frac{FY_1}{Z - Z_2} = \frac{F \times P \times \cos(\theta)}{Z - P \times \sin(\theta)} \quad \text{and} \quad e_2 = \frac{FY_1}{Z + Z_2} = \frac{F \times P \times \cos(\theta)}{Z + P \times \sin(\theta)} \tag{12}$$

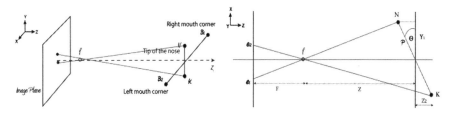

Fig. 4 Representation of the noise tip and its projection relative to (B_1B_2) axis in 3D coordinates (*top*) and their 2D projection in the image plane (*bottom*)

The distance Z is determined as follows

$$Z = \frac{P \times (F \times \cos(\theta) + e_1 \times \sin(\theta))}{e_1} \text{ and } Z = \frac{P \times (F \times \cos(\theta) - e_2 \times \sin(\theta))}{e_2}$$

(13)

Equation 13 give

$$\frac{P \times (Q + W)}{e_1 \times e_2} = 0$$

(14)

then

$$Q = e_2 \times (F \times \cos(\theta) + e_1 \times \sin(\theta))$$

(15)

and

$$W = e_1 \times (e_2 \times \sin(\theta) - F \times \cos(\theta))$$

(16)

Since the distance $P \neq 0$ and the product $(e_1 \times e_2) \neq 0$. It is possible to conclude

$$Q + W = 0$$

(17)

After the development of Eq. 17 the angle Θ is estimated according to the following equation

$$\theta = arctg \left(\frac{F \times (e_2 - e_1)}{-2e_2 e_1} \right)$$

(18)

3 Experimental Study

We describe in this section the experimental study. We used an RBG camera with 640×480 of resolution and 30 fps. In order to evaluate our approach of head orientation, we tested our algorithm on five subjects.

We limit the degree of rotation angles value between $[-35°, 35°]$ for all axes. Our work is not planned for the exact estimate of the 3D pose of installed subject. Indeed, we can tolerate in the error rates in order to know the face orientation. If rotation exceed $[-20°, 20°]$ in the pitch and the yaw and $[-15°, 15°]$ in the roll, we alert the driver of his drowsiness state. We divide these states in four categories: widely right, widely left, widely top and widely bottom. We note that there are states which are not detected. Figure 5 shows an average rate of recall which is equal to 84.2, 73.4, 90 and 85.2 % for four classes widely right, left, top and bottom respectively.

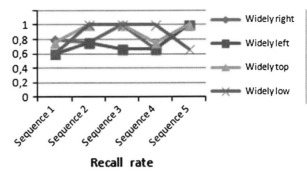

Fig. 5 Curve recall rates for different head orientation

Recall rate

3.1 Conclusion and Future Work

This paper presents a new approach for 3D driver pose estimation. This method makes it possible to determine the states of drowsiness if the driver directs his head in four positions: widely right, left, top and bottom, compared to the vision angle. The choice of the two mouth corners and the nose tip is improved by their visibility in the majority of rotation angles. The average recall of orientation classes is near 85 %. Our method proves a success under various light conditions. On the other hand, it presents limits if one of the feature points is not detected correctly. These errors are explained by the noises or blurs effects in the recorded videos. We propose in our next work to resolve this kind of problem.

References

1. Garcia I, Bronte S, Bergasa LM et al (2012) Vision-based drowsiness detector for real driving conditions. In: IEEE intelligent vehicles symposium, Spain
2. Horng W, Chen C, Chang Y (2004) Driver fatigue detection based on eye tracking and dynamic template matching. In: Proceeding of the IEEE international conference on networking sensing and control, New York, pp 7–12
3. Hongbiao M, Zehong Y, Yixu S, Peifa J (2008) A fast method for monitoring driver fatigue using monocular camera. In: Proceedings of the 11th joint conference on information sciences, China
4. Picot A, Caplier A, Charbonnier S (2009) Comparison between EOG and high frame rate camera for drowsiness detection. In: Proceedings of the IEEE workshop on applications of computer vision, USA
5. Akrout B, Mahdi W, Ben hamadou A (2013) Drowsiness detection based on video analysis approach. In: Proceedings of the 8th international conference on computer vision theory and applications (VISAPP), Spain
6. Murray J, Andrew T, Robert C (2005) A new method for monitoring the drowsiness of drivers. In: Proceedings of the international conference on fatigue management in transportation operations, USA
7. Takuhiro O, Fumiya N, Takashi K (2008) Driver drowsiness detection focused on eyelid behavior. In: Proceedings of the 34th congress on science and technology of Thailand, Thailand

8. Lee JJSJ, Jung HG, Park KR, Kim J (2011) Vision-based method for detecting driver drowsiness and distraction in driver monitoring system. In: Proceedings of the optical engineering
9. Viola P, Jones M (2001) Rapid object detection using a boosted cascade of simple features. In: Proceedings of the computer vision and pattern recognition, USA
10. Harris C, Stephens M (1988) A combined corner and edge detector. In: Proceedings of the 4th Alvey vision conference

Potentiality for Executing Hadoop Map Tasks on GPGPU via JNI

Bongen Gu, Dojin Choi and Yoonsik Kwak

Abstract Hadoop has good features for storing data, task distribution, and locality-aware scheduler. These features make Hadoop suitable to handle Big data. And GPGPU has the powerful computation performance comparable to super-computer. Hadoop tasks running on GPGPU will enhance the throughput and performance dramatically. However the interaction way between Hadoop and GPGPU is required. In this paper, we use JNI to interact between them, and write the experimental Hadoop program with JNI. From the experimental results, we show the potentiality GPGPU-enabled Hadoop via JNI.

Keywords: Hadoop · GPGPU · Map Task · Map/Reduce · CUDA · JNI · Cluster

1 Introduction

Hadoop [1] is suitable to handle Big Data. The reason of using Hadoop is that it has good features for handling Big Data as following: MapReduce programming model, HDFS, data locality-aware scheduler for multiple nodes on cluster. MapReduce is a programming model to simply express tasks which are concurrently executed for handling data, and developed by Google, Inc [2, 3]. HDFS (Hadoop Distributed File System) is a distributed file system on Hadoop cluster. It

---

B. Gu · D. Choi · Y. Kwak (✉)
Department of Computer Engineering, Korea National University of Transportation,
ChungJu-Si 380-702, Chungbuk-Do, South Korea
e-mail: yskwak@ut.ac.kr

B. Gu
e-mail: bggoo@ut.ac.kr

D. Choi
e-mail: mycdj91@gmail.com

J. J. (Jong Hyuk) Park et al. (eds.), *Multimedia and Ubiquitous Engineering*,
Lecture Notes in Electrical Engineering 240, DOI: 10.1007/978-94-007-6738-6_7,
© Springer Science+Business Media Dordrecht(Outside the USA) 2013

partitions files off, and stores each partition on multiple nodes redundantly for fault-tolerant data accessing service. Hadoop scheduler takes into account data locality to efficiently assign tasks to nodes [4], and reassigns abnormally terminated or delayed tasks to other nodes to prevent them from delaying job completion due to the abnormal execution state task. So Hadoop programmer can make his/her code without consideration of data storage, task/data assignment and migration, etc.

GPGPU is used to enhance computing throughput and performance [5]. Graphics Processing Units (GPU) is designed to process a huge number of graphics objects such as points, polygons, etc. To get enough graphics performance, GPU has many processing elements which can operate in parallel manner. These processing elements on recent GPU became to have additional functions suitable to perform general operations, and can be used to perform computation executed by CPU [6]. General Purpose computing on Graphics Processing Units (GPGPU) is using GPU to perform general computation handled by CPU. To efficiently handle a huge number of data in parallel manner, GPU has many processing elements with large number of register.

To make use of the computing power of GPU while MapReduce tasks are executed, there are many researches. Mars [7] is GPGPU-based MapReduce framework. Mars partitions data stored in local disk, and assigns it to threads executed on GPGPUs in parallel manner. This framework can enhance the computing throughput and performance. However it cannot handle data whose size is larger than the available capacity of local disk. And it cannot be operated on multiple nodes. DisMaRC [8] is GPGPU-based MapReduce framework for multiple nodes. However it cannot handle data whose size is larger than the available disk capacity of a node because it does not support distributed file system. And it does not have any mechanism to resolve fault state generated by GPGPU nodes due to hardware failure and data transferring problem, etc.

We think that Hadoop is suitable to resolve the problems previously described. To the best of our knowledge, the meaningful result about GPGPU-enabled Hadoop has not be reported yet. In this paper, we show the potentiality that GPGPU can be used by Hadoop framework. Our approach to show potentiality for executing Hadoop Map tasks on GPGPU is using Java Native Interface (JNI). JNI is the interaction mechanism between Java and other programming language like C. Java is the basic language for Hadoop MapReduce program. However Java is not the suitable language for programming GPGPU until now. Therefore it is necessary to use JNI for interaction between Hadoop MapReduce and GPGPU code.

This paper is organized as follows: Section 2 describes interaction between Hadoop Mapper and GPGPU code, Sect. 3 describes implementations of our approach. Section 4 concludes and describes further studies.

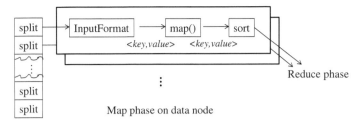

Fig. 1 Simplified data flow of Hadoop map phase

2 Interaction Between Mapper and GPGPU Code via JNI

In HDFS, data file is partitioned into blocks whose default size is 64 MB, and blocks are distributed among nodes. One Map task is scheduled for each block by Hadoop scheduler. To maximize parallelism for handling big data, Hadoop assigns data blocks to all available Map tasks on cluster.

Figure 1 shows the data flow for Map task in Hadoop. Block assigned to a task is called 'split'. Hadoop schedules a Map task executed on data node for handling a split in HDFS. Map task loads split and partitions it into <key,value>-pair records. Each record fetched from split is passed to *map()* method to handle it. *Map()* method makes <key,value> pair as a result. And then, this pair is passed to sort step to sort all <key,value> pairs generated by Map task, and all sorted <key,value> pairs are passed to reduce phase.

Our approach in this paper to use GPGPU for GPGPU-enabled Hadoop is changing the function of *map()* in Map phase. The function of *map()* in original Hadoop handles data expressed as <key,value> pair. However *map()* in our approach transfers data to GPGPU, initiates GPGPU code, fetches results from GPGPU, and converts the result into <key,value> format. Figure 2 shows the simplified data flow of GPGPU-enabled Hadoop Map phase.

Normally Hadoop Map/Reduce task is written in Java. Hadoop also has mechanisms, such as stream and pipeline, to write Map task with other language. But Java is standard language for programming Hadoop task. However C/C++ is currently standard language for GPGPU, and the famous GPGPU programming framework such as CUDA and OpenCL is based on C/C++. Therefore we use JNI for interaction between *map()* and GPGPU code. The JNI enables Java code running in JVM to call, and to be called by, native program written in other languages like C. JNI is normally used to call a hardware and operating system function written in a native language from Java application.

The procedure for executing Hadoop Map task on GPGPU is as following: At first, *map()* is called by Hadoop framework with <key,value> record. *Map()* transfers record to GPGPU via a system bus in host. The record transferred by *map()* is stored in a memory for GPGPU. NVIDIA calls this memory as *global memory*. And then *map()* initiates GPGPU code, and waits until the execution of GPGPU code is complete. GPGPU code processes data stored in its memory, and

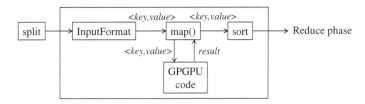

Fig. 2 Simplified data flow of GPGPU-enabled Hadoop map phase

stores the result in the memory. Of course, the GPGPU code is previously prepared by programmer to process the record. When GPGPU code is done, *map()* fetches the result from the memory of GPGPU, and make it *<key,value>* record for the following step such as sort, merge, etc.

Using JNI to execute Hadoop task on GPGPU has the advantages as following: In Hadoop cluster consisted of multiple GPGPU-enabled node, the computation throughput and performance are dramatically enhanced without the consideration about how to store data, how to distribute tasks, and how to recover the faulty task. The feature of the throughput and performance enhancement is due to GPGPU. And the feature of programmability without the consideration about data store, task distribution, and fault tolerant task processing is due to Hadoop. So Hadoop application developers only focus on his/her algorithm to handle record, and get the maximized throughput and performance via GPGPU computing power at the same time.

3 Implementation of Hadoop Map Task on GPGPU via JNI

To show the potentiality for executing Hadoop Map task on GPGPU via JNI, we configure the small Hadoop cluster, and implement very simple experimental Hadoop program. The configuration of Hadoop cluster for our approach in this paper is as Table 1. The Hadoop cluster consists of three nodes, and two GPGPU add-on boards are installed in two data nodes.

Table 1 Configuration of Hadoop cluster to show the potentiality of our approach

The number of nodes: 3	Name node	1 (GPGPU is not installed)	
		CPU	Intel Xeon
		Memory	4G
	Data node	2 (GPGPU is installed in all data node)	
		CPU	Intel Core2Duo/AMD Phenom II x6
		Memory	2 GB/8GM
OS		CentOS 6.3	
GPGPU		Nvidia GeForce GTX 670	
Hadoop		1.0.1	

To write the simple experimental Hadoop program executed on GPGPU, we use CUDA developed by Nvidia. The Compute Unified Device Architecture (CUDA) is a parallel computing platform and programming model. It enables dramatic increases in computing performance by using the computing power of GPU. The experimental program simply adds two numbers in each record. To do this, we create about 110 MB data file which consists of about seventeen million records. The 110 MB data file is partitioned into two splits by HDFS, and assigned two Map tasks.

The execution time of this experimental program is about 110 min even though GPGPUs installed in nodes are up-to-date devices, and each node has the good performance shown in other experiments. It is very long execution time in our cluster configuration. The reason of very long execution time is as follows: As shown in Fig. 2, *map()* is called once for each record in split. And each split averagely has about 8.5 million records. So *map()* is called about 8.5 million times, and for each call repeatedly executes the processing steps: transferring record to GPGPU, initiating GPGPU code, fetching the result. This repetition is very big overhead.

However the result due to the overhead cannot obscure our approach to use JNI for executing Hadoop task on GPGPU. Using JNI enables GPGPU to execute Hadoop Map tasks. And this shows the potentiality for executing Hadoop task on GPGPU via JNI. We think that our result is the first step for implementing GPGPU-enabled Hadoop.

4 Conclusion

Hadoop is known to be suitable platform for handling Big data because it has good features such as simple MapReduce programming model, distributed file system, and data locality-aware scheduling policy, etc. Therefore many researchers study on Hadoop and application fields. And recently many researchers are also interested in GPGPU because it has a powerful computation power comparable to supercomputer.

If Hadoop application can use the computation power of GPGPU, the throughput and performance will be dramatically enhanced. To the best of our knowledge, the GPGPU-enabled Hadoop is not reported yet. Some researchers reported GPGPU-enabled MapReduce frameworks. But they didn't target to Hadoop. To realize the GPGPU-enabled Hadoop, the interaction mechanism between Hadoop tasks and GPGPU code is required.

In this paper, we used the JNI to interact between Hadoop tasks and GPGPU code via JNI. And we experimentally implemented Hadoop Map tasks running on GPGPU. The execution time is very long due to the previous described reasons. However we showed the potentiality for executing Hadoop tasks on GPGPU via JNI.

In the future, we will revise Hadoop framework. In the current version of Hadoop, *map( )* is called for each record in split. This strategy is good for normal Hadoop cluster. But this makes very big overhead in Hadoop cluster with GPGPU. So the additional *map( )* calling strategy is required that *map( )* is called for one split or splits in Hadoop cluster with GPGPU.

Acknowledgments This research was supported by a grant from the Academic Research Program of Chungju National University in 2010. And this research was partially supported by Technology Development Program for 'Bio-Industry Technology Development', Ministry for Food, Agriculture, Forestry and Fisheries, Republic of Korea.

References

1. Tom W (2011) Hadoop: The definitive guide. O'relilly: 1–13
2. Dean J, Ghemawat J (2004) MapReduce: Simplified data processing on large cluster. In: '04: Sixth symposium on operating system design and implements (OSDI '04), SanFrancisco, pp 137–150
3. Jorda P, David C, Yolanda B, Jordi T, Eduard A, Malgorzata S (2010) Performance-Driven task co-scheduling for MapReduce environments. In: IEEE network operations and management symposium (NOMS), pp 373–380
4. Matei Z, Dhruba B, Joydeep SS, Khaled E, Scott S, Ion S (2010) Delay scheduling: a simple technique for achieving locality and fairness in cluster scheduling. In: Proceedings of the 5th European conference on computer systems (EuroSys' 10), New York, pp 265–278
5. GPGPU http://en.wikipedia.org/wiki/GPGPU
6. Cayrel PL, Gerhard H, Michael S (2011) GPU implementation of the Keccak Hash function family. IJSA 5:123–132
7. He B, Fang W, Govindaraiu N, Luo Q, Yang T (2008) Mars: a MapReduce framework on graphics processors. In: PACT '08: Proceedings of the 17th international conference on Parallel architectures and compilation techniques, New York, pp 260–269
8. Mooley A, Murthy K, Singh H (2008) DisMaRC: A distributed map reduce framework on CUDA.TechRep, The University of Texas, Austin, pp 65–66

An Adaptive Intelligent Recommendation Scheme for Smart Learning Contents Management Systems

Do-Eun Cho, Sang-Soo Yeo and Si Jung Kim

Abstract This study aims to provide personalized contents recommendation services depending on a learner's learning stage and learning level in the learning management system using open courses. The intelligent recommendation system proposed in this study selects similar neighboring groups by performing user-based collaborative filtering process and recommends phased learning contents by using prior knowledge information between contents and considering the relevance and levels of learning contents. The proposed learning contents recommendation is applied flexibly according to a user's learning situation and situation-specific contents recommendation link is created by performing the intelligent learning process of recommendation system. This service allows a variety of industrial classification learners using open course to effectively choose more accurate curriculum.

Keywords Personalization · Recommendation system · Collaborative filtering · E-learning · Learner's preference

---

D.-E. Cho
Innovation Center for Engineering Education, Mokwon University, Daejeon, Korea
e-mail: decho@mokwon.ac.kr

S.-S. Yeo
Division of Computer Engineering, Mokwon University, Daejeon, Korea
e-mail: sangsooyeo@gmail.com

S. J. Kim (✉)
Center for Teaching and Learning, Hannam University, Daejeon, Korea
e-mail: sjkim6183@gmail.com

J. J. (Jong Hyuk) Park et al. (eds.), *Multimedia and Ubiquitous Engineering*,
Lecture Notes in Electrical Engineering 240, DOI: 10.1007/978-94-007-6738-6_8,
© Springer Science+Business Media Dordrecht(Outside the USA) 2013

1 Introduction

Recently entering the era of lifelong learning, E-learning, which is the new education paradigm, provides a variety of learning contents. The advantages of E-learning are to promote learner-centered education and enable customized education for individual learners. Therefore, a variety of lifelong learning programs or learning contents using it are provided. Generally, most of E-learning is conducted based on web and learning is carried out by choosing materials posted by instructors and materials posted by learners. In these circumstances, it is very hard for learners themselves to select necessary matters from a variety of learning contents and determine the learning process. Therefore, personalization strategy is needed in order for learners to obtain academic efficiency and learning effect [1–3]. For this personalization strategy, the recommendation system identifying learners' learning objectives and automatically filtering differentiated information by individual is required [4].

As representative recommendation research classification, there is a Content-based recommendation classification method by considering a user's previous preferences first and recommending first items. The content-based classification method is the method using the fact that the preferences of the past are highly likely to choose the future. In addition, there are other methods such as Demographic-based recommendation method recommending items by referring to the use form of learners showing similar patterns with using demographic information, Rule-based recommendation method which is the recommendation method according to several rules with existing data and Collaborative filtering recommendation method using approach value of groups with similar contents access data [5]. Recently, these recommendation methods are used variously for movies, music, video and other services.

This paper attempts to present contents recommendation services for providing personalized contents depending on the learning step and learning level of a learner in the learning management system using open courses. The method proposed in this paper is to gather individual learning information first based on learning information performed by learners in the learning management system. And then, it recommends learning contents deemed to be best suited for learners by using prior knowledge information between contents and considering the relevance and levels of learning contents. This paper is organized as follows. First, Sect. 2 learns about the existing recommendation method of the recommendation system and E-learning system and Sect. 3 describes the proposed intelligent recommendation system and service model. Section 4 makes conclusions.

2 Related Work

2.1 Recommendation System

2.1.1 Content-Based Recommendation Schemes

Content-based recommendation is based on information retrieval and recommendations are made by comparing the user profile and the contents to improve performance. Information about the user's tastes, preferences, need is included in the user profile. Profile information can be obtained in the explicit way by asking questions to the user or in the implicit way by observing the user's behaviors. The content-based recommendation has the characteristics that the user's attention on specific areas can be reflected and recommendation is available when new areas of interest occur.

However, it is difficult to independently use it in multimedia information such as music, photos, pictures which are hard to define the characteristics of contents and the user's potential interests cannot be indicated by solely relying on the user profile. Also, in order to improve the accuracy of recommendations, it is important to accurately extract the characteristics of contents well reflecting users' intention. Therefore, sufficient prior information about a user such as contents preferred by a user in the past, feedback etc. is required.

2.1.2 Collaborative Filtering Recommendation Schemes

Collaborative filtering recommendation methods used in the recommendation system are classified into user-based collaborative filtering method and item-based collaborative filtering method [6]. The user-based collaborative filtering method is the method to recommend contents that a particular user may prefer based on contents evaluated by other users with similar preferences by measuring the similarity between users. The techniques to select neighbors with similar preferences and any particular user based on the association between users include clustering, best N-neighbor, Bayesian network etc. The item-based collaborative filtering method is the method to recommend by predicting which items a specific user prefers by measuring the similarity, that is, similarity between existing items that a user entered preferences and items to be recommended. If using the collaborative filtering method, when sufficient preference information of users showing a similar tendency, contents can be recommended actively to those who accessed to the system for the first time. Eventually, in case of the collaborative recommendation method, if the number of users who have the similar preferences with them is less, the selection probability of the recommended list is lowered. Also, the disadvantage is that the evaluation on specific contents is not made, the system cannot be applied.

2.2 E-Learning Recommendation System

The methods of E-learning recommendation system currently ongoing include the recommendation system using contents-based collaborative filtering, recommendation system using user based collaborative filtering and automatic recommendation system using hybrid filtering [7, 8]. The recommendation system using contents-based collaborative filtering is the method used in currently active online education sites and is the method to recommend the courses of similar themes based on the courses that learners have taken in the past. Like this, if other courses similar to those that a learner has taken are uploaded newly, recommended courses can be offered easily. However, in case of a new learner, recommendation is impossible because any information does not exist indicating in which course he/ she is interested. First, the recommendation system using user-based collaborative filtering calculates similarity with neighbor learners with the same idea by using the Pearson correlation coefficient based on what a learner evaluated after listening to a course. At this time, courses are recommended by extracting the list that the learner did not take from the courses taken by neighbor learners with high similarity value. Like this, if recommended by similar neighbor learners, the recommendation of unnecessary courses is reduced so the reliability of learners can be improved. However, its disadvantage is that if a learner's learning activities are not active, it is not easy to configure neighbor learners and therefore, correct recommendation is difficult. Hybrid recommendation system creates the learner profile by using the log of learners and identifies neighbor learners with similar interests through collaborative filtering. And it recommends a new list by text-mining courses and applying content-based filtering method to created contents profile. By mixing the list recommended through these two filtration methods, it finally recommends top N lists to learners.

3 Intelligent Learning Content Recommendation System Design and Service Model

Intelligent learning recommendation system proposed in this study performs user-based collaborative filtering process and selects similar neighboring groups and then recommends phased contents according to levels by considering prior knowledge information between contents. Learning content recommendation is applied flexibly depending on the learner's learning situation and creates context-sensitive contents recommendation links by performing intelligent learning process of the recommendation system.

3.1 System Structure for Intelligent Learning Content Recommendation

In the recommendation system proposed in this study, learners are classified into learners taking learning and pre-learners who already took learning. The manager enters the contents profile for contents registered in the system.

The contents profile includes learning difficulty and prior knowledge information. Learning contents are saved in knowledge save location in the form of complex knowledge considering basic information as well as information occurring during learning etc. Through information gathering handler, information entered extracts contents use frequency and contents preference information of similar learners. By using values generated in information gathering handler, the candidate content recommendation engine provides the recommendation list by group selected by learners. A candidate content recommendation list is finally recommended to learners through weighting and ranking. The following Fig. 1 shows the overall configuration of proposed intelligent learning-based content recommendation system.

3.2 Information Creation and DB Configuration

When using the system for the first time, a learner must enter learner's personal profile information such as his/her interest parts, log-in information etc. An manager performs the contents registration process for users who set content

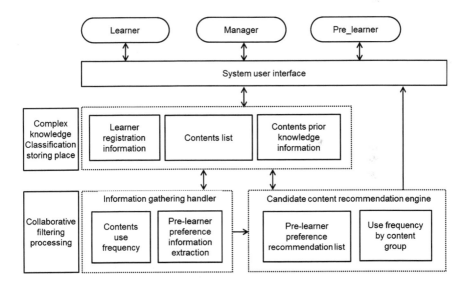

Fig. 1 Recommendation system configuration

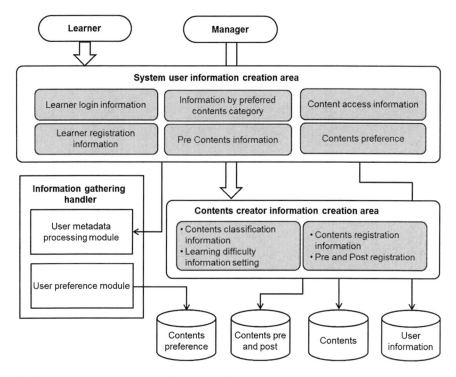

Fig. 2 Recommendation system DB configuration

provision. In this process, the manager classifies the contents into by category and each content is saved in content DB by setting the difficulty rating (high, medium, low). Also, the manager creates pre/post course links of all contents by specifying pre-course for learning contents by the course. This information is saved in pre/post DB. And learners who completed learning and are classified into pre-learners perform the process of entering course information on each learning content and information on content satisfaction after learning. This information is used for creating a recommendation list corresponding to learning content category of next learners. The following Fig. 2 is the configuration on information creation of each DB and configuration items of the proposed system. Pre contents information receives values entered when registering learning contents based on association degree between learning topics and learning contents of learners.

3.3 Candidate Content Recommendation Engine

Contents attribute data on initial data are collected through information gathering handler and analyzed and the value of created metadata is passed to the candidate content recommendation engine. Based on prior knowledge information of

An Adaptive Intelligent Recommendation Scheme

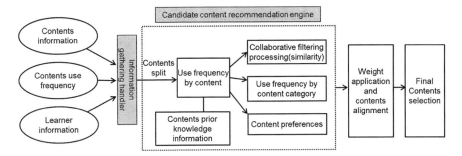

Fig. 3 Configuration of candidate content recommendation engine

learning level information and learning contents of a learner, the candidate content recommendation engine derives best n-neighborhood and derives recommendation lists by calculating learners and similarity. Also, by applying weight according to contents registration time, it recommends recently registered contents first. Figure 3 shows candidate contents recommendation method, candidate contents recommendation process and necessary elements.

The similarity of learners is found by calculating Pearson correlation coefficient [9]. And by using the evaluation value and similarity within Best n-neighborhood, the evaluation value of the contents that learners did not learn is predicted. At this time, by applying the mean values and similarity of the learning results of each learner as weight, evaluation predictive value for items of the learners should be calculated [10].

The calculated candidate contents list performs the functions of applying weight to select contents best meeting learners' preferences and selecting final recommendation contents through ranking. For weight, w, which is the weight value according to contents registration time, is applied. For weight, the method of subtracting 0.1 depending on year is used. Recently registered contents will have 1 and older contents 0.1. As Top-N technique, it creates and provides top N recommendation lists and learners study at least one learning content.

3.4 Content Recommendation Service Model

In order to gain access to the system, each learner basically performs the registration process creating his/her registration information. The system provides services by classifying accessors into learners and contents mangers. A leaner enters basic information at the initial access and searches lists of 1st contents category of learning process that he/she wants. After checking the results, he/she searches contents lists with corresponding difficulty and receives the results. And then, to search for recommendation lists, the system determines whether there are students who take the contents and if so, creates recommendation lists by using

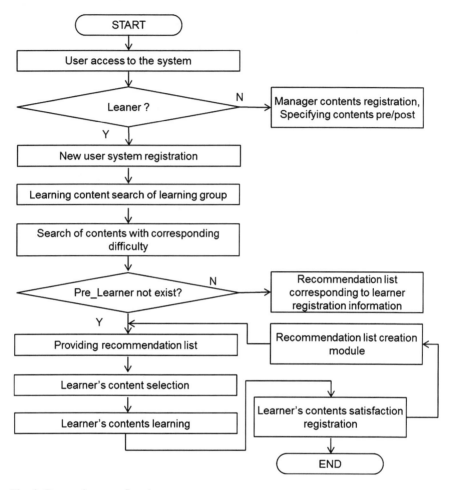

Fig. 4 Proposed system flowchart

input data for pre-learners and if there are no pre-learners, it provides recommendation lists based on contents registration information and learner basic registration information. Figure 4 shows the flow of the system service provision.

4 Conclusion

This study aims to provide personalized contents recommendation services depending on a learner's learning stage and learning level in the learning management system using open courses. Intelligent learning recommendation system proposed in this study performs user-based collaborative filtering process and selects similar neighboring groups and then recommends phased contents according to levels by considering prior knowledge information between contents. Also, by applying weight according to contents registration time, it recommends latest contents first. Also, even when a learner accesses for the first time, the contents are recommended by each interest part category by using user profile information provided at the beginning of registration. In the system, learners receive next learning contents by using prior knowledge information set by the manager and receive recommendation list top information on the corresponding contents category. Currently, a variety of contents are provided in many learning management systems but a lot of content information is required in order for learners to select contents appropriate for them. Also, even if a lot of information is provided, it is not easy for learners themselves to select contents appropriate for learning progress. The existing various recommendation systems use the method of recommending based on how many students chose different contents or simply based on learners' marks.

This study provides recommendation lists appropriate for the personal environment by using a variety of complex knowledge such as prior knowledge information of learning contents, selection frequency of pre-learners and learner preferences without learners' contents selection information in the open course learning management system of various learning areas commonly utilized in the lifelong learning environment. System implementation of contents recommendation service and performance and use assessment of models proposed as future research project will be carried out.

Acknowledgments This research was supported by Basic Science Research Program through the National Research Foundation of Korea (NRF) funded by the Ministry of Education, Science and Technology (2011-0014394).

References

1. Cho D-E, Kim S-J, Kwak Y (2011) A study of personalized contents recommendation method based on user preference learning. J Korean Inst Inf Technol 9(9):229–235
2. Schafer et al (1999) Recommender system in E-Commerce. In: Proceedings of the ACM E-Commerce 1999 conference
3. Arazy O, Kumar N, Shapira B (2010) A theory-driven design framework for social recommender systems. J Assoc Inf Syst 11(9):455–490
4. Liang T, Lai H, Ku Y (2007) Personalized content recommendation and user satisfaction: theoretical synthesis and empirical findings. J Manag Inf Syst 23(3):45–70
5. Burke R (2002) Hybrid recommender systems: survey and experiments. User Model User Adapt Interact 12(4):331–370
6. Konstan J, Miller B, Maltz D, Herlocker J, Gordon K, Riedl J (1997) GroupLens: applying collaborative filtering to usenet news. Commun ACM 40(3):77–87
7. Kang Y-J, Sun C-Y, Park K-S (2010) A Study of IPTV-VOD program recommendation system using hybrid filtering. J Institute Electron Eng Korea 47(4):9–19
8. Inay Ha, Song G-S, Kim H-N et al (2009) Collaborative recommendation of online video lectures in e-learning system. J Korean Soc Comput Inf 9(14):87–94

9. Herlocker JJ, Konstan A, Borschers R et al (1999) An algorithmic framework for performing collaborative filtering. In: Proceedings of the 22th ACM SIGIR conference on research and development in information retrieval, pp 230–237
10. Sarwar B, Karypis G, Konstan J et al (2000) Item-based collaborative filtering recommendation algorithm. WWW10, pp 285–295

Part II
Ubiquitous and Pervasive Computing

An Evolutionary Path-Based Analysis of Social Experience Design

Toshihiko Yamakami

Abstract Service engineering is quickly moving forward to social services. Social service engineering is one of the most promising arenas of service engineering in the 2010s. The term social experience represents the analogy of user experience in a social service context. The author proposes an evolutionary path model of social experience design in order to highlight the design principles of social experience design.

1 Introduction

Social interaction is difficult to design, manage and measure. These difficulties prevent social service engineering from being analyzed in a scientific manner.

The term user experience gained visibility in the 1990s as the computing power enabled human-centric affective user interface design with rich-media capabilities. This transition from user interface to user experience provides the basic insight for this research.

The concept of social experience design was proposed in a previous paper by the author. In this paper, the author examines the social experience using the transition paths and changes invoked by each transition.

The concept of social experience design was coined in order to provide an umbrella concept to guide social service designs. In this paper, the author extends the concept of social experience design using a transition view model from user interface to social experience.

T. Yamakami (✉)
ACCESS, Software Solution, 1-10-2 Nakase, Mihama-ku, 261-0023 Chiba-shi, JAPAN
http://www.access-company.com/

J. J. (Jong Hyuk) Park et al. (eds.), *Multimedia and Ubiquitous Engineering*,
Lecture Notes in Electrical Engineering 240, DOI: 10.1007/978-94-007-6738-6_9,
© Springer Science+Business Media Dordrecht(Outside the USA) 2013

2 Backgrounds

The aim of this research is to identify the unique characteristics of, and guidelines for social experience design.

The term User Experience Design was coined by Don Norman while he was Vice President of the Advanced Technology Group at Apple Computer in the 1990s. The term User Experience has been impacting user interaction design for two decades with the departure from computer–human interface toward high-level interaction design. He also discussed emotional design and mentioned that emotion is a necessary part of life, affecting how we feel, how we behave and think. He mentioned that usability and pleasure should go hand in hand.

Grudin presented eight challenges for groupware from social dynamics [1]. Social aspects of information technology research focused organizational ones.

The originality of this paper lies in the examination of key factors of social experience design using evolutionary-path-based analysis.

3 Definition and Method

3.1 Definition

The definitions of user interface, user experience, social interface, and social experience, are depicted in Table 1. In these definitions, the social interface is similar to the multi-user interface in this paper. The definition of social experience is coined by the author. Examples of each term are depicted in Table 2.

In the early stages, computers were precious assets and their ability to deal with human interactions was limited. This leads to a design where human beings were limited to following the computer-side restrictions. The drop in computer hardware prices and the increase of computing power brought an increase in the human-

Table 1 Definitions

Term	Description
User interface	Design of human–machine interaction where interaction between humans and machines takes place. It aims at effective operation and control of the machine with usable feedback from the machine
User experience	Design of how a person feels about using a product, system or service. It highlights valuable aspects of human–computer interaction and product ownership
Social interface	User-computer interface that deals with human–human interactions. User-computer interface that deals with Multi-user interactions. (This is the definition used in this paper. Social interface may represent human-like computer interface in other contexts)
Social experience	Design of the way a person feels about other humans through computer-user interface

Table 2 Examples of each term

Term	Examples
User interface	Artifacts to provide interfaces to a computer, network, or system. Artifacts that consist of each modality computer–human interaction. Menus, icons, command sequences, command parameters, and so on
User experience	Total experience aspect that governs multiple aspects of user interface. Holistic aspect of space-dimensional and time-dimensional integration of multiple components of user interface. For example, creating architecture or interaction models that affect the user's perception of computer, device or system. It deals with the improvement of perception of total systems in order to satisfy both technical needs and business needs
Social interface	User interface that deals with multi-user factors. User interface to deal with roles, role-taking, conflict-resolution, collective culture, social awareness, and so on
Social experience	Total experience aspects that deal with different social roles with a single user interface

Fig. 1 Shift of key concepts of design through transitions

interaction capabilities of computers. This provided a challenge to the legacy concept of user interface with the implication that humans should have to follow computer ways. The word "user experience" was coined to provide the best experience for users in terms of human–computer interaction.

Improved network capabilities brought opportunities for multi-user interactions. Multi-user interactions were encumbered with conflicts between human and computers, as well as conflicts among humans. Early multi-user interfaces needed to address control arbitration and other exclusive control matters.

Further advances of the Internet brought the new infrastructure of world-scale real-time human interactions. With this transition, we have to re-focus on the importance of user experience in social contexts. The shifts in key concepts of design through transitions are shown in Fig. 1.

3.2 Method

The research method is as follows:

- identify transition paths towards social experience,
- for each path, the characteristics of transition are examined,
- using the transition semantics, the key aspects of social experience are parsed.

4 Evolutionary Path-Based Analysis

An evolutionary path model of social experience design is depicted in Fig. 2.

The analysis of Path 1 is depicted in Table 3.

From this analysis, the transition from user interface to social interface takes place where the entity to be designed accepts simultaneous operations from multiple users. Then, the transition from social interface to social experience takes place where social emotion is invoked or where social relationship is built up over a span of time.

From this consideration, the transition to the final social experience takes place where a time-dimensional long-term approach is taken or where social semantics such as social emotion is taken in interaction models.

The important factors in social experience in path 1 are depicted in Table 4.

A typical example of this transition is the structure of knowledge-sharing as depicted in Fig. 3.

There are two layers in the knowledge sharing structure. One is an inner core layer, where core members exchange their expert knowledge. In this layer, information sharing is bidirectional. Experts actively engage in sharing the knowledge of other experts.

The other is an outer follower layer, where follower members connect to a core member. A follower member actively makes use of the knowledge of an expert. In this layer, information mainly flows from an expert to followers.

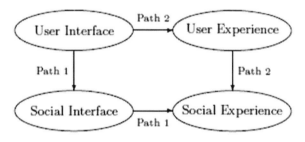

Fig. 2 Design path model of social experience

Table 3 Path 1 analysis

Path	Changes
From user interface to social interface	Architecture shifts toward for multi-user interface, interpersonal interaction, and mutual exclusion. Interaction model deals with multi-use interference, synchronization, arbitration, role-taking and conflict resolution
From social interface to social experience	Design deals with socially-leveraged experience. It deals with satisfaction of role distribution. Simultaneous satisfaction with different roles. It aims at building collective experience with social causes. It covers collective cultural factors in satisfaction. It deals with collective satisfaction with a variety of skills and experience

Table 4 Important social experience factors in path 1

Factor	Description
Time dimensional factors	Long-term relationships. High-level social roles
Fits with business goals	Satisfaction with overall experience. Maintaining high level satisfaction in the social contexts
Integrating social interface into positive social experience	Creating positive social experience with social rewards with extending the social interface

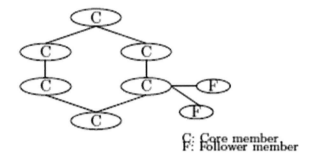

Fig. 3 Knowledge sharing structure

C: Core member
F: Follower member

Generally, knowledge sharing system do not pay attention to the values of information depending upon which layer a user belongs to. Social experience design deals with this social structure which stands for a relatively long-term.

The path 2 analysis is depicted in Table 5.

In this path, the transition from user interface to user experience takes place where high level requirements of user satisfaction or business goals are created.

Then, the transition from user experience to social experience takes place where the user satisfaction or business goals are tightly bound to multi-user interaction.

The important factors in social experience in path 2 are depicted in Table 6.

A typical example of this takes place in the beginner-veteran collaboration in a social game as depicted in Fig. 4.

In a mobile social game, veterans have knowledge, experience, have accumulated game points, and have paid premium items. It is difficult for a beginner to match these veterans. Mobile social game design has to deal with generating

Table 5 Path 2 analysis

Path	Changes
From user interface to user experience	Architecture and interaction models to deal with higher level satisfaction and affective aspects rather than individual aspects of computer–human interaction
From user experience to social experience	Experience shifts from computer–human interaction to interpersonal interaction. Satisfaction has origins in social interactions such as support, gifts, thanks, greetings, and so on. It deals with collective experience of achievement and shared excitement. It also deals with bidirectional interactions, such as reciprocity and mutual education

Table 6 Important social experience factors in path 2

Factor	Description
Satisfaction portfolio	Different satisfaction for different types of users in a shared context
Emotion engineering	Engineering of social-context-based positive emotions. Creating rewarding social experiences (group achievement, collaboration, reciprocal support, being acknowledged as a member, greetings, and so on)
Improving user experience in social contexts	Upgrading user experience into socially-positive experience
Short-term time management	Human beings are asynchronous except in cases of strict time-keeping and real-time meeting. This asynchronicity makes people gradually accept asynchronous triggers

satisfaction among users with multiple skill levels. When we examine the relationship between helping and being helped in battles in mobile social games, it becomes clear that each battle brings different kinds of satisfaction and affective factors. A beginner receives support from a veteran. One feels like a princess, receiving protection and services from a guardian. A veteran helps a beginner, with the feeling of knight. One also exercises skills and premium items, as feeling like a "knight in shining armor." The same game scene serves as different affective aspects depending on user experience and skills. It is a typical example of social experience design.

The term experience is broad. In order to further examine the best practices of social experience design, it is necessary to parse multiple layers of social experience design. The experience can be used in the total perception of long-term use of social systems, or in the individual socially-leveraged emotion. The detailed analysis of this spectrum remains for further research.

Considering the above analysis, the author presents the design components of social experience design, as depicted in Table 7.

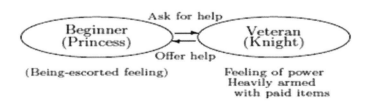

Fig. 4 Beginner-veteran collaboration in a social game

Table 7 Components of social experience design

Time-scale	Component	Description
Long-term	Total experience	Total design of each interface with integrated delivery of social experience
	Culture	Design fits with social interaction with culture of corresponding group
Mid-term	Design of mid-term social experience	Creating social experience over a span of time. Experience utilizing reciprocity principle
	Awareness experience	Creating socially-leveraged awareness, we-feeling
Short-term	Emotion design	Creating positive social experience, greetings, thanks, gifts, social acknowledgement, feeling as a member, social achievement, and so on
	Shared emotion	Success of group action, social achievement

5 Discussion

5.1 Advantages of the Proposed Approach

Social experience design is a natural extension of user experience design in the domain of social service engineering. Social service engineering increases in importance according to the increasing stay time of people in social services. The natural extension from single-user to multi-user is not easy because it involves multi-faceted challenges that take a long time to identify and resolve.

There are three approaches that identify the unique characteristics of social experience design as depicted in Table 8.

The author takes the transition analysis approach. This approach focuses on what takes place at the time of transition. The author proposes an evolutionary path model where two evolutionary paths toward social experience design are identified.

Table 8 Approaches that identify the uniqueness of social experience design

Approach	Description
Model-based approach	Highlighting the high level aspects of social factors, roles, social task, role-taking, conflicts, coordination, collective culture, and so on
Bottom-up approach	Collect examples of social experience in the socially-connected system examples
Transition analysis	Observe transitions from single user experience to multi-user experience from the design perspective to collect the distinguishing factors that separate single-user experience and social experience

5.2 Limitations

This research is a qualitative study. The quantitative measures for identifying multiple aspects of social experience design discussed in this paper remain for further study.

User acceptance of social experience design in the real world environment is beyond the scope of this paper. Quantitative analysis of performance and user satisfaction of social experience design requires future research. The concrete design methodology of social experience design is beyond the scope of this paper.

6 Conclusion

Social service engineering is increasing in importance as the Internet changes its primary role from information access to social interaction. Facebook reached one billion active users this year. This demonstrates that the social aspect of human lives is penetrating into the virtual world.

As Donald Norman conceived the concept of user experience, the increased capabilities of social interaction in the virtual world can lead to a new concept "social experience" coined by the author. This is analogous to how increased power from dealing with rich-media user interfaces led to the concept of user experience with the departure from the concept of the user interface.

Social experience design is different from user experience design with emphasis on satisfaction in social interactions. Social experience design aims at creating different types of user satisfaction depending upon each user's expectation.

There are multiple approaches to social experience design. One is to focus on the social aspects of user experience, such as roles, role-taking, collective culture, conflicts, shared values, and so on. Another is to collect socially-connected design examples of user experience design. Another is to highlight the uniqueness of the social experience design in comparison with other approaches in design.

The author takes an approach that deals with the analysis of the evolutionary paths toward social experience design. The author examines two paths from user interface to social experience. With the examination of these two paths, the author clarifies the unique characteristics of social experience design.

References

1. Grudin J (1994) Groupware and social dynamics: eight challenges for developers. CACM 37(1):92–105

Block IO Request Handling for DRAM-SSD in Linux Systems

Kyungkoo Jun

Abstract This paper proposes a method to improve the performance of DRAM-SSD in Linux systems by modifying the block device driver. Currently, it processes requests in a segment-by-segment way. However it involves overheads because it needs to perform overlapped works repetitively when finishing one segment and starting next one. It prevents DRAM-SSD from running in full speed. The proposed method reduces the overheads by grouping multiple segments into one request, removing unnecessary duplicated steps. But, the grouping also involves overhead. Thus we propose to determine adaptively whether to do grouping or not according to the number of segments contained in requests. From the evaluation results, the throughput of the proposed method improved compared with the segment-by-segment way.

Keywords SSD · DISK IO · Throughput · Block device

1 Introduction

As cloud-based storage services are growing, demands for high performance storage are increasing. However, I/O performance of hard disks is still relatively slower than processors because hard disks depend on mechanical operations. Recently, Solid State Drives (SSD) [1, 2] are widely employed as high performance storages. Because SSD does not have mechanical operations, it is superior in reliability and I/O performance.

SSDs are categorized depending on the type of memory; flash-SSD and DRAM-SSD. Flash-SSD is already widely used in diverse computing devices due

---

K. Jun (✉)
Department of Embedded Systems Engineering, University of Incheon, Incheon, Korea
e-mail: kjun@incheon.ac.kr

J. J. (Jong Hyuk) Park et al. (eds.), *Multimedia and Ubiquitous Engineering*,
Lecture Notes in Electrical Engineering 240, DOI: 10.1007/978-94-007-6738-6_10,
© Springer Science+Business Media Dordrecht(Outside the USA) 2013

to high speed read and low power characteristic. DRAM-SSD inherits all the advantages of flash-SSD. In addition, it is faster in read/write than flash-SSD and more reliable, thus more suitable for handling massive data.

The DRAM-SSD that we consider in this paper is configured as follows. It consists of DRAM modules, DRAM controller, and PCIe [3] controller. The DRAM-controller is FPGA-based and it controls read/write to the DRAM modules, which is performed via DMA.

Figure 1 shows a set of registers that controls DMA data transfer between DRAM-SSD and a host to which DRAM-SSD is installed. DMA start register signals the beginning of transfer, direction register determines whether transfer is read or write, and DMA result register is a flag indicating whether transfer succeeds or fails. The region of DMA transfer is specified by using BCT base address and BCT counter register. One BCT defines one memory area for which DMA transfer performs and usually a set of BCTs are defined for one read or write operation. The BCT base address is the start address from which BCTs are stored consecutively. In detail, a BCT specifies the start address of memory, the start block address of DRAM-SSD, and a length. The addresses are 64-bit long.

Figure 2 shows the procedure of DMA transfer in Linux. Firstly, a set of BCTs are configured according to given read or write request. Then the direction register and the BCT counter register are set. On writing to the DMA start register, DMA transfer starts. And, an interrupt signals the end of the transfer and whether it is completed or not can be found by reading the DMA result register.

Considering such DMA transfer procedure, it should be noted that the number of DMA transfer has more influence on I/O performance than the transfer size. Due to high speed of DRAM, most of time is wasted in configuring BCTs and waiting for the completion interrupt. Therefore, it is obvious to minimize the number of

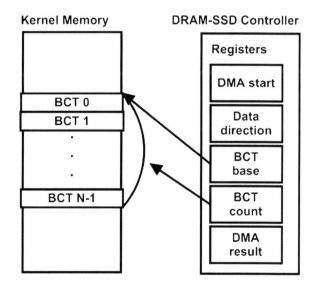

Fig. 1 Registers that control DRAM-SSD

transfers while increasing per-transfer size in order to maximize I/O performance of DRAM-SSD.

However, in reality, Linux operating system which is widely adopted for storage systems takes the opposite action; decreasing per-transfer size while increasing the number of transfers. It is because Linux block layer is designed in a hard-disk oriented way. In the sense of spinning time of hard disks, reduced per-transfer size is more desirable for Linux; it divides one IO requests into multiple segments, resulting in increased number of transfers and decreased size. However, such behavior is not adequate for DRAM-SSD.

Regarding IO performance of SSD, scheduling without queue [4] for reducing scheduling overhead and SSD-oriented IO scheduler [5, 6, 7] are proposed. However, these methods are designed for flash-SSD.

In this paper, we propose a method that adaptively determines transfer size according to the whole IO size in order to optimize the number of transfers. This paper is organized as follows. Section 2 proposes an adaptive method with the explanation about the operation of block device drivers. Section 3 compares the performance of the proposed method with existing methods and Sect. 4 concludes this paper.

2 Adaptive Block Handling considering sizes

When Linux kernel performs the read or write on DRAM-SSD, a request queue and a block device driver are used as shown in Fig. 2. The block device driver processes the requests by fetching them from the request queue in sequence. Requests consist of a set of segments and each segment can specify maximum 4 KB transfer. The number of segment is different depending between the requests.

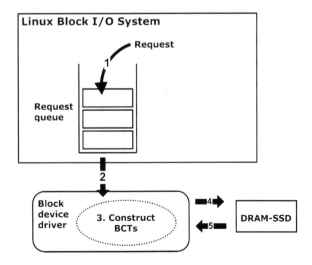

Fig. 2 Procedure to handle IO requests from Linux kernel

The block device driver processes a request segment by segment. Read/write on DRAM-SSD is also handled in this way. Given a request, the block device driver configures a BCT only for a first segment and begins DMA transfer, and then proceeds to a next segment after the completion interrupt. If it is successful, the device driver releases DMA mapping information regarding the transfer. As a result, each segment processing repeats the step 3, 4, and 5. It involves overheads to wait for the interrupt and release the mapping information. Such overheads increase linearly as the transfer size of a request increases because of the limitation of the maximum 4 KB segment.

The overheads can be easily reduced by performing only one DMA transfer for all of the segments, namely request-by-request. However, it requires hardware support. The controllers that we used in this paper can support up to 1024 DMA transfers in sequence at once. Also, it has another type of overheads to save DMA mapping information separately for the segments because the mapping should be freed after transfer completion. The segment-by-segment processing is free from this overhead.

Another way to reduce the overheads is to perform only one DMA transfer for multiple requests, but it is impractical because of the need to modify kernel. Current kernel, in some cases, is not allowed to proceed to next requests until previous requests complete. Another reason is that some requests cannot be combined together, for example, a read and a write.

This paper proposes a method to alternate between the segment-by-segment way and the request-by-request way depending on the number of contained segments. If the number of segments is less than a threshold N, the segment-by-segment is preferable because the overhead of the request-by-request way is larger. On the contrary, if the number is more than N, the request-by-request way is adopted. The request-by-request way requires that the BCTs for each segment be configured in advance which is different from the segment-by-segment. Note that the DMA mapping information should be saved separately to be freed after completion.

3 Performance Evaluation

The performance of the proposed method is evaluated and compared with the case when only the segment-by-segment is used and also with the case of the request-by-request. For the evaluation, we used the system running the Linux kernel version of 2.6.31. A benchmark program IOMeter [8] is used to generate four types of loads; sequential read/write, random read/write. Also the transfer size can vary ranging from 512 byte up to 256 KB. As a performance metric, throughput (MB/s) is measured.

We firstly perform a set of experiment to determine an optimal N. We measure the throughput as we increase the request size step by step from 512 byte. When $N = 2$, its throughput is superior to other cases. Particularly when the request size is 8 KB, the throughput gap between $N = 2$ and $N = 3, 4, 5$ is the largest. Since 8 KB is a multiple of the maximum segment size of 4 KB, it is one of the perfect conditions for the request-by-request processing. The results of 16 KB can be explained in the same way. However, the case of 32 KB shows similar throughput for all N because it is large enough to use the request-by-request for any N. Not only for random ready, but also for the other work loads, similar performance was observed. Since $N = 2$ shows the best performance, the following experiments set $N = 2$.

Figure 3 shows the throughput of the sequential read and the sequential write when the request sizes are large such as 16 KB or larger. The throughput of the segment-based processing is lower than the request-based and the proposed method in all the cases. As the request sizes increases, the number of the included segments in a request also increases. Therefore the segment-based processing incurs more overheads than in the case of smaller sizes of requests. And the increased overheads lower the throughput. On the other hand, the request-based processing shows the similar performance as that of the proposed method. It is because it operates in the same way as the proposed method when the request size is bigger than 4 KB.

Figure 4 shows the throughput of the sequential read and the sequential write when the request sizes are small such as less than 16 KB. Different from the results of Fig. 3, the throughput of the segment-based processing increases as the request sizes increases. It is because its overhead does not increase as the number of the included segments in a request does not increase. However, the throughput when the request size is 8 KB shows differences. It can be explained by the effect of the overheads.

Generally hard disks show different performance between random access and sequential access. However, DRAM-SSD is not affected by access pattern. Therefore, the performance of random access is similar to Figs. 3 and 4. We do not present the results in this paper because of the limitation of space.

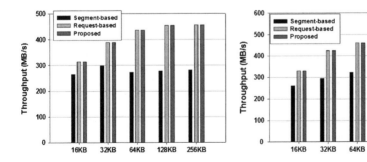

Fig. 3 Throughput of sequential read (*left*) and sequential write (*right*) according to large request sizes

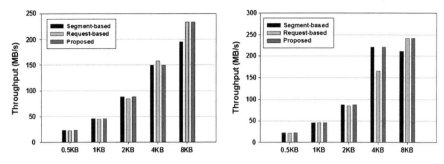

Fig. 4 Throughput of sequential read (*left*) and sequential write (*right*) according to small request sizes

4 Conclusions

This paper proposed a method to improve the throughput of DRAM-SSD by modifying the request handling procedure of Linux block device driver. It adaptively decides whether to use the request-by-request handling or the segment-by-segment according to the number of the contained segments in a request size. The number of the segments increases as the request sizes increases. If more than one segment is included, the request-based handling is more advantageous than the segment-based way. It is because of the overheads concerning the processing of the segments in sequence. On the other hand, if the number of the segments is less than two, the segment-based way is better. The request-based handling has its own overhead. Depending on the number of the segments, our method chooses a proper handling method. We evaluated the performance of our proposed method and observed that it is effective in improving the throughput.

References

1. Takeuchi K (2013) Flash signal processing and NAND/ReRAM SSD. In: Inside solid state drives, vol 37. Springer Series in Advanced Microelectronics, pp 357–374
2. Zambelli C, Olivo P (2013) SSD reliability, In: Inside solid state drives, vol 37. Springer Series in Advanced Microelectronics, pp 203–231
3. PCI Express Base Specification (2010) PCI SIG
4. Seppanen E, OKeefe M, Jilja D (2010) High performance solid state storage under Linux. In: The 26th IEEE symposium on MSST, pp 1–12
5. Hui S, Rui Z, Jin C, Lei L, Fei W, Sheng X (2011) Analysis of the file system and block IO scheduler for SSD in performance and energy consumption. In: 2011 IEEE Asia Pacific services computing conference, pp 48–55
6. Zhang X, Davis K, Jiang S (2012) iTransformer: using SSD to improve disk scheduling for high-performance I/O. In: 2012 IEEE parallel and distributed processing symposium, pp 715–726
7. Kang S, Park H, Yoo C (2011) Performance enhancement of I/O scheduler for solid state devices. In: 2011 IEEE ICCE, pp 31–32
8. http://www.iometer.org

Implementation of the Closed Plant Factory System Based on Crop Growth Model

Myeong-Bae Lee, Taehyung Kim, HongGeun Kim, Nam-Jin Bae, Miran Baek, Chang-Woo Park, Yong-Yun Cho and Chang-Sun Shin

Abstract The paper proposed the Closed Plant Factory System (CPFS) applied the crop growth model. The CPFS monitors climate data in a closed building or room and the actuator's status for control devices, and provides optimized operations for controlling growth environments. The CPFS monitors environmental data, plant growth data and the control devices' status data. This system can analyse the optimal growth environment and the correct control environment. We implemented the system and applied it to a testbed, also confirmed that the CPFS operated real-time monitoring service and controlling service correctly.

Keywords Vertical farm · USN · Growth monitoring · Plant factory

---

M.-B. Lee · T. Kim · H. Kim · N.-J. Bae · M. Baek · C.-W. Park · Y.-Y. Cho
C.-S. Shin (✉)
Department of Information and Communication Engineering, Sunchon National University,
Sunchon, South Korea
e-mail: csshin@sunchon.ac.kr

M.-B. Lee
e-mail: lmb@sunchon.ac.kr

T. Kim
e-mail: taehyung@sunchon.ac.kr

H. Kim
e-mail: khg_david@sunchon.ac.kr

N.-J. Bae
e-mail: bakkepo@sunchon.ac.kr

M. Baek
e-mail: tm904@sunchon.ac.kr

C.-W. Park
e-mail: jwpark@sunchon.ac.kr

Y.-Y. Cho
e-mail: yycho@sunchon.ac.kr

J. J. (Jong Hyuk) Park et al. (eds.), *Multimedia and Ubiquitous Engineering*,
Lecture Notes in Electrical Engineering 240, DOI: 10.1007/978-94-007-6738-6_11,
© Crown Copyright 2013

1 Introduction

Recently, environmental pollution and climate change cause concern for many in terms of a future food production system. Additionally, consumers' change in demand (easier access to cleaner and organic foods) raised the need for local farms. CPFS is created to address such demands and needs [1].

A CPFS is a new farming format that maximizes the production by optimizing light, temperature, humidity, nutrients and moisture, etc. in a controlled environment. It enables the highly optimized environmental control and more accurate estimation of production through active monitoring than existing greenhouses. In short, it is an agricultural IT technology combined with BT technology that seeks to identify the most optimized growth points in areas such as light source technology such as LED, automated manufacturing process, USN and integrated control, etc.

A CPFS artificially controls the growth environment enabling planned faming during anytime of the year, while eliminating external elements such as climate, pollution or geographic limitation, etc. Therefore, a development of an optimized crop growth model is an essential field of study to maximize the benefits of the CPFS. The crop optical growth model will be a basis for the development of research in the field of standardization of crop growth process and quality, and automation control framework for the plant factory [2, 3].

Generally, plant factories can be categorized as plant production factories and vertical plant production factories. The most important task of such facilities is how to monitor and control the growth environment. In particular, to develop a crop growth model requires a continuous monitoring service and accumulated data crop growth cycle.

In this paper, we construct a testbed for the CPFS. Through this, we are able to research a monitoring about a variety of environment elements in the CPFS. Section 2 explains the actualization of the suggested system and results of its performance in Sect. 3; and then draws conclusions and provides additional topics for future studies in Sect. 4.

2 Design of Closed Plant Factory System

2.1 Structure of CPFS

A CPFS can be categorized into; a physical level that consists of controlling devices that sense and adjust the factory environment accordingly and an application that processes gathered data and makes necessary adjustments to the system.

The physical level transmits all obtained data that sensors receive around the facility to the server's middleware which identifies any abnormalities in the data,

Implementation of the Closed Plant Factory System

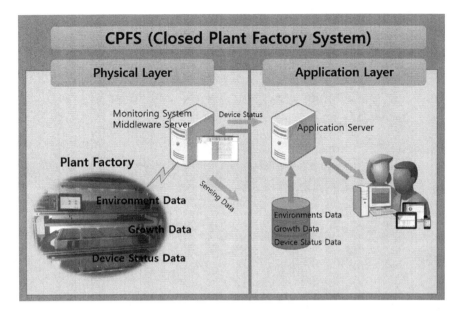

Fig. 1 System structure

then stores them in the database. The application server provides the factory necessary monitoring services so users can have real-time monitoring capabilities through PCs or smart terminals. Figure 1 describes the monitoring system of a CPFS.

2.2 Components of a CPFS

Unlike a typical glasshouse, a CPFS is rarely influenced by external elements therefore eliminating a concern for weather related issues. On the other hand, it must satisfy all necessary conditions that crops require in a controlled environment; therefore a continuous monitoring of internal environment is critical. Such monitoring must involve factory's environment data, a crop's growth data and conditions of controlling devices [4–6]. Table 1 describes each of the elements.

Figure 2 identifies all elements that form the monitoring system. The physical level consists of a climate sensor, an integrated sensor node, a manual controller

Table 1 Monitoring elements

Division	Elements
Environment info	Temperature., humidity, illumination, CO_2
Growth info	Leaf temp., nutrient solution EC, PH
Control device info	Irrigation pumps, LED, heater, fan, humidifier, CO_2 generator

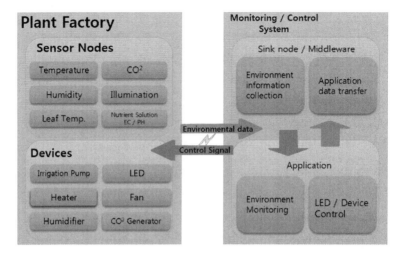

Fig. 2 System integration

that can override equipment's that are stationed within a factory (water pumps, lighting, CO_2 generator, ventilator, humidifier, etc.).

The application level consists of an application server that monitors the factory environment and monitoring software for users.

Figure 3 provides pictures of an integrated controller and sensor node that were used at the physical level. Sensor nodes are equipped with a sensor board that is equipped with a climate sensor and they are stationed around the factory to measure different conditions, they then transmit the data wirelessly. The middleware receives data from the sink node through USB communication and stores them to the database. Then, the stored data become available to users through the application server.

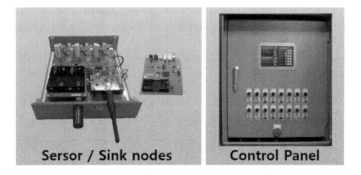

Fig. 3 Sensor node and control panel

Fig. 4 CPFS testbed

3 Realization of CPFS

3.1 Test Settings

In order to confirm the functionality of the suggested system, this study constructed a test bed as described in Fig. 4. A vertical plant production factory that is installed at the Rural Development Administration served as a role model and we modified to fit the purpose of this study. The study also chose leaf lettuce as a test crop because it has a relatively short growth span and less impacted by its environment [7].

The testbed has two parts that are laid side by side and each part consists of four layers. The first part utilized florescent lighting and it was used to germinate seeds, while the second part used red/blue LED light to grow the germinated seeds. Each light source was to be exchanged to another if future tests required doing so. Any data that were obtained from the test bed became available users through a monitoring program, so users can easily analyze data to optimize the factory environment for the crop.

3.2 Plant Factor's Monitoring Program

The factory's supporting software which was used to monitor its environment was developed in three different types; a middleware, an application server and a monitoring software. The middleware not only stores data that sensors gather but also transmits each device's control signals to the integrated controller.

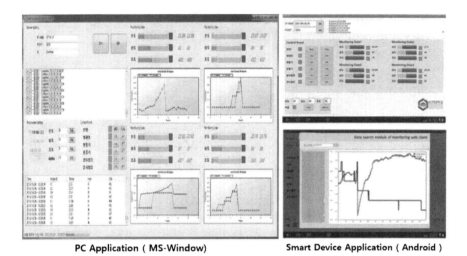

Fig. 5 Monitoring applications

The application server transmits data that the middleware gathers and stores to the client program, in addition to transmitting back to the middleware its decision on appropriateness of control signals.

Data from the factory are sent to users through an application server and users can monitor such data through their PCs or Smartphone applications. Figure 5 shows the monitoring software for PC and Android devices that was specifically developed for this study.

Each type of software shows the real-time state of the factory environment and controlling devices, and users can utilize such information to actively monitor and adjust the environment accordingly to ensure an optimized condition for the crops.

4 Conclusion

This study designed a monitoring system of a CPFS, constructed a test environment while enabling a continuous monitoring and analyzing of the plant growth. This study is to serve as a basis for future studies. Data that were retrieved in this research shall be further reviewed and analyzed in order to identify the most optimized conditions for crops, and to develop an automated controlling system that utilizes the most efficient algorithm. Through this, we will define the optical control set point. In addition, we will design an optimal control algorithm and develop an optimal control system throughout the crop growth cycle.

Acknowledgments This work was supported by the Industrial Strategic technology development program, 10040125, Development of the Integrated Environment Control S/W Platform for Constructing an Urbanized Vertical Farm Funded by the Ministry of Knowledge Economy (MKE, Korea).

This research was supported by Basic Science Research Program through the National Research Foundation of Korea (NRF) funded by the Ministry of Education, Science and Technology (2011-0014742).

References

1. Despommier D (2012) Advantages of the vertical farm. Sustainable environmental design in architecture. Springer optimization and its application 2012. pp 259–275
2. Gim BG, Lee WJ, Heo SY (2010) Construction of a testbed for ubiquitous plant factory monitoring system using artificial lighting. Korea Institute of Information Technology, pp 272–275
3. Lee EJ, Lee KL, Kim HS, Kang BS (2010) Development of agriculture environment monitoring system using integrated sensor module. Korea Contents Soc 10(2):63–71
4. Yiming Z (2007) A design of green house monitoring & control system based on ZigBee wireless sensor network. In: Proceedings of wireless communications, networking and mobile computing. pp 2563–2567
5. Song Y, Ma J, Zhang X, Feng Y (2012) Design of wireless sensor network-based greenhouse environment monitoring and automatic control system. J Network 7(5):838–844
6. Cha MK, Lee SH, Cho YY (2012) Selection of leaf vegetables and set-up of planting density and light intensity in the plant factory. J Asian Agric Biotechnol 28(1):17–23
7. Park DH, Park CY, Cho SE, Park JW (2010) Greenhouse environment monitoring and automatic control system based on dew condensation prevention. In: Proceedings of EMC 2010: embedded and multimedia computing 2010, pp 1–5

Part III
Ubiquitous Networks and Mobile Communications

An Energy Efficient Layer for Event-Based Communications in Web-of-Things Frameworks

Gérôme Bovet and Jean Hennebert

Abstract Leveraging on the Web-of-Things (WoT) allows standardizing the access of things from an application level point of view. The protocols of the Web and especially HTTP are offering new ways to build mashups of things consisting of sensors and actuators. Two communication protocols are now emerging in the WoT domain for event-based data exchang, namely WebSockets and RESTful APIs. In this work, we motivate and demonstrate the use of a hybrid layer able to choose dynamically the most energy efficient protocol.

Keywords Web-of-things · RESTful services · WebSockets

1 Introduction

In the last few years, a vision of inter-connected sensors and actuators attached to physical objects has emerged, leading to the concept of Internet-of-Things (IoT) [14]. This idiom includes the concept of Wireless Sensor Networks (WSN) and goes beyond with all kind of physical objects able to communicate. The field of building automation is a potential target for IoT approaches where numerous communicating sensors and actuators are in use [2]. In such smart-buildings, new communicating objects are also appearing, for example to provide the user with feedback on the energy consumption [7]. The IoT has since then been extended from the IP usage towards the inclusion of well-known Web patterns to ease the

G. Bovet (✉)
LTCI, Telecom ParisTech, Paris, France
e-mail: gerome.bove@telecom-paristech.fr

J. Hennebert
ICT Institute, University of Applied Sciences of Western Switzerland, Fribourg, Switzerland
e-mail: jean.hennebert@hefr.ch

J. J. (Jong Hyuk) Park et al. (eds.), *Multimedia and Ubiquitous Engineering*,
Lecture Notes in Electrical Engineering 240, DOI: 10.1007/978-94-007-6738-6_12,
© Springer Science+Business Media Dordrecht(Outside the USA) 2013

the integration and communication with things at the application level, leading to the concept of Web-of-Things (WoT) [10]. One of the main problems of the IoT is certainly in the management of the energy consumption of this multitude of communicating nodes. Although new low-power standards like 6LoWPAN, IEEE802.15.4 and RPL are being established at the network layer, the WoT framework is actually not energy aware at the application level. We believe that the protocol and data structure used for communicating with things at the highest layers could contribute to a significant reduction of the energy consumption.

In this paper, we show the feasibility of using an additional layer at the application level able to select the most suitable communication method in order to reduce the energy consumption of things connected to the Internet through Wi-Fi. We rely on the *Web-of-Things* paradigm proposing to use WebSockets or RESTful APIs for event-based data exchange. Instead of forcing application developers choosing a communication method, they can rely on an hybrid layer dynamically selecting which method is less energy consuming depending on how much and how frequently data should be sent. This represents a meaningful advantage letting developers focus on other tasks than thinking about costs. Sections 2 and 3 summarize related work and the principles of event-based WoT communications. In Sect. 4, we present our proposal for improving the event-based communication. In Sect. 5, we present the experimental measurements and their analysis. Section 6 provides details on the implementation of our hybrid layer and energy consumption measurements. Section 7 concludes our paper and provides insights on further research.

2 Related Work

The Cooltown project [13] is one of the early projects considering people, places and things as Web resources, using HTTP GET and POST requests for manipulating things. The recent progresses in embedded devices are now enabling the integration of Web servers on things. The tendency is clearly shown with, for example, the WebPlug WoT framework where sensors and actuators used to build so-called mashups [16]. An important step towards a standardization of the communication at the application level for web services was the introduction of the SOAP protocol. However, SOAP is not optimized in terms of energy consumption due to the large overhead of XML and of the protocol itself [8]. Much lighter, RESTful APIs provided a clear answer to this problem, with an increased adoption for many IS, especially in the domain of IoT and WoT [1, 11]. Recently, persistent TCP connections called WebSockets have been proposed for the communication between things [17]. Preliminary comparisons between HTTP and WebSockets in terms of energy consumption have been reported in [3]. This previous research shown differences between these protocols in terms of energy consumption, with complex variations as a function of the payload and frequency of the communication. Motivated by this previous work, the research presented in this paper focuses on the analysis of the optimal choice between RESTful APIs and persistent TCP connections targeting energy efficiency. More specifically, we open the question if rules may be implemented on things for choosing automatically the most efficient way of communicating.

3 WoT Event-Based Communications

Sensor and actuator data can vary in quantity and frequency according to the context of use. For example a power outlet will continuously notify about the electricity consumption when a device is plugged in, while on the other side a presence sensor will only signal a change of state. This kind of behavior is leading to so-called event-based communications. The WoT proposes two fundamentally different approaches for managing event-based communication: HTTP callbacks and persistent TCP connections [9]. Both approaches are detailed below.

Registration. The first step for event-based system is the registration of the consumer at the producer. Using things with REST, we can simply expand the API with a service dedicated to registration [9]. A thing interested in being notified by change of states of another object will announce itself by providing the required callback information. For example, a lamp actuator will register a door contact sensor to be notified when someone enters or leaves the room. The lamp sends a HTTP POST request to http://door.office.home/register. This request can be of two types: (1) REST service—containing a JSON message indicating a REST service as callback, (2) WebSocket—containing the HTTP upgrade header field for switching to WebSocket, keeping open the connection.

HTTP requests. The WoT relies on REST for exposing things as resources to the Web [5]. Unlike SOAP, REST uses HTTP as application protocol for interacting with things and not only as transport protocol. The advantages of REST over SOAP are in having less overhead, and being resource oriented, which fits naturally with physical objects. With WoT, every object is embedding a built-in Web server exposing an API for interacting with its sensing, actuating and configuration capabilities. Self-descriptives URLs are used through common HTTP requests, like GET, PUT, POST and DELETE. For example, reading a sensor value is done using the GET verb and actuating using POST. For event-based communications, POST is actually the only necessary operation. A "consumer" object typically provides a REST service to be notified of changes in another object. The service URL is provided as callback at the registration on the producer. This is a significant aspect of our approach as we can link sensors with actuators.

Persistent TCP connections. The second way of managing event-based communications proposed in the Web-of-Things framework is using persistent TCP connections also known as WebSockets [4]. This kind of communication is mostly used in push scenarios where data has to be sent from a server to a client not running a Web server, as for example Web browsers. The channel is kept open on both sides as long as possible.

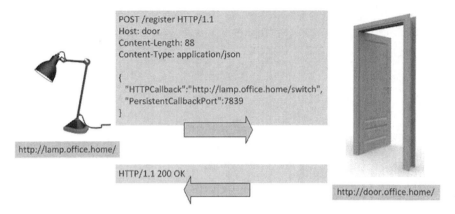

Fig. 1 Example of the registration process

4 Proposal for Energy Efficient Communications

The main idea of our proposal is to let the producer decide the most energy efficient way to communicate, either through REST HTTP or through WeSockets. As it will be shown later in Sect. 5, either mode become optimal as a function of the frequency and payload of the messages exchanged. To enable dynamic switching between modes, we explain here the modifications that are requested. It concerns mainly the registration process and persistent TCP connections concept explained above. In our vision, both modes are supported and therefore, the registration JSON message has to include the available callbacks for REST and persistent TCP connection. The producer will further select automatically which method is best suited for exchanging data from an energy efficiency point of view. This is illustrated in Fig. 1 with an example involving a lamp and a door.

For persistent TCP connections, our proposal slightly differs from what is currently done with WoT. Indeed, WoT approaches suppose that the consumer initiates the persistent connection, keeping the channel open while the producer sends its data. In our approach, the producer has to select between HTTP or TCP and therefore initiates the connection. If the connection is lost due to network faults, the producer will retry to open a connection on the same port, unless the consumer registers with another one.

5 Experimental Measurements and Analysis

HTTP requests and persistent TCP connections have different impact on energy consumption. This is especially true for objects connected to the Internet with a Wi-Fi transceiver. We show here how each method can influence the energy consumption of things.

5.1 Test Environment

We used the openPICUS FLYPORT programmable Wi-Fi module. This tiny module (35 × 48 × 11 mm), is Wi-Fi IEEE 802.11 certified and embeds a full TCP/IP stack, able to connect to IEEE 802.11b/g/n networks. It supports 1 or 2 Mbit/s rates as well as security protocols such as WEP, WPA-PSK and WPA2-PSK. The FLYPORT can be powered either at 5 V or at 3.3 V and drains 128 mA current at 3.3 V when connected to Wi-Fi. An IDE is available for developing application in C [15]. We set up an isolated test environment composed of a FLYPORT module acting as the producer, an access point and a PC acting as the consumer. The wireless network set up is 802.11 g, no encryption and long preamble. We also used a Hameg HM8115-2 for measuring the energy consumption of the FLYPORT [12]. Having a dedicated test bench ensures that no other device will be disturbing the proper running of the experiment as it would be in a public network.

5.2 Power Consumption Measurements

We describe here our measurement campaign for both TCP and HTTP. During each test of 30 s, the producer sent packets with a fixed payload size at a specific interval. For measuring precisely the consequence of each method on the power consumption, the FLYPORT was only running a minimal program sending events. The values of payload and interval are chosen to match the behavior of some specific devices and therefore to perform more realistic measurements. We made the payload size in bytes vary from 1 to 400 and the intervals between packets in milliseconds from 50 to 800, which correspond to certain devices one can find in smart buildings. The combination of the payload sizes and intervals gives us a campaign of 30 measurements as illustrated in Fig. 2.

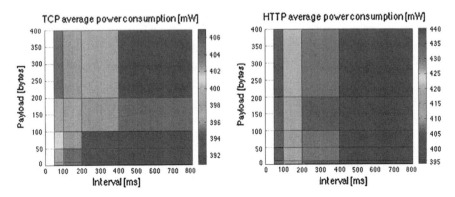

Fig. 2 Results of the average power consumption measurements for TCP and HTTP

From Fig. 2, we see that TCP is overall less energy consuming than HTTP. TCP appears to be on average 4 % less consuming than HTTP with a maximal gain of 9.5 %. The quantity of transmitted packets is indeed lower for TCP than HTTP. With TCP, once the connection is established, only one packet is necessary to send the JSON message. HTTP is more complex as a connection has to be established every time a JSON message must be sent. An HTTP connection includes the potential TCP window negotiation, the HTTP header, the HTTP response, and finally the connection closing. All this overhead causes an increase in consumption. The measurements also show that the amount of payload data plays a less important role in the power consumption. This is especially true for HTTP consumption. On the other hand, a factor influencing the consumption is clearly the sending interval. The main observation is that both modes are overlapping in terms of efficiency, with TCP is becoming less optimal than HTTP in some conditions.

5.3 Consumption Approximation for TCP and HTTP

HTTP. With TCP, the variable is the necessary time needed to send data, including all underlying protocols. With Wi-Fi (802.11 g) frame composition is taken from [19]. The energy consumption for one packet of data can be computed with the following function: E(payload) = {PLCP preamble + (MAC header + IP header + TCP header + payload) * ByteRate} * TransmitPower with IPheader, TCPheader and ByteRate known from [6, 18] and TransmitPower previously measured. When comparing the theoretical values of the approximation to the measurements of the FLYPORT in Fig. 2, it comes out this function is accurate enough with an average error of 0.86 %.

TCP. As explained earlier, the HTTP case is more complicated as for TCP. Instead of using a theoretical model, we opted for a parametric model where the parameters are fit to the observations. We converged to an exponential function, approximated as in P(interval) = a*exp(b*interval) + c*exp(d*interval). Through a numerical fitting algorithm, we computed the parameters a, b, c and d for every case of payload (1, 10, 50, 100, 200 and 400), ending up with 6 functions for the different payload sizes. The computed parameters allow the functions to be quite precise with an average error of 0.05 %.

6 Implementation and Evaluation of the Hybrid Layer

We had first to develop a **REST server** library in C for the FLYPORT. The services are registered by indicating a URL scheme corresponding to the Web service, and providing a pointer to a callback function that will be called when the server receives a request for this particular service. We then implemented the hybrid layer in charge of dynamically choosing the appropriate method between TCP and HTTP when sending events to registered consumers. We first implemented a history structure for recording the past events sent to consumers. An instance of this structure is created for every consumer registered. This allows computing the energy consumed to send the previous events. According to this result, the layer then switches to the most efficient method and this every time a new event must be sent. For computing the TCP mode energy consumption, we implemented the function described in Sect. 5.3. The implementation includes a rule for intervals higher than 10 s to consider the keep-alive packets (specific to the FLYPORT as it may differ on other modules). The final value is computed as follows: *energy of each packet sent in history + energy at idle between the shipments + energy of keep-alive packets*. For HTTP, we implemented the function as in Sect. 5.4. Using the history, we know the interval and the average payload. Those values are then used as parameters for our approximation function. Linear interpolations are used in the case of payload different as our reference values (1, 10, 50, 100, 200 or 400). The obtained power value is then converted in energy by knowing the time duration of the history.

Table 1 shows the energy measurements of our hybrid layer where some relevant saves were achieved. For comparison purposes, we had to rerun the campaign for each TCP and HTTP modes as our REST server running on the module is also consuming some energy. The column *Gain* shows the percentage of energy saved relative to the highest value between TCP and HTTP. The column *Loss* shows the percentage of energy lost relative to the lowest value between TCP and HTTP. The negative values in the *Gain* can be explained by the consumption due to the hybrid layer. Nevertheless, our hybrid layer clearly shows its usefulness

Table 1 Power consumption comparison between TCP, HTTP and the hybrid layer

Payload (bytes)	Interval (ms)	TCP (mW)	HTTP (mW)	Hybrid (mW)	Gain (%)	Loss (%)
1	50	406	429	407	5.41	0.25
1	100	404	420	406	3.45	0.49
1	200	402	409	403	1.49	0.25
10	50	405	430	405	6.17	0.00
10	100	405	422	405	4.20	0.00
10	200	403	411	404	1.73	0.25
50	50	408	432	409	5.62	0.24
50	100	404	422	404	4.46	0.00
50	200	402	409	403	1.49	0.25
100	50	407	430	408	5.39	0.25
100	100	403	422	404	4.46	0.25
100	200	402	411	402	2.24	0.00
200	50	411	431	414	4.11	0.72
200	100	406	423	406	4.19	0.00
400	50	415	429	418	2.63	0.72
400	100	410	423	412	2.67	0.49

allowing saving 6.2 % of energy in the best case and 2.1 % on average. The hybrid layer also chooses the best method for higher intervals above 10 s as it selects HTTP, which is theoretically the best one for higher intervals.

7 Discussion and Conclusion

Our measurements showed that TCP and HTTP are not equivalent in terms of energy, even if their purpose is the same. By offering a hybrid layer, we expect to globally reduce the energy consumption and lengthen battery life of Web-of-Things. Although our hybrid layer allows energy savings for sensors sending at a fixed interval, the behavior remains open for varying intervals. The number of records saved in the history will play a role on how the layer will respond to changes of interval. Another unresolved issue concerns the rate of symbols sent over Wi-Fi. The approximation function for TCP requires knowing at which rate the module sends its data. Due to changes in the surrounding environment, traffic congestions and other reasons, this rate may be changing. In our case, we forced a rate of 2 Mb/s in our test infrastructure.

In this paper, we explored a new way on how to reduce the energy consumption of things working inside the WoT framework. Instead of giving the responsibility of choice between TCP and HTTP for event notifications to developers, we introduce an hybrid layer doing the job for them. Our results show that energy savings can be achieved by selecting the most appropriate transport protocol. Further to this, we believe that our approach simplifies callbacks between things. Future work includes addressing the varying interval of events and finding the best history size to conciliate reaction time and filtering of outlier intervals. While the measured energy savings are relatively limited, we believe our hybrid layer has further potentials, for example if used as caching method of events by considering time penalties to limit the radio's use.

References

1. Aijaz F, Chaudhary M, Walke B (2009) Performance comparison of a SOAP and REST mobile web server. In: Proceeding of the 3rd international conference on open-source systems and technologies, Lahore, Pakistan
2. Bovet G, Hennebert H (2012) The web-of-things conquering smart buildings. Bulletin 10s:15–19
3. Bovet G, Hennebert H (2012) Communicating with things: an energy consumption analysis. In: Proceeding of the 10th International conference on pervasive computing, Newcastle, UK
4. Fette I, Melnikov A (2011) The WebSocket protocol. RFC
5. Fielding R, Taylor R (2002) Principled design of the modern Web architecture. ACM Trans Internet Technol 2:115–150
6. Gast M (2005) 802.11 wireless networks: the definitive guide, 2nd ed. O'Reilly Media

7. Gisler C, Barchi G, Bovet G, Mugellini H, Hennebert J (2012) Demonstration of a monitoring lamp to visualize the energy consumption in houses. In: Proceedings of the 10th international conference on pervasive computing, Newcastle, UK
8. Groba C, Clarke S (2010) Web services on embedded systems: a performance study. In: Proceeding of the 8th IEEE international conference on pervasive computing and communications, Mannheim, Germany
9. Guinard D (2011) A web of things application architecture: integrating the real-world into the web. ETHZ, Zurich
10. Guinard D, Trifa V, Mattern F, Wilde E (2011) From the internet of things to the web of things: resource oriented architecture and best practices In: Uckelmann D, Harrison M, Michahelles F (eds) Architecting the internet of things. Springer, Heidelberg, p 97
11. Hamad H, Saad M, Abed R (2010) Performance evaluation of RESTful web services. Comput Eng 2:72–78
12. Hameg (2012) HM8115-2 power meter description. http://www.hameg.com/0.147.0.html
13. Kindberg T et al (2002) People, places, things: web presence for the real world. Mobile Netw Appl 7:365–376
14. Mattern F, Floerkemeier C (2010) From the internet of computers to the internet of things. In: Sachs K, Petrov I, Guerrero P (eds) From active data management to event-based systems and more. Springer, Heidelberg, p 242
15. OpenPicus (2012) FLYPORT datasheet. http://space.openpicus.com/u/ftp/datasheet/flyport_wifi_datasheet_rev8.pdf
16. Ostermaier B, Schlup F, Römer K (2010) WebPlug: a framework for the web of things. In: Proceedings of the first IEEE international workshop on the web of things (WOT2010), Mannheim, Germany
17. Priyantha N, Kansal A, Goraczko M et al (2008) Tiny web services: design and implementation of interoperable and evolvable sensor networks. In: Proceeding of the 6th ACM conference on embedded network sensor systems, Raleigh, USA
18. Stevens R (1993) TCP/IP illustrated: the protocols. Addison-Wesley Longman Publishing Co, Boston
19. Vassis D, Rouskas A, Maglogiannis I (2005) The IEEE 802.11 g standard for high data rate WLANs. IEEE Netw J 9:21–26

A Secure Registration Scheme for Femtocell Embedded Networks

Ikram Syed and Hoon Kim

Abstract Recently, femtocell received a signification interest to improve the indoor coverage and provide better voice and data services. Lots of work has been done to improve the femtocell security, but still there are some issues which need to be addressed. Our contribution to the femtocell security is to protect secure zone (femtocell coverage area within macrocell) from unauthorized (non-CSG) users. In this paper, we propose a secure registration scheme for femtocell embedded network. In this scheme, only Closed Subscriber Group (CSG) users are allowed to access both the femtocell and macrocell services within the secure zone. By prioritizing the femtocell over macrocell within the secure zone, every user will try to camp on femtocell and invoke location registration to the femtocell as the user enters to the femtocell coverage area. If the user is within the allowed users list, the femtocell will allow the user otherwise femtocell will send a reject message to the user and also send the user information to the core network.

Keywords Femtocell · Macrocell · Closed subscriber group · Location area update · Secure zone

1 Introduction

Femtocell is small base station, connected with the service provider network through broadband (DSL, cable modem), it typically designs for home use and office use, they are short range, low cost and low power base stations that provides

---

I. Syed · H. Kim (✉)
University of Incheon, Incheon, Korea
e-mail: hoon@incheon.ac.kr

I. Syed
e-mail: ikram@incheon.ac.kr

J. J. (Jong Hyuk) Park et al. (eds.), *Multimedia and Ubiquitous Engineering*,
Lecture Notes in Electrical Engineering 240, DOI: 10.1007/978-94-007-6738-6_13,
© Springer Science+Business Media Dordrecht(Outside the USA) 2013

better coverage, better indoor voice and data services [1, 2]. Recent research shows that more than 50 % of voice calls and more than 70 % of data traffic are generated indoors [3]. Femtocell improves indoor coverage and capacity of the cellular providers with very low cost compare to the traditional macrocell base station [4]. The femtocell operates in licensed spectrum and may use the same or different frequency from the macrocell [5].

Recently, femtocells received a significant interest in the telecommunications industry. According to the ABI Research, 5.3 million femtocells will be deployed by end of 2012 [6]. Many of the major issues in femtocell have been studied, especially in security, lots of work has been done on femtocell security [7–9], but still there are some issues which needs to be addressed, especially on embedded networks security. In this paper, we focus on the protection of specific coverage area of femtocell within the macrocell coverage area, we called it secure zone. Our contribution is to protect the secure zone from unauthorized (non-CSG) users. Only authorized users are allowed to access both the femtocell and macrocell services within the secure zone.

There are some restrictions in the femtocell network for unauthorized users, when femtocell is working in CSG mode, but there is no restriction in macrocell for unauthorized (non-CSG) users of accessing the macrocell services. When the non-CSG users are in femtocell coverage area, they can access the macrocell services if the macrocell service is available. Our main contribution is to protect the femtocell coverage within the macrocell coverage area from non-CSG users. The secure zone would be used for security purpose in security agencies and military organizations.

The remainder of this paper is organized as follows. In Sect. 2 Access Control methods in femtocell, Location Area Update and cell selection and reselection criterion are discussed, Sect. 3 presented the proposed scheme for the secure zone and conclusion are presented in Sect. 4.

2 Background

2.1 Access Control Methods in Femtocell

There are basically three types of access control in femtocell, namely Closed Access, Open Access and Hybrid Access.

1. *Closed Access*: In closed access mode, the femtocell doesn't want to share their resources with other users due to limited resources or security reasons. Only authorized users are allowed to access the CSG cell [10]. In 3GPP the closed access is known as the CSG cell.
2. *Open Access*: In open Access, all users are allowed to access the femtocell. There is no restriction on any user [10].

3. *Hybrid Access*: In Hybrid access mode, both CSG users and non-CSG users are allowed to camp on femtocell but there are some exceptions for non-CSG users. In this mode CSG users are given more priority over non-CSG users [10].

2.2 Location Area Update

The Location Area Update (LAU) normally performs when a user turn on the power or enter to a new location area different from the old location area of the user. The access control need to be invoked, when a user move from one LAI to another LAI. In macrocell network, access control is normally invoked during the LAU [10]. Each femtocell has assigned a femtocell specific LAI different from Macrocell LAI [10]. Figure 1 shows the deployment of femtocells within the macrocell coverage area. In this approach, each femtocell has assigned different and unique LAI from the other femtocell and also from the macrocell, whenever a user enters to the femtocell coverage area and try to camp on femtocell. If a user is not allowed to a specific femtocell and try to camp on it, the user will received negative response in location update procedures.

2.3 Cell Selection and Reselection

Cell selection and reselection are still more complex problem in femtocell network, during the cell selection and reselection, the users need to carry out cell measurement parameters for intra-frequency, inter-frequency and inter-RAT

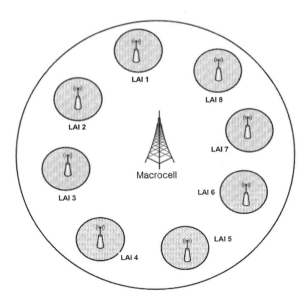

Fig. 1 Femtocells deployment

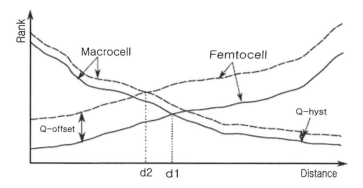

Fig. 2 Prioritizing femtocell over the macrocell in cell reselection [10]

neighbor cells, and rank the cells base on the policy used for specific cell [10]. Searching criterion (S-criterion) invokes the users to do monitoring and measurements for the intra-frequency, inter-frequency and inter-RAT, for the corresponding cell [10]. A lower S-criterion with the macrocell can help the users to start searching for the femtocell as soon it reached to the femtocell coverage area.

Further we also consider cell Ranking Criterion (R-Criterion) in which the Hysteresis values for the serving cell (Q-hyst) and offset values for neighboring cells (Q-offset) jointly affect the cell ranking. Figure 2 shows that by setting the Qoffset values negative and low Q-hyst in the macrocell neighbor cell list (NCL), the cell searching point d1 moves toward d2. It prioritizes the femtocell over the macrocell within the macrocell NCL. The main advantage of these methods is that all users in femtocell coverage area will prioritize the femtocell over the macrocell and will try to camp on femtocell. The non-CSG users will receive a negative response from the femtocell.

3 Proposed Scheme

Our proposed scheme focused on the secure zone protection from non-CSG users. We proposed a secure registration scheme for femtocell embedded network. It is assumed that the femtocell will operate in CSG mode in the secure zone. The authorized user information is store in CSG list. The CSG members list is located in the core network (CN) [11]. The CN entities such as Mobile Switching Centre (MSC)/Visitor Location Registration (VLR), Serving GPRS Support Node (SGSN), Mobility Management Entity (MME) and Home Subscriber Server (HSS) [10].

In normal case non-CSG users can access the macrocell service if the macrocell coverage is available. Due to Security reasons, we want to block non-CSG users of accessing both the femtocell and macrocell services within the secure zone. Figure 3 shows the basic scenarios of the secure zone within the macrocell coverage. U1, U3 are CSG members while U2, U4 are non-CSG members. U1 and U3

A Secure Registration Scheme for Femtocell Embedded Networks

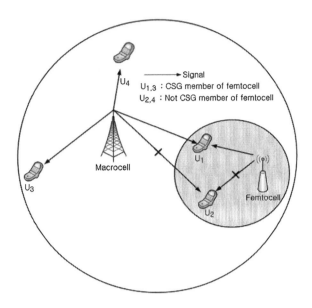

Fig. 3 Secure zone in macrocell coverage area

can access both the femtocell and macrocell services. U2 and U4 can access the macrocell service out of secure zone, but within secure zone U2 and U4 can't access the femtocell and macrocell services. In the Cell selection and reselection, femtocell will be prioritized over macrocell in Searching Criterion by setting the lower S-criterion within the macrocell NCL. Every user will start searching femtocell as soon as they reached to the femtocell coverage area. We also prioritized the femtocell over macrocell in Ranking Criterion by setting lower value of Q-offset for femtocell within macrocell, by doing this, every user will try to camp on femtocell as the user enter femtocell coverage area and initiates the initial Non Access Stratum (NAS) procedure by establishing the RRC Connection with the femtocell.

The user capabilities are reported to the Femtocell as a part of the RRC connection establishment procedure [12]. The RRC connection message includes user identity (IMSI or TMSI) and establishment cause etc. by sending the RRC connection procedure to femtocell, the femtocell will check the user capabilities. If no context ID exists for the user, the femtocell will initiate user registration request to the Femtocell gateway (FGW). The FGW will check the user capabilities provided in RRC Connection message. If the user is a CSG user, the FGW may accept the user registration and allocate a context ID for the user [12]. If the user is non-CSG user the FGW will send reject message to the femtocell and will also send the user information to the CN to block the user within the specific femtocell coverage area.

The user information includes IMSI or TMSI and the serving femtocell location area identity. Figure 4 shows the flow chart of the proposed scheme, and the signal flow diagram of the proposed scheme is shown in Fig. 5.

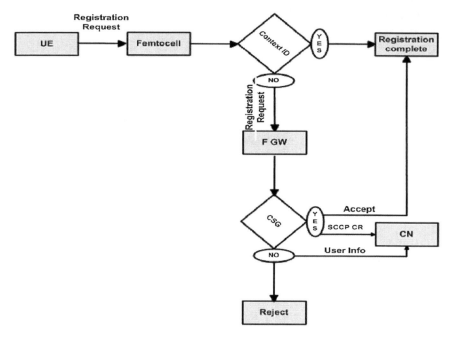

Fig. 4 Flow chart of the proposed scheme

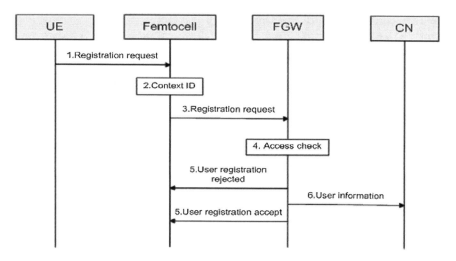

Fig. 5 Signal flow chart of user registration process

After receiving a rejected response from the femtocell, the user will try to camp on macrocell, but femtocell already sent the user information to the macrocell to block this user within the serving femtocell coverage area. The femtocell sends this information to the macrocell in the user information massager. When the user try to camp on macrocell, the macrocell will first check the user capabilities (the user location area and user membership), If the user is CSG user then the macrocell will allow the user to access his services, but if the user is not in the CSG member list then check the LA of the user. If the user is in the secure zone, the macrocell will block the user.

4 Conclusions

This paper focused on the protection of secure zone, to make the secure zone secure from non-CSG users. We proposed a new user registration scheme for both CSG and non-CSG users, by prioritizing the femtocell over macrocell within the secure zone. By doing this, every user will try to camp on femtocell and invoke location registration to the femtocell as soon it enter the secure zone. The femtocell will reject the non-CSG users and send the user information to the CN.

References

1. Namgeol O, Han S, Kim H (2010) System capacity and coverage analysis of femtocell networks. In: WCNC, Sydney, pp 1–5
2. Chandrasekhar V, Andrews J, Gatherer A (2008) Femtocell networks: a survey. IEEE Commun Mag 46(9):59–67
3. Sang YJ, Hwang HG, Kim KS (2009) A self-organized femtocell for IEEE 802.16e system. In: GLOBECOM, pp 1–5
4. Chowdhury MZ, Trung BM, Jang YM (2011) Neighbor cell list optimization for femtocell-to-femtocell handover in dense femtocell networks. In: ICUFN, pp 241–245
5. Cao F, Fan Z (2010) The tradeoff between energy efficiency and system performance of femtocell deployment. ISWCS, UK, pp 315–319
6. Femtocell shipments flatline due to slow inventory burn rate. ABI research, Research report (2012)
7. Han C-K, Choi H-K, Kim I-H (2009) Building femtocell more secure with improved proxy signature. In: GLOBECOM, pp 1–6
8. Vanek T, Rohlik M (2011) Perspective security procedures for femtocell backbone. In: ICUMT, pp 1–4
9. Brassil J, Manadhata PK, (2012) Securing a femtocell-based location service. Mobile and wireless networking, pp 30–35
10. Zhang J, de la Roche G (2009) Femtocells: technologies and deployment
11. 3GPP TS 25.304 (2009) User equipment (UE) procedures in idle mode and procedures for cell reselection in connected mode. 3GPP-TSG RAN, 8.5.0
12. 3GPP TS 25.467 (2010) UTRAN architecture for 3G Home Node B (HNB). 3GPP-TSG RAN, 9.3.0

Part IV
Intelligent Computing

Unsupervised Keyphrase Extraction Based Ranking Algorithm for Opinion Articles

Heungmo Ryang and Unil Yun

Abstract Keyphrase extraction is to select the most representative phrases within a given text. While supervised methods require a large amount of training data, unsupervised methods can perform without prior knowledge such as language. In this paper, we propose a ranking algorithm based on unsupervised keyphrase extraction and develop a framework for retrieving opinion articles. Since the proposed algorithm uses an unsupervised method, it can be employed to multi-language systems. Moreover, our proposed ranking algorithm measures the importance in three aspects, the amount of information within articles, representativeness of sentences, and frequency of words. Our framework shows better performance than previous algorithms in terms of precision and NDCG.

Keywords Opinion article · Ranking algorithm · Unsupervised keyphrase extraction

1 Introduction

With the expanse of the e-commerce, people are using the Internet to check opinion article of products written by other people before buying them. The rapid growth of online stores not only leads to explosive increasing of opinion information but also presents new challenges to Information Retrieval (IR) field. As the result of the expanding information, it becomes difficult to find helpful opinion articles. Thus, appropriate ranking algorithms are required for searching and retrieving meaningful opinion articles. Ranking algorithms used in the IR measure

---

H. Ryang · U. Yun (✉)
Department of Computer Science, Chungbuk National University, Chungbuk, South Korea
e-mail: yunei@chungbuk.ac.kr

H. Ryang
e-mail: riangs@chungbuk.ac.kr

J. J. (Jong Hyuk) Park et al. (eds.), *Multimedia and Ubiquitous Engineering*,
Lecture Notes in Electrical Engineering 240, DOI: 10.1007/978-94-007-6738-6_14,
© Springer Science+Business Media Dordrecht(Outside the USA) 2013

the importance of targets such as web pages and words in documents, and many algorithms [1, 3, 5] have been proposed. Although these traditional ranking algorithms have played an important role, it is hard to retrieve relevant opinion articles since the algorithms do not consider characteristics of them. To address this issue, opinion ranking algorithms [2, 6, 7] have been proposed, and they can be divided into two approaches: (1) sentimental and semantic analyze opinions using dictionaries which contain opinion words; (2) measure the importance of opinion articles based on additional information. The former approaches [2, 7] need the prior data such as dictionary, and thus these approaches can be called supervised methods. Because they demand sentimental and semantic words list, it is difficult for multi-language system to apply the methods. In view of this, the latter approaches [6] can be easily employed to the multi-language system due to ability of computing rankings without any prior knowledge. Thus, these approaches can be called unsupervised methods. However, they do not consider the importance of sentences. In this paper, therefore, motivated by the above, we propose a ranking algorithm, called ROU (Ranking Opinion articles based on Unsupervised keyphrase extraction), for reflecting the importance of sentences to rankings. The proposed algorithm computes ranking scores in three aspects, the amount of information within articles, representativeness of sentences, and frequency of words. The remainder of this paper is organized as follows. In Sect. 2, we introduce the related work. In Sect. 3, we describe the proposed algorithm and framework for retrieving opinion articles in detail. In Sect. 4, we show and analyze experimental results for performance evaluation. Finally, conclusions are given in Sect. 5.

2 Related Work

Various ranking algorithms were proposed for IR. TF-IDF [1] is one of the most popular and important algorithms. In this algorithm, terms are given more weights when they appear frequently in a single document (Term Frequency) or they are included smaller set of documents in the corpus (Inverse Document Frequency). Although TF-IDF can measure the importance of words, it cannot know how important sentences or documents are. Nevertheless, it still can be used as effective measurement to words. Meanwhile, ranking algorithms [3, 5] for web pages were also proposed, such as PageRank [5] adopted by Google (http://www.google.com), and they have played significant role. Recently, with the rapid growth of the e-commerce, the amount of opinion information is increasing explosively and appropriate ranking algorithms are required since the general purpose algorithms cannot reflect the characteristics of opinion articles. To address this issue, opinion ranking algorithms [2, 6, 7] were proposed. There are two types of work, sentimental analysis based and additional information based ranking algorithms. The former methods [3, 5] identify nouns, adjectives, and inversion words using dictionary which contains positive, negative, and neutral words and calculate ranking based on the found words. It means that they need prior knowledge for certain language, and thus they can be called supervised methods. Therefore, it is difficult for multi-language systems to apply the algorithms. In contrast, the latter measures the importance of opinion articles without any prior knowledge, and thus they can be easily adopted by multi-language systems. RLRank [6] is one of the latter algorithms and employs weight of words and the amount of information. However, they do not reflect the importance of sentences. For these reasons, this study aims to develop framework for opinion article retrieval and reflect importance of sentences. On the other hand, TextRank [4] is one of algorithms for unsupervised keyphrase extraction. It calculates ranking score of sentences using both similarity and other graph based ranking algorithms such as PageRank and HITS [3]. Since TextRank is an unsupervised method, it can be performed regardless of language. Thus, our proposed ranking algorithm applies TextRank for measuring the importance of sentences based on representativeness.

3 Unsupervised Keyphrase Extraction Based Algorithm

The proposed and developed framework in this study is a keyword based retrieval system for opinion articles, and it searches important opinion articles. In addition, the framework consists of three steps, data preprocessing, ranking, and indexing. In the first step, data of the collected dataset is preprocessed for applying the proposed ranking algorithm; at the same time, information for the algorithm such as title is extracted. In the second step, ranking scores are computed using the preprocessed data. Especially, the ranking algorithm measures the importance of keywords in three aspects. Let *key* be a certain keyword. The first aspect is about how useful an opinion article includes *key*, and the proposed algorithm uses the amount of information as the measurement. The second aspect is how representative sentences containing *key* are. The last aspect is about how important *key* is in both the opinion article and dataset.

Figure 1a is a diagram of the three aspects. In the last step, index is constructed by creating inverted index files including index information such as ranking scores. Figure 1b shows the system architecture of the framework. It consists of not only the three modules described in the above but also searching module for providing search results. The searching module first accepts queries and extracts keywords from the queries. Then, it finds inverted index files containing each extracted keyword, and calculates ranking score by employing the information in the files. Finally, the module provides results by sorting in ranking score descending order.

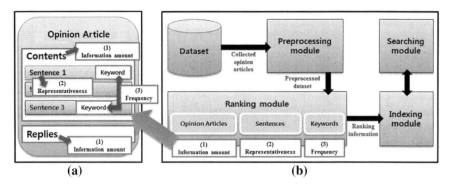

Fig. 1 The proposed framework

3.1 Measuring the Amount of Information in Opinion Articles

The proposed ranking algorithm, ROU first analyzes the usefulness of each opinion article in the dataset by measuring the amount of information with two assumptions. The first assumption is that if opinion articles include more contents, they contain more information related to products. It can be measured using the length of contents. The second assumption is that reply involves information about not only products but also opinion articles. Thus, for this purpose, preprocessing module extracts information of reply and the number of words. Ranking module computes the importance of opinion articles using the extracted information through following equation.

$$Importance(OA) = L(OA) + R(OA) \times (1 + RO(OA)/R(OA)) \quad (1)$$

In the equation, OA is a certain opinion article, $L(OA)$ is the length of contents in OA, $R(OA)$ is the number of reply contained by OA, and $RO(OA)$ is the number of reply written by other users.

3.2 Measuring Representativeness of Sentences

After calculating of *Importance(OA)*, ranking module measures the importance of each sentence in *OA* based on representativeness, and TextRank [4] is used for this purpose. Most of global online stores such as Amazon (http://www.amazon.com) provide multi-language systems. Thus, ROU algorithm applies the unsupervised method so as to be performed without prior knowledge of languages. In this stage, ranking module first divides *OA* into sentences (i.e. keyphrases) and computes similarity between the divided sentences. The similarity is measured by a function which analyzes their content overlap [4]. Then, by using the computed similarity,

Unsupervised Keyphrase Extraction

TextRank score, that is representativeness of the sentences is obtained based on PageRank [5]. In TextRank, the more representative sentences, the more scores are assigned. The obtained scores are normalized as [0, 1] for calculating the importance of keywords in *OA*, and the details are described in the following section.

3.3 Ranking Technique Based on Unsupervised Keyphrase Extraction

Once both *Importance(OA)* and the normalized TextRank score of sentences in *OA* are computed, ranking module calculates final ranking scores of keywords based on the results obtained through previous steps. First, TF-IDF [1] of each keyword is computed to measure how the each keyword is important in both *OA* and dataset. After that, the average importance of sentences containing the each keyword is calculated for reflecting representativeness of the sentences to ranking. Following equation is used for this purpose.

$$Representative(key) = tf \cdot Idf \times \left(1 + \sum NTextRank(sentence, key)/S(key)\right)$$

(2)

In the equation, *tf·Idf* is TF-IDF value of a certain keyword, *key*, *sentence* is a sentence which has *key*, *NTextRank(sentence, key)* is a normalized TextRank score

---

Construct_Index(*dataset, ratio*)
1. **For each** opinion article *OA* in *dataset*
2.   **Extract** information of title, contents, and reply from *OA*
3.   **Analyze** the extracted information and **Save** the analyzed data into files *PF*
4. **For each** preprocessed data *POA* in *PF*
5.   *L* ← the length of contents
6.   *R* ← the number of reply
7.   *RO* ← the number of reply written by other users
8.   *Importance* ← *L* + *R* × (1 + *RO* / *R*) /* Eq. (1) */
9.   **Divide** contents of *POA* into sentences
10.   *Scores*[][] ← ϕ
11.   **For each** sentence *S* in the sentences
12.   **Calculate** TextRank of *S* and **Normalize** TextRank as [0, 1]
13.   *TR* ← the normalized TextRank score
14.   **Extract** keywords from *S*
15.   **For each** keyword *key* of the extracted keywords
16.   *Scores*[*key*][*score*] ← *Scores*[*key*][*score*] + *TR*
17.   *Scores*[*key*][*count*] ← *Scores*[*key*][*count*] + 1
18.   **For each** *Scores*[*key*] in *Scores*
19.   **Calculate** TF-IDF of *key*
20.   *tf·Idf* ← the calculated TF-IDF value
21.   *Representative* ← *tf·Idf* × (1 + *Scores*[*key*][*score*] / *Scores*[*key*][*count*]) /* Eq. (2) */
22.   *Ranking* ← (*Importance* × *ratio*) + (*Representative* × (1 − *ratio*)) /* Eq. (3) */
23.   **Save** information of *Ranking* and *OA* into inverted index file for *key*

Fig. 2 Algorithm for constructing index with the ranking algorithm

of all *sentence* including *key*, and $S(key)$ is the number of all *sentence*. Then, final ranking score of keywords is calculated through following equation based on Eqs. (1) and (2).

$$Ranking(key) = (Importance(OA) \times r) + (Representative(key) \times (1-r)) \quad (3)$$

In the equation, r is the adoption rate of two ranking factors. After this process of ranking module, indexing module creates inverted index files with respect to each keyword with information such as the ranking scores for construction of index. The constructed index is used by searching module to provide service of keyword based opinion article retrieval. Figure 2 shows our ranking algorithm for constructing index with ROU algorithm of the framework.

4 Performance Evaluation

In performance evaluation, all experiments were performed on 3.3 GHz Intel processor with 8 GB main memory, and run with Microsoft Windows 7 operating system. Algorithms are implemented in C++ language. In addition, about 58,000 opinion articles have been collected from Amazon. To evaluate performance of algorithms, we compare our ROU algorithm with RLRank [6] and TF-IDF [1]. Common settings for performance evaluation are as follows. First, top-50 searching results are extracted in respect to sampled keywords, which are related to product names selected from categories in the Amazon. Second, the number of relevance articles is counted from the searched results. In the experiments, relevance article is an article containing contents closely related to a given keyword. In addition, it is an article having no less than the average number of evaluations such as helpful in regard to the all articles in the collected dataset, and the average number is 5.414 in our collected dataset. Third, performances of algorithms are measured according to precision and NDCG.

We first perform precision test of ROU with RLRank and TF-IDF. For given retrieved results, precision is defined as the percentage of retrieved relevant articles to the results. In the experiment, five sampled keywords are used. The left figure of Fig. 3 shows the results of precision evaluation. From the figure, we can observe that our ROU algorithm outperforms RLRank and TF-IDF in the sampled keywords.

Next, we evaluate performance of compared algorithms in terms of Normalized Discounted Cumulate Gain (NDCG). NDCG is used to measure performance of IR systems using graded relevance and ranking order since users usually refer to the top results as important, not the bottom results. In the right figure of Fig. 3, the proposed algorithm, ROU shows better NDCG results than the previous algorithms in the sampled keywords. Although previous algorithms show better performance with respect to a keyword "jewelry", all performances of compared algorithms is almost the same. It means that the more retrieved relevant articles appear in the top of the results.

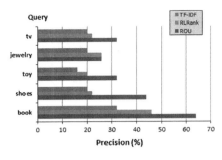 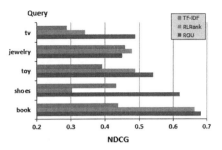

Fig. 3 Precision and NDCG evaluations

5 Conclusions

In this paper, we proposed unsupervised keyphrase extraction based ranking algorithm, ROU which measures the importance in three aspects, the amount of information in articles, representativeness of sentences, and frequency of words. To reflect the representativeness to ranking, ROU calculates TextRank scores of sentences. Moreover, we conducted precision and NDCG experiments for performance evaluation. The experimental results showed that our ranking algorithm, ROU outperformed previous ranking algorithms. In addition, our framework can provide service of keyword based opinion article retrieval.

Acknowledgements This research was supported by the National Research Foundation of Korea (NRF) funded by the Ministry of Education, Science and Technology (NRF No. 2012-0003740 and 2012-0000478).

References

1. Aizawa AN (2003) An information-theoretic perspective of Tf-idf measures. J Info Process Manage 39(1):45–65
2. Eirinaki M, Pisal S, Singh J (2012) Feature-based opinion mining and ranking. J Comp Syst Sci 78(4):1175–1184
3. Kleinberg JM (1999) Authoritative sources in a hyperlinked environment. J ACM 46(5):604–632
4. Mihalcea R, Tarau P (2004) TextRank: bringing order into text. In: Proceedings of EMLNP 2004, Barcelona, pp 404–411
5. Page L, Brin S, Motwani R, Winograd T (1999) The PageRank citation ranking: bringing order to the Web. Technical report, Stanford InfoLab
6. Yun U, Ryang H, Pyun G, Lee G (2012) Efficient opinion article retrieval system. Lecture Note in Computer Science. In: Proceedings of ICHIT 2012, Daejeon. pp 566–573
7. Zhang L, Liu B, Lim SH, O'Brien-Strain E (2010) Extracting and ranking product features in opinion documents. In: Proceedings of COLING 2010, Beijing, pp 1462–1470

A Frequent Pattern Mining Technique for Ranking Webpages Based on Topics

Gwangbum Pyun and Unil Yun

Abstract In this paper, we propose a frequent pattern mining technique for ranking webpages based on topics. This technique shows search results according to selected topics in order to give users exact and meaningful information, where we use an indexer with the frequent pattern mining technique to comprehend webpages' topics. After mining frequent patterns related to topics (i.e. frequent topics) in collected webpages, the indexer compares new webpages with the generated patterns and calculates degree of topic proximity to rank the new ones, where we also propose a special tree structure, named RP-tree, to compare the new webpages to the frequent patterns. Since our technique reflects topic proximity scores to ranking scores, it can preferentially show webpages which users want.

Keywords Frequent pattern mining · Ranking · RP-tree · Topic search

1 Introduction

Information retrieval algorithms find and show users webpages related to their needs. Previous information retrieval algorithms generally use a method that finds webpages through keywords inputted by users, and the found results contain a number of webpages with unrelated topics. A topic search is a method which provides webpages related to specific topics selected by users. To conduct the topic search, topic analysis of webpages is needed. FIIR [1], one of topic analysis

---

G. Pyun · U. Yun (✉)
Department of Computer Science, Chungbuk National University,
Chungbuk, Republic of Korea
e-mail: yunei@chungbuk.ac.kr

G. Pyun
e-mail: pyungb@chungbuk.ac.kr

J. J. (Jong Hyuk) Park et al. (eds.), *Multimedia and Ubiquitous Engineering*,
Lecture Notes in Electrical Engineering 240, DOI: 10.1007/978-94-007-6738-6_15,
© Springer Science+Business Media Dordrecht(Outside the USA) 2013

methods, uses a word dictionary associated with a specific topic and assigns weights to webpages according to the number of food-related words. However, this method has a limitation that does not consider word combinations since it analyzes topics in terms of the number of related words. In this paper, we propose a frequent pattern mining technique for precisely analyzing webpages' topics and reasonably assigned weights. The technique mines frequent word patterns from topic-related webpages and calculates topic scores for new webpages through the mining results. Then, the topic score is applied as a weight for ranking, and our algorithm computes exact ranking scores by using the weight. As related works, DTM [2] preferentially shows webpages for frequently accessed topics by considering queries inputted by users so far. In LDA [3], a method for finding hidden topics through webpage's domain information was proposed. LDA can discover the hidden information by calculating relations between users' queries and webpages. As a food information retrieval algorithm, FIIR [1] computes the number of words matched with its food dictionary and sets weights to webpages. In contrast to FIIR which only uses word quantity information derived from the food dictionary. FP-growth [4], one of the frequent pattern mining methods, finds meaningful patterns.

2 Frequent Pattern Mining Technique for Analyzing Topics of Webpages

2.1 Extracting Patterns of Webpages Related to Topics

To analyze topics of webpages, appropriate analysis criteria are required, and we use patterns of webpages associated with topics, where the patterns are composed of special combinations of words in webpages. Therefore, patterns, which occur frequently among patterns derived from a number of topic-related webpages, become important data representing topics. We can determine whether new webpages are associated with a current topic or not, according to the topic score. To find frequent patterns for any topic, an indexer searches them through the frequent pattern mining technique, and for this purpose, we propose a preprocessing method regarding webpage's words.

Definition 1 (*Preprocessor*) A preprocessor generates unique IDs reflected to words consisting of webpages.

Assuming that any webpage, P has n words, i, P is denoted as $P = \{i_1, \ldots i_n\}$, where IDs are uniquely assigned for each i. For $k < n$, $r < n$, $k \neq r$, any ID, Q is expressed as $Q = k$, iff $Q \neq r$, and vice versa. After the preprocessor converts words of any webpage as IDs, the webpage is denoted as a set of IDs, and the set from one webpage is considered as one transaction. Our indexer extracts frequent patterns by pattern mining approach after constructing a database from the webpages. However, general pattern mining approach generates too many frequent

A Frequent Pattern Mining Technique for Ranking Webpages Based on Topics

Fig. 1 RP-tree structure

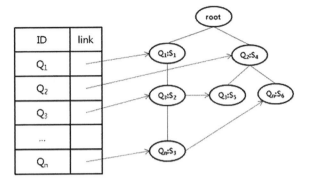

patterns. To solve the disadvantage, we propose an efficient comparison method using a new tree structure, named RP-tree.

Definition 2 RP-tree (Result Pattern–tree) is a tree structure for storing frequent patterns.

RP-tree's basic structure is shown in Fig. 1, where the tree has a link table and a tree, and the link table has an ID list composing RP-tree. Note that sorting order of the list is the same as that of frequent pattern mining (i.e. support descending order). Each ID in the link table has a pointer, and the pointer is connected to all of the nodes with the same ID. RP-tree's pattern storage procedure is as follows. We first create a link table according to sort order of frequent patterns, and then insert frequent patterns' IDs into a tree, starting from a root. When IDs are inserted in the tree, we assign 0 if a new node is generated, and we only change the support for not middle nodes but the end node after any transaction is inserted. Thereby, the node where the last ID of the frequent pattern is entered has support information for the pattern. Thus, RP-tree represents the characteristics as shown in Lemma 1.

Lemma 1 *Support of all nodes in RP-tree does not interfere in other supports of frequent patterns.*

Proof Given a frequent pattern, $F = \{i_1, i_2, \ldots i_k\}$ with a length, k, and its support, S, the only node with i_k has S in RP-tree. Let us assume that a sub-pattern of F, $F' = \{i_1, i_2, \ldots i_{k-1}\}$ and a super pattern of F, $F'' = \{i_1, i_2, \ldots i_{k+1}\}$ are inserted into the tree after F. Then, the nodes with i_{k-1} and i_{k+1} are set as their corresponding supports. However, these two patterns do not have any effect in the support of F. The reason is that RP-tree only updates the last node's support when any transaction is inserted. Thus, since F' is a sub-pattern of F, there is no change for F. In the case of F'', F'' has an effect on the node with i_k, since F'' is a super pattern of F and has all items in F. However, any problem is not caused as in the cases of F' since RP-tree only updates a support for the last node with i_{k+1}, not changing the node with i_k. Namely, the information for F is preserved. As a result, RP-tree does not cause any interference among all of the patterns. □

2.2 Analyzing Topics of Webpages by RP-Tree

If a web robot collects new webpages, our indexer compares the new webpages with the frequent patterns in RP-tree and computes topic scores, where our algorithm converts words of the new ones to ID forms to calculate topic scores. Thereafter, the generated IDs are sorted according to RP-tree's sort order. Then, the algorithm selects IDs from the bottom one by one and finds nodes with the current ID as using the link table containing ID and link information.

Definition 3 (*Matching List*) A matching list stores a set of paths matched with new webpages as traversing RP-tree and supports corresponding to the paths.

When the algorithm visits nodes linked to the selected ID, IDs and supports for the nodes are stored into the matching list. After that, a pointer moves to the parent of the current node, and its ID is added in the matching list if the parent has ID of the new webpage. These operations are iterated by the root. For all of the IDs from the link table, the above steps are performed. Then, the finally generated matching list has match information between the new webpages and RP-tree's frequent patterns. If any webpage is related to the current topic, the patterns in the matching list have long lengths and high supports. The topic score, W is computed as

$$W = maximal\ length\ *\ (1 + sum\ of\ maximal\ patterns'\ supports)$$

where *maximal length* means a length value of the pattern with the longest length in the matching list and *sum of maximal patterns' supports* are to add all of the patterns' supports with *maximal length*. We add 1 to the sum to increase an effect by *maximal length*. In the equation, *sum of maximal patterns' supports* has a value between 0 and 1, and we assign higher scores to frequent patterns with 2 or more lengths rather than 1 length. Topic scores and words' supports are saved into an inverted file, and they are used as important factors when we consider ranking scores. As an example, the left list in Fig. 2 represents that any new webpage is converted to its IDs. Then, we first select the last item in the ID list, G, and search the tree by using the link table as shown in the figure. After searching the tree from each node with G to the root, {A, E, G} and {G} are included in the matching list, where the corresponding *maximal length* and *sum of maximal patterns' supports* are 3 and 0.042 respectively. Thus, its topic score, W is calculated as W = 3*(1 + 0.042) = 3.126. Figure 3 shows an algorithm for analyzing webpages' topics and generating an inverted file. The algorithm mines frequent patterns related to the current topic as using the FP-growth method [4], and then RP-tree is generated though the mined patterns. If new webpages are collected, the algorithm removes a few words unrelated to the topic in the webpages and converts them as IDs. After comparing the ID list to the RP-tree, the score of corresponding topic is computed, and finally, an inverted file for P is generated.

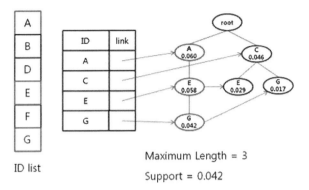

Fig. 2 A webpage pattern and its RP-tree

Algorithm : Analysis topic of webpage

Input : Topic dictionary DIC Database of topic webpages **Output** : inverted file 1. pattern ← **mining process**(Database) 2. RP-tree ← **build RP-tree**(pattern) 3. for each collected webpage, P 4. ID list ← **Preprocessor**(P, DIC) 5. Topic score ← **Compare patterns**(ID list, PR-tree) 6. create inverted file **End procedure** **Procedure Preprocessor**(P, DIC) 7. for each i_n ← P 8. if(i_n = one of the DIC's words) 9. insert ID list(i_n) 10. return ID list **End procedure**	**Compare patterns**(ID list, PR-tree) 11. sort ID list according to RP-tree's ID order 12. for each ID list, i_n 13. find i_n in the RP-tree's link table 14. for each link 15. now ← link 16. while(now ≠ root) 17. traverse RP-tree in bottom up manner 18. if(ID of now = one of the ID list's IDs) 19. insert temporary list(ID) 20. insert Matching list(temporary list) 21. for each Matching list 22. find max length in Matching list 23. max_score ← sum of max length patterns' supports 24. Topic score ← max length * (1+max_score) **End procedure**

Fig. 3 Algorithm for topic analysis

2.3 Ranking Technique by Frequent Pattern Mining

In general information retrieval, a ranking score is computed by multiplying a frequency by idf [5]. However, the above calculation is not suitable for the topic search in this paper. Therefore, we calculate ranking scores as reflecting topic scores computed through frequent patterns, and the proposed ranking score, RS is as follows.

$$RS = tf * idf * ln(W + e)$$

Here, tf, idf, and W are a relative frequency reflecting how many a searched word is included in a webpage, an inverted document frequency related to a searched word, and a topic score comparing frequent patterns for topics with the current webpage respectively. W is utilized as a weight condition changing ranking scores of webpages, and we need to adjust this value to obtain better scores. If W is too low, a role as a weight condition for topics is lost. In contrast, if it is too high, webpages simply related to topics are presented regardless of user-inputted keywords. To solve these problems, we use the natural logarithm, where this is a logarithm which has a natural constant, e as a base and is used as a normal distribution value in statistics. Accordingly, we adjust W by using the natural logarithm to compute reasonable ranking scores which both contain as many queries as possible and consider topics. However, the ranking score using the adjusted W has a limitation, if $0 \leq W < e$. Topic scores of webpages unrelated to topics become 0 while those highly related to topics are more than 0. However, if $W < e$, the ranking score is lower than the score when W is not included. Therefore, our ranking technique computes the scores as tf * idf, if $W = 0$, while it reflects W to RS after adding e, if $W > 0$.

3 Performance Evaluation

In this section, we compare our FPR (Frequent Pattern mining for Ranking webpages by topics) with the other topic search algorithm, FIIR [1], in terms of precision, recall, and NDCG. Webpages used in these experiments were gathered from www.washingtonpost.com between 01/01/2011 and 12/31/2011, where the number of webpages is 32,248. In addition, "A Dictionary of food [6]" was utilized for the evaluation. FPR first performs frequent pattern mining process to construct RP-tree, and then calculates ranking scores based on the food dictionary and the comparison technique by RP-tree. Queries used in the experiments are {Travel, Friday, Beach, Young, Water, Chief, Car, Train Apple, Meat, Food, Corn}. X-axis in Figs. 4 and 5 denotes the above 12 queries in sequence. To evaluate precision [7], we measure ratios of food-related webpages in the top-30 webpages gained from the two algorithms. In Fig. 4, FPR shows outstanding precision compared to FIIR [1] in all of the cases except for the 9th query, "Apple". In the recall test [8], given webpages including both the current query and food-related information, we calculate ratios for how many food-related webpages exist in the top-30 webpages found by the algorithms. Figure 5

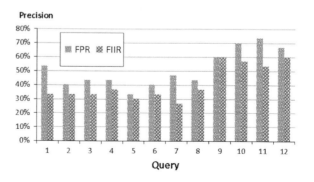

Fig. 4 Precision test

Fig. 5 Recall test

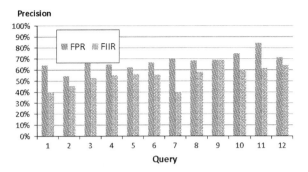

Fig. 6 NDCG test

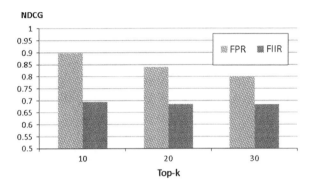

represents recall results, where FPR guarantees higher recall ratios than those of FIIR in most cases. NDCG [8] has a high value if any webpage with food information is located in upper ranks while it has a low value if this is in lower ranks. A food-related score, relk is set as 0 when there is no information related to food, 1 when there is food-related information in the other topics, 2 when the topic is only contained to food, or 3 when the topic belongs to food and there are many food-related contents. Figure 6 shows average values of NDCG for the 12 queries, where top-k is set as 10–30. In this result, FPR also guarantees outstanding NDCG performance compared to FIIR in every case.

4 Conclusion

In this paper, we proposed a frequent pattern mining technique and the corresponding algorithm which can rank webpages based on topics. To analyze topics exactly, we conducted frequent pattern mining operations regarding topic-related webpages, and thereafter we calculated topic scores by comparing webpages with the mined frequent patterns and applied the topic scores to ranking scores. In the various experiments, it was observed that our algorithm outperforms the previous topic-based algorithm in terms of precision, recall, and NDCG. Through the proposed techniques and algorithm, we expect that they will contribute to improving the level of both information retrieval and frequent pattern mining fields.

Acknowledgment This research was supported by the National Research Foundation of Korea (NRF) funded by the Ministry of Education, Science and Technology (NRF No. 2012-0003740 and 2012-0000478).

References

1. Pyun G, Yun U (2011) Efficient food retrieval techniques considering relative frequencies of food related words. Lect note Comput Sci 368–375
2. Chen KY, Wang HM, Chen B (2012) Spoken document retrieval leveraging unsupervised and supervised topic modeling techniques. IEICE Trans 1195–1205
3. Andrzejewski D, Buttler D (2011) Latent topic feedback for information retrieval. Knowl Discovery Data Min 600–608
4. Han J, Pei J, Yin Y, Mao R (2004) Mining frequent patterns without candidate generation: a frequent pattern tree approach. DMKD 8(1):53–87
5. Donald M (2008) Generalized inverse document frequency. Paper presented at conference on information and knowledge management, pp 399–408
6. Charles S (2005) A dictionary of food: international food and cooking terms from A to Z, 2nd edn. A&C Black Publishers Ltd
7. Kato MP, Ohshima H, Tanaka K (2012) Content-based retrieval for heterogeneous domains: domain adaptation by relative aggregation points. Paper presented at ACM SIGIR conference on Research and development in information retrieval, pp 811–820
8. Croft WB, Metzler D, Strohman T (2010) Search engines: information retrieval in practice. Addison-Wesley, Boston

Trimming Prototypes of Handwritten Digit Images with Subset Infinite Relational Model

Tomonari Masada and Atsuhiro Takasu

Abstract We propose a new probabilistic model for constructing efficient prototypes of handwritten digit images. We assume that all digit images are of the same size and obtain one color histogram for each pixel by counting the number of occurrences of each color over multiple images. For example, when we conduct the counting over the images of digit "5", we obtain a set of histograms as a *prototype* of digit "5". After normalizing each histogram to a probability distribution, we can classify an unknown digit image by multiplying probabilities of the colors appearing at each pixel of the unknown image. We regard this method as the baseline and compare it with a method using our probabilistic model called Multinomialized Subset Infinite Relational Model (MSIRM), which gives a prototype, where color histograms are clustered column- and row-wise. The number of clusters is adjusted flexibly with Chinese restaurant process. Further, MSIRM can detect *irrelevant* columns and rows. An experiment, comparing our method with the baseline and also with a method using Dirichlet process mixture, revealed that MSIRM could neatly detect irrelevant columns and rows at peripheral part of digit images. That is, MSIRM could "trim" irrelevant part. By utilizing this trimming, we could speed up classification of unknown images.

Keywords Bayesian nonparametrics · Prototype · Classification

---

T. Masada (✉)
Nagasaki University, 1-14 Bunkyo-machi, Nagasaki-shi, Nagasaki 852–8521, Japan
e-mail: masada@nagasaki-u.ac.jp

A. Takasu
National Institute of Informatics, 2-1-2 Hitotsubashi, Chiyoda-ku, Tokyo 101–8430, Japan
e-mail: takasu@nii.ac.jp

J. J. (Jong Hyuk) Park et al. (eds.), *Multimedia and Ubiquitous Engineering*,
Lecture Notes in Electrical Engineering 240, DOI: 10.1007/978-94-007-6738-6_16,
© Springer Science+Business Media Dordrecht(Outside the USA) 2013

1 Introduction

This paper considers image classification. While there are a vast variety of methods, we focus on *prototype*-based methods. We construct a prototype for each image category and classify an unknown image to the category whose prototype is the most similar. In this paper, we describe prototypes with probability distributions and classify an unknown image to the category whose prototype gives the largest probability.

We assume that all images are of the same size, say N_1 by N_2 pixels, because we consider *handwritten digit images* in this paper. We can obtain a set of color histograms by counting the number of occurrences of each color at each pixel. Consequently, we obtain $N_1 N_2$ color histograms, each at a different pixel. When we conduct this counting over the images of the same category, e.g. the images of handwritten digit "5", the resulting set of histograms gives a color configuration specific to the category and can be regarded as a *prototype* of the category. Formally, we describe each prototype with parameters $\left\{ g^h_{ijw} \right\}$, where g^h_{ijw} is a probability that the w th color appears at the 2D pixel location (i, j) for $w = 1, \ldots, W$, $i = 1, \ldots, N_1$, $j = 1, \ldots, N_2$, where W is the number of different colors. The superscript h is a category index. By using the parameters, we can calculate the log probability of an unknown image as $\sum_{i,j,w} n_{ijw} \ln g^h_{ijw}$, where n_{ijw} is 1 if the image has the w th color at the pixel (i, j) and 0 otherwise. Based on the obtained probabilities, a category for the image can be determined by $\arg\max_h \sum_{i,j,w} n_{ijw} \ln g^h_{ijw}$. We regard this method as the baseline method and would like to improve it in terms of *efficiency*.

Any prototype the baseline gives has $N_1 N_2 W$ parameters, whose number can be reduced by *clustering* histograms. We propose a new probabilistic model called Multinomialized Subset Infinite Relational Model (MSIRM) for clustering. MSIRM clusters histograms column- and row-wise. Denote the numbers of column and row clusters as K_1 and K_2, respectively. In MSIRM, each pixel is assigned to a pair of column and row clusters, and the colors of the pixels assigned to the same pair of column and row clusters are assumed to be drawn from the same distribution. Consequently, we can reduce the complexity of prototypes from $N_1 N_2 W$ to $K_1 K_2 W$. MSIRM determines K_1 and K_2 flexibly with Chinese restaurant process (CRP) [4]. MSIRM has an important feature: it detects columns and rows *irrelevant* for constructing prototypes. This feature can speed up classification, because we can reduce execution time of classification by skipping irrelevant columns and rows. The skipping technically means giving probability one to all pixels in irrelevant columns and rows. We will show that the skipping lead to only a small degradation in classification accuracy.

The rest of the paper is organized as follows. Section 2 gives preceding proposals important for us. Section 3 provides details of MSIRM. Section 4 presents the results of our comparison experiment. Section 5 concludes the paper with discussions.

2 IRM and SIRM

We propose MSIRM as an extension of Subset Infinite Relational Model (SIRM) [1], which is, in turn, an improvement of Infinite Relational Model (IRM) [2].

Assume that we have N_1 entities of type T_A and N_2 entities of type T_B, and that a binary relation R is defined over a domain $T_A \times T_B$. The relation can be represented by an N_1 by N_2 binary matrix as the left most panel of Fig. 1 shows. IRM discovers a column- and row-wise clustering of entities so that the matrix gives a relatively clean block structure when sorted according to the clustering as the center panel of Fig. 1 shows. IRM can determine the numbers of column and row clusters flexibly with CPR. Further, IRM associates each block, enclosed by thick lines in the center panel of Fig. 1, with a binomial distribution determining the probability that the pairs of entities in the block fall under the relation R. While IRM can be extended to the cases where there are more than two types of entity, we do not consider such cases.

When constructing a prototype from the images of the same category, we can consider a relation between pixel columns and rows by taking an IRM-like approach, because contiguous pixels are likely to give similar color distributions. However, IRM is vulnerable to noisy data. Therefore, a mechanism for detecting *irrelevant* columns and rows is introduced by SIRM, where clustering is conducted only on a *subset* of columns and rows, i.e., on columns and rows other than irrelevant ones. In SIRM, we flip a coin at each column and row to determine whether relevant or not. Irrelevant pixels are bundled into the same group, as the right most panel of Fig. 1 depicts with red cells. However, both IRM and SIRM can only handle binary data. Therefore, we "multinomialize" SIRM for handling multiple colors and obtain our model, MSIRM.

3 MSIRM

We below describe how observed data are generated by MSIRM.

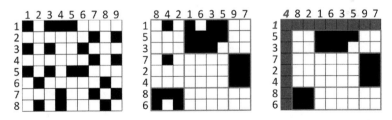

Fig. 1 Clustering of a binary relation (*left*) by IRM (*center*) and by SIRM (*right*). Each cluster is enclosed by *thick lines*. *Red cells* in the *right* most panel correspond to irrelevant pixels

1. Draw the parameter λ_1 of a binomial $Bi(\lambda_1)$ from a Beta prior $Be(a_1, b_1)$. Then, for each column, draw a 0/1 value from $Bi(\lambda_1)$. Let r_{1i} denote the value for the ith column, which is irrelevant if $r_{1i} = 0$ and is relevant otherwise.
2. Draw the parameter λ_2 of a binomial $Bi(\lambda_2)$ from a Beta prior $Be(a_2, b_2)$. Then, for each row, draw a 0/1 value from $Bi(\lambda_2)$. Let r_{2j} denote the value for the jth row, which is irrelevant if $r_{2i} = 0$ and is relevant otherwise.
3. Draw the parameters $\phi_{kl1}, \ldots, \phi_{klW}$ of a multinomial distribution $Mul(\phi_{kl})$ from a Dirichlet prior $Dir(\beta)$. $Mul(\phi_{kl})$ is the multinomial for generating colors of the pixels belonging to the kth column cluster and, at the same time, to the lth row clusters.
4. Draw the parameters $\psi_1, \ldots, \psi_W$ of a multinomial $Mul(\psi)$ from a Dirichlet prior $Dir(\gamma)$. $Mul(\psi)$ is the multinomial for generating colors of the irrelevant pixels.
5. For each relevant column, draw a cluster ID based on a Chinese restaurant process $CRP(\alpha_1)$. We introduce a latent variable z_{1i}, which is equal to k if the ith column is relevant and belongs to the kth column cluster.
6. For each relevant row, draw a cluster ID based on a Chinese restaurant process $CRP(\alpha_2)$. We introduce a latent variable z_{2j}, which is equal to l if the jth row is relevant and belongs to the lth column cluster.
7. For each pixel (i, j), draw a color from the multinomial $Mul(\psi)$ if $r_{1i}r_{2j} = 0$ (i.e., the pixel is irrelevant) and from the multinomial $Mul\left(\phi_{z_{1i}z_{2j}}\right)$ otherwise.

We adopt Gibbs sampling technique for inferring posterior distribution of MSIRM. For each column, we update the relevant/irrelevant coin flip r_{1i} and the cluster assignment z_{1i} based on the following posterior probability:

$$p\left(z_{1i}, r_{1i} | X, Z^{\backslash 1i}, R^{\backslash 1i}\right) \propto p\left(X^{+1i} | z_{1i}, r_{1i}, X^{\backslash 1i}, Z^{\backslash 1i}, R^{\backslash 1i}\right) p\left(z_{1i}, r_{1i} | Z^{\backslash 1i}, R^{\backslash 1i}\right),$$

where the details of the two terms on the right hand side are omitted due to space limitation. The derivation is almost the same with that for SIRM [1]. We also conduct a similar update for each row. Additionally, we update the hyperparameters of the Dirichlet and the Beta priors by Minka's method.[1]

4 Experiment

Our experiment compared MSIRM with the baseline and also with a method using Dirichlet process mixture of multinomial (DP-multinomial) [3] for clustering histograms. Our target was MNIST data set,[2] consisting of 60,000 training and

---

[1] http://research.microsoft.com/en-us/um/people/minka/papers/dirichlet/

[2] http://yann.lecun.com/exdb/mnist/

Trimming Prototypes of Handwritten Digit Images

10,000 test images. All images are 28×28 pixels in size, i.e., $N_1 = N_2 = 28$. We quantized 8 bit gray-scaled colors into 4 bit colors uniformly, i.e., $W = 16$.

Let n_{ijw}^h be the number of times the wth color appears at the pixel (i,j) in the training images of category h, where h ranges over ten digit categories. The baseline method calculates a probability g_{ijw}^h that the wth color appears at the pixel (i,j) in the images of category h as $g_{ijw}^h = (n_{ijw}^h + \eta_w)/\sum_w(n_{ijw}^h + \eta_w)$, where a Dirichlet smoothing with the parameters $\eta_1,\ldots,\eta_W$ is applied. We determine the category of an unknown image by $\arg\max_h \sum_{i,j} \hat{n}_{ijw} \ln g_{ijw}^h$, where $\hat{n}_{ijw}$ is 1 if the wth color appears at the pixel (i,j) in the unknown image and is 0 otherwise. For the DP-multinomial method, we set g_{ijw}^h to a posterior probability estimated by a Gibbs sampling [3] and determine the category of an unknown image in the same manner with the baseline method. For MSIRM, we determine the category of an unknown image as follows:

$$\arg\max_h \sum_{i=1}^{N_1} \sum_{j=1}^{N_2} \hat{n}_{ijw} \left\{ r_{1i}r_{2j} \ln \frac{n_{z_{1i}z_{2j}w}^h + \beta_w^h}{\sum_w(n_{z_{1i}z_{2j}w}^h + \beta_w^h)} + (1 - r_{1i}r_{2j}) \frac{q_w^h + \gamma_w^h}{\sum_w(q_w^h + \gamma_w^h)} \right\},$$

where n_{klw}^h is the number of times the wth color appears at the pixels belonging to the kth column and the lth row clusters in the training images of category h. q_w^h is the number of times the wth color appears at the irrelevant pixels of the training images of category h. β_w^h and γ_w^h are the hyperparameters for category h.

We give results of the experiment. Let K be the number of clusters given by DP-multinomial. With respect to the complexity of prototypes, DP-multinomial was superior to MSIRM. While $K_1 \approx 20$ and $K_2 \approx 20$ for MSIRM, $K \approx 85$ for DP-multinomial with respect to all digits. That is, $K < K_1 K_2$ for all digits.

Classification accuracies of the baseline, DP-multinomial, and MSIRM were 0.840, 0.839 and 0.837 respectively. While the accuracies were far from the best reported at the Web site of the data set, they were good enough for a meaningful comparison. MSIRM and DP-multinomial gave almost the same accuracies with that of the baseline. Therefore, it can be said that we could reduce the complexity of prototypes by clustering histograms without harming clustering accuracy.

While DP-multinomial and MSIRM showed no significant difference in terms of accuracy, MSIRM could neatly detect irrelevant columns and rows at peripheral part of images. Figure 2 visualizes the prototypes obtained by MSIRM. We mixed grayscale colors linearly by multiplying their probabilities at each pixel and obtained the visualization in Fig. 2. The red-colored area corresponds to irrelevant columns and rows. Astonishingly, MSIRM precisely detected the area irrelevant for digit identification.

By utilizing this "trimming" effect, we could speed up classification. We skipped irrelevant columns and rows when we calculated the log probabilities of unknown images. Technically, this means that we assigned probability one to all irrelevant pixels. For the prototypes in Fig. 2, we could skip 32.2 % of the $28^2 \times 10 = 7,840$ pixels. This led to 32.2 % speeding up of classification.

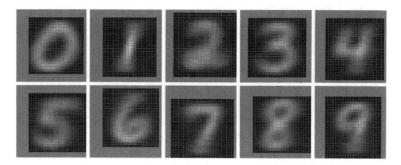

Fig. 2 Irrelevant portion (*red*-colored pixels) trimmed out by MSIRM from each prototype

The achieved accuracy was 0.819, which shows a small degradation from 0.837. Therefore, it can be said that MSIRM could speed up classification with only a small degradation in accuracy.

5 Conclusion

This paper proposes a prototype-based image classification using MSIRM. We could neatly detect irrelevant pixels and could speed up classification by skipping the irrelevant pixels. This led to only a small degradation in accuracy. While DP-multinomial was comparable with MSIRM in accuracy and was superior to MSIRM in reduction of the complexity of prototypes, it has no mechanism for detecting irrelevant pixels.

We have a future plan to extend MSIRM for realizing a clustering of training images by introducing an additional axis aside from the column and the row axes.

References

1. Ishiguro K, Ueda N, Sawada H (2012) Subset infinite relational models. In: Proceedings of AISTATS 2012, JMLR W&CP 22, pp 547–555
2. Kemp C, Tenenbaum JB, Griffiths TL, Yamada T, Ueda N (2006) Learning systems of concepts with an infinite relational model. In: Proceedings of AAAI'06. p 381–388
3. Neal RM (2000) Markov chain sampling methods for Dirichlet process mixture models. J Comput Graph Stat 9(2):249–265
4. Pitman J (2002) Combinatorial stochastic processes. Notes for Saint Flour Summer School

Ranking Book Reviews Based on User Influence

Unil Yun and Heungmo Ryang

Abstract As the number of people who buy books online increases, it is becoming a way of life for people. Recently, algorithms to retrieve book reviews have been proposed for searching meaningful information since ranking algorithms for general purpose are not suitable. Although the previous algorithms consider features of book review, they calculate ranking scores without reflecting user influence. In this paper, thus, we propose a novel algorithm for ranking book reviews based on user influence. To apply user influence, the proposed algorithm uses recommendations by other users. For performance evaluation, we perform precision and recall tests. The experimental results show that the proposed algorithm outperforms previous algorithms for searching book reviews.

Keywords Book review · Information retrieval · User influence

1 Introduction

Since World Wide Web has emerged, the number of documents on the Internet has been increasing exponentially. In face of the overwhelming information volumes, people focus on solving information overload rather than its shortage, and it is a difficult task to find meaningful information on the Internet. Thus, the importance of searching relevant and useful documents has risen. IR (Information Retrieval) systems look for the information by measuring its importance. If given a query of user, general search engines evaluate the relation between the query and indexed

---

U. Yun (✉) · H. Ryang
Department of Computer Science, Chungbuk National University, Chungcheongbuk-do, South Korea
e-mail: yunei@chungbuk.ac.kr

H. Ryang
e-mail: riangs@chungbuk.ac.kr

J. J. (Jong Hyuk) Park et al. (eds.), *Multimedia and Ubiquitous Engineering*, Lecture Notes in Electrical Engineering 240, DOI: 10.1007/978-94-007-6738-6_17, © Springer Science+Business Media Dordrecht(Outside the USA) 2013

documents on the Internet, and it lists results based on the relation in descending order of ranking scores. This process is called ranking documents, and algorithms used for such retrieval are called ranking algorithms. Various ranking algorithms [1, 2, 4–7] have been proposed, and some of them are based on references or quotations between documents through hyperlinks [1, 2, 5]. One of the most famous algorithms is PageRank [5] adopted by Google (http://www.google.com) and it becomes a fundamental algorithm of IR. Although PageRank has played an important role in IR area, it is not suitable for the system finding meaningful book reviews since there are few references or quotations between book reviews. To address this issue, algorithms, LengthRank and ReplyRank [6], were proposed, and the algorithms reflect features of book review. Nevertheless, ranking scores are calculated without reflecting influence of users in the algorithms. That is, only the features of book review are considered in the ranking scores. For this reason, we propose a novel algorithm for ranking book reviews based on user influence. Major contributions of this paper are summarized as follows: (1) a novel algorithm, called IRRank (Influence based Review Rank), is proposed and (2) various experiments are conducted for performance evaluation of the proposed algorithm.

The remainder of this paper is organized as follows. In Sect. 2, we introduce the influential related work. The proposed ranking algorithm is described in Sect. 3. In Sect. 4, we show performance of the proposed algorithm through various experimental evaluations. Finally our contributions are summarized in Sect. 5.

2 Related Work

In the IR field, extensive studies [1–3, 7] have been proposed to search and retrieve relevant information, and ranking algorithms are divided into content-based and inbound link-based methods. Content-based ranking algorithms determine the relevant degrees between user queries and documents through inside information such as the frequency and distance. TF-IDF [9] is a popular weighting scheme and a numerical statistics which is reflected how important words are to documents in the collection, and it is calculated by multiplying tf by idf, where tf refers to the frequency of words in documents and idf is a measure as general importance of the words. In this paper, we employ TF-IDF for measuring the importance of words in book reviews. On the other hand, inbound link-based ranking algorithms improve the quality of search results based on external information. They measure the importance of documents based on concept of casting votes such as inbound hyperlinks, as votes from some documents are regarded as having a greater ranking score. RageRank [5] estimates the ranking score of documents using hyperlinks between the documents. Although PageRank has become the most famous method after it was adopted by Google, it is not suitable for the systems searching meaningful book reviews. The reason is that there are few references or quotations between book reviews by hyperlinks. To address this issue, LengthRank and ReplyRank [6] were proposed for evaluating the importance of book reviews, and the algorithms consider features of book review regarding the length of contents and the number of reply, respectively. In the algorithms, book reviews which have more length or contain a larger number of replies receive higher ranking. However, they cannot consider user influence, and thus this study aims to evaluate the relevance of book reviews by reflecting user influence.

3 Ranking Algorithm Based on User Influence

The framework of the proposed method consists of three steps. In the first step, influence of each user is analyzed. In the second step, features of book review are extracted, and then morphemic analysis is performed. In the last step, ranking scores are calculated using information with the analyzed influence and extracted features. In the following subsections, we first describe and define user influence. Then, we illustrate the proposed ranking algorithm in detail.

3.1 User Influence

The proposed ranking algorithm computes ranking scores by reflecting user influence as well as features of book review. The user influence of a certain review used in the proposed algorithm indicates that the average number of recommendations in reviews written by a user. On the other hand, reply means additional information as well as more interesting in the reviews. That is, people can gain more information about books from reviews having more reply. Especially, reply written by users with high influence can be more useful. The user influence is defined as follows.

$$Influence(user) = \sum Rec(rv)/Cnt(user) \tag{1}$$

In the equation, $user$ is a certain user, rv is a review written by $user$, $Rec(rv)$ is the number of recommendations of rv such as $helpful$ evaluation in Amazon (http://www.amazon.com) which is a well-known online book store, and $Cnt(user)$ is the total number of book reviews written by $user$. For example, if there is a certain user who wrote n reviews and the total number of recommendations of the reviews is r, influence of the user, $Influence(user) = r/n$. Figure 1 is an example of user influence.

3.2 Ranking Book Reviews

For calculating ranking score, it is needed to compute influence of each user. Thus, in the first step of the framework, the influence is calculated with Eq. (1). First, each user who wrote book review or reply is extracted from the collected dataset.

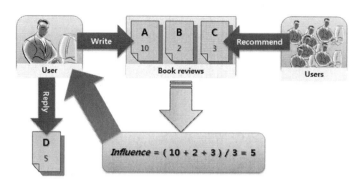

Fig. 1 Example of user influence

Then, the number of reviews written by each user is counted; at the same time, the total number of recommendations with respect to the reviews is computed. After that, values of user influence are calculated. The importance of book reviews can be represented by two aspects. The first is influence of users who wrote reviews or reply. This aspect indicates how an influential user wrote the reviews and how many influential other users participate in the reviews through reply. The other aspect is how much information is contained in the reviews. This aspect can be measured easily by checking the number of replies. On the other hand, keywords involved in each review have to be analyzed and extracted for keyword-based searching. To extract the keyword information, morphemic analysis is performed. The analyzed information is saved to indexed files for employing in the stage of calculating ranking scores and searching book reviews. In addition, each indexed file is created with respect to each keyword except for stop words. In the last step, ranking scores are obtained using the extracted and analyzed information in the indexed files. In this stage, the importance of book reviews is first computed based on user influence. Since the proposed framework is a keyword-based retrieval system, the relevance of keywords is also calculated. Basically, TF-IDF [9] is used for this purpose, and thus we also apply TF-IDF in the proposed ranking algorithm. TF-IDF represents multiplication of the term frequency and the inverse document frequency which is a measure of the general importance of the term as following formula.

$$tf \cdot Idf = log_2 N - -log_2 d_k + 1 \qquad (2)$$

After that, the relevance of keywords is adjusted using the importance of book reviews containing the keywords. To compute the importance of a review rv, influence of users who are involved in the review and the number of replies are used through the following equation.

$$Irv(rv) = \sum Influence(u_{rep}) + Influence(u_{rv}) + Reply(rv) \qquad (3)$$

In the equation, $Irv(rv)$ is the importance of a review rv, u_{rv} and u_{rep} refer to a writer of rv and each user who wrote reply in regard to rv except for u_{rv}, respectively, and $Reply(rv)$ is the number of replies. $Reply(rv)$ is used as the factor for reflecting the amount of additional information in rv. Figure 2 shows an algorithm for ranking book reviews based on user influence.

First, we calculate TF-IDF of each keyword in the indexed files (lines 3–10). That is, the importance of keywords is calculated first. After that, the relevance of each review is computed using Eq. (3) (lines 11–20), and then ranking scores of each keyword is obtained by employing both TF-IDF and $Irv(rv)$ as the following Eq. (4) (line 21).

$$IRRank(keyword, \ rv) = tf \cdot Idf(keyword) + (tf \cdot Idf(keyword) \times Irv(rv)) \quad (4)$$

4 Performance Evaluation

In this section, we provide performance evaluation of IRRank with LengthRank, ReplyRank [6], and Google. Especially, we perform the test with Google by using search operator to limit target dataset as book reviews in online bookstores. We have collected about 114,409 reviews from GoodReads (http://www.goodreads.com) which is a collecting book review site and Amazon. Algorithms are written in Microsoft Visual C++ 2010. In addition, they run with the Windows 7 operating system on an Intel Pentium Quad-core 3.2 GHz CPU with 8 Giga bytes main memory.

Function *IRRank(dataset, indexedFiles)*
1. $IRL \leftarrow \phi$ /* a list of the importance of each review in *dataset* */
2. $N \leftarrow$ the total number of reviews in *dataset*
3. **For each** indexed file *idx* in *indexedFiles*
4.   **Let** *kw* be a keyword of *idx* and *info* be information stored in *idx*
5.   **Count** the number of reviews contained *kw* using *info*
6.   **Set** d_k as the counted value
7. $idf \leftarrow log_2N - log_2d_k + 1$ /* Equation (2) */
8.   **For each** review rv in *idx*
9. $inf \leftarrow 0$
10.   **Let** tf be the frequency of *kw* in rv
11. $TF\text{-}IDF \leftarrow tf \times idf$
12.   **Check** a value of the importance of rv is stored in the *IRL*
13.   **If** $Irv(rv)$ is not stored in the *IRL* **then**
14. $inf \leftarrow Influence(u_{rv})$ /* Equation (1) */
15.   **For each** user u_{rep} in rv
16.   **If** u_{rep} is u_{rv} **then** continue
17.   **Increase** inf by $Influence(u_{rep})$
18.   **Add** inf to *IRL*
19. $Irv(rv) \leftarrow inf + Reply(rv)$ /* Equation (3) */
20.   **Else Set** $Irv(rv)$ as the value stored in *IRL*
21. $IRRank \leftarrow TF\text{-}IDF + (\ TF\text{-}IDF \times Irv(rv) \ )$ /* Equation (4) */

Fig. 2 IRRank algorithm

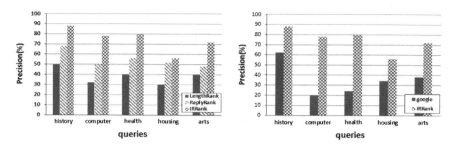

Fig. 3 Precision evaluation

4.1 Precision Evaluation

We first compare performance of IRRank with LengthRank, ReplyRank, and Google by evaluating precision. Precision refers to the rate of retrieved relevant reviews by the searched K reviews. We used five keywords selected from categories in the Amazon and K is set to 50. Figure 3 is the precision results of IRRank with LengthRank and ReplyRank which use the same dataset and Google, respectively. In this paper, we define the important review as a review having no less than the average value of evaluations such as *helpful* in the collected dataset. From Fig. 3, we can observe that IRRank outperforms other algorithms in the sampled keywords. Especially, the reason in regard to result compared with Google is that ranking technique adopted by Google measures the importance based on references or quotations by hyperlinks while there are little references or quotations between book reviews.

4.2 Recall Evaluation

Second, we evaluate performance through recall which is the fraction of relevant reviews to the retrieved reviews with the five queries which are used in the previous experiment. Note that answer dataset is required for performing

Fig. 4 Recall evaluation

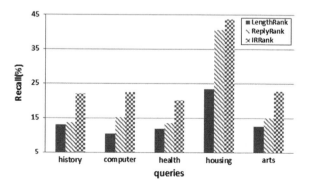

evaluation of recall. Thus, we only perform recall experiment with LengthRank and ReplyRank so that they can use the same answer dataset. Figure 4 shows results of recall evaluation, and we can know that IRRank shows better performance than the previous algorithms in the sampled five keywords.

5 Conclusions

In this paper, we proposed a ranking algorithm, IRRank based on user influence for evaluating the importance of book reviews. Moreover, we conducted precision and recall experiments for performance evaluation. The experimental results showed that the proposed ranking algorithm, IRRank outperformed previous ranking algorithms in terms of both precision and recall. We expect that our research will take effects to not only searching book reviews but also retrieving area based on user influence.

Acknowledgments This research was supported by the National Research Foundation of Korea (NRF) funded by the Ministry of Education, Science and Technology (NRF No. 2012-0003740 and 2012-0000478).

References

1. Aktas MS, Nacar MA, Menczer F (2004) Using hyperlink features to personalize web search. WebKDD. Seattle, pp 104–115
2. Alyguliev RM (2007) Analysis of hyperlinks and the ant algorithm for calculating the ranks of web pages. Autom Control Comput Sci 41(1):44–53
3. Chen MY, Chu HC, Chen YM (2010) Developing a semantic-enable information retrieval mechanism. Expert Syst Appl 37(1):322–340
4. Kritikopoulos A, Sideri M, Varlamis I (2009) BLOGRANK: ranking weblogs based on connectivity and similarity features. CoRR
5. Page L, Brin S, Motwani R, Winograd T (1999) The PageRank citation ranking: bringing order to the Web. Technical report, Stanford InfoLab
6. Ryang H, Yun U (2011) Effective ranking techniques for book review retrieval based on the structural feature. Lecture note in computer science, ICHIT, pp 360–367
7. Tayebi MA, Hashemi S, Mohades A (2007) B2Rank: an algorithm for ranking blogs based on behavioral features. Web intelligence. Silicon Valley, pp 104–107
8. Weng J, Lin EP, Jiang J, He Q (2010) TwitterRank: Finding topic-sensitive influential Twitterers. WSDM. New York, pp 261–270
9. Xia T, Chai Y (2011) An improvement to TF-IDF: term distribution based term weight algorithm. J Softw 6(3):413–420

Speaker Verification System Using LLR-Based Multiple Kernel Learning

Yi-Hsiang Chao

Abstract Support Vector Machine (SVM) has been shown powerful in pattern recognition problems. SVM-based speaker verification has also been developed to use the concept of sequence kernel that is able to deal with variable-length patterns such as speech. In this paper, we propose a new kernel function, named the Log-Likelihood Ratio (LLR)-based composite sequence kernel. This kernel not only can be jointly optimized with the SVM training via the Multiple Kernel Learning (MKL) algorithm, but also can calculate the speech utterances in the kernel function intuitively by embedding an LLR in the sequence kernel. Our experimental results show that the proposed method outperforms the conventional speaker verification approaches.

Keywords Log-Likelihood ratio · Speaker verification · Support vector machine · Multiple kernel learning · Sequence kernel

1 Introduction

The task of speaker verification problem is to determine whether or not an input speech utterance U was spoken by the target speaker. In essence, speaker verification is a hypothesis test problem that is generally formulated as a Log-Likelihood Ratio (LLR) [1] measure. Various LLR measures have been designed [1–4]. One popular LLR approach is the GMM-UBM system [1], which is expressed as

$$L_{\text{UBM}}(U) = \log p(U|\lambda) - \log p(U|\Omega), \tag{1}$$

Y.-H. Chao (✉)
Department of Applied Geomatics, Chien Hsin University of Science and Technology, Taoyuan, Taiwan
e-mail: yschao@uch.edu.tw

J. J. (Jong Hyuk) Park et al. (eds.), *Multimedia and Ubiquitous Engineering*, Lecture Notes in Electrical Engineering 240, DOI: 10.1007/978-94-007-6738-6_18,
© Springer Science+Business Media Dordrecht(Outside the USA) 2013

where λ is a target speaker Gaussian Mixture Model (GMM) [1] trained using speech from the claimed speaker, and Ω is a Universal Background Model (UBM) [1] trained using all the speech data from a large number of background speakers. Instead of using a single model UBM, an alternative approach is to train a set of background models $\{\lambda_1, \lambda_2,..., \lambda_N\}$ using speech from several representative speakers, called a cohort [2], which simulates potential impostors. This leads to several LLR measures [3], such as

$$L_{\mathrm{Max}}(U) = \log p(U|\lambda) - \max_{1 \leq i \leq N} \log p(U|\lambda_i), \tag{2}$$

$$L_{\mathrm{Ari}}(U) = \log p(U|\lambda) - \log\left(\sum_{i=1}^{N} p(U|\lambda_i)/N\right), \tag{3}$$

$$L_{\mathrm{Geo}}(U) = \log p(U|\lambda) - \left(\sum_{i=1}^{N} \log p(U|\lambda_i)\right)/N, \tag{4}$$

and a well-known score normalization method called T-norm [4]:

$$L_{\mathrm{Tnorm}}(U) = L_{\mathrm{Geo}}(U)/\sigma_U, \tag{5}$$

where σ_U is the standard deviation of N scores, $\log p(U|\lambda_i), i = 1, 2,...,N$.

In recent years, Support Vector Machine (SVM)-based speaker verification methods [5–8] have been proposed and successfully found to outperform traditional LLR-based approaches. Such SVM methods use the concept of sequence kernels [5–8] that can deal with variable-length input patterns such as speech. Bengio [5] proposed an SVM-based decision function:

$$L_{\mathrm{Bengio}}(U) = a_1 \log p(U|\lambda) - a_2 \log p(U|\Omega) + b, \tag{6}$$

where a_1, a_2, and b are adjustable parameters estimated using SVM. An extended version of Eq. (6) using the Fisher kernel and the LR score-space kernel for SVM was investigated in [6]. The supervector kernel [7] is another kind of sequence kernel for SVM that is formed by concatenating the parameters of a GMM or Maximum Likelihood Linear Regression (MLLR) [8] matrices. Chao [3] proposed using SVM to directly fuse multiple LLR measures into a unified classifier with an LLR-based input vector. All the above-mentioned methods have the same point that must convert a variable-length utterance into a fixed-dimension vector before a kernel function is computed. Since the fixed-dimension vector is formed independent of the kernel computation, this process is not optimal in terms of overall design.

In this paper, we propose a new kernel function, named the LLR-based composite sequence kernel, which attempts to compute the kernel function without needing to represent utterances into fixed-dimension vectors in advance. This kernel not only can be jointly optimized with the SVM training via the Multiple Kernel Learning (MKL) [9] algorithm, but also can calculate the speech utterances in the kernel function intuitively by embedding an LLR in the sequence kernel.

2 Kernel-Based Discriminant Framework

In essence, there is no theoretical evidence to indicate what sort of LLR measures defined in Eqs. (1)–(5) is absolutely superior to the others. An intuitive way [3] to improve the conventional LLR-based speaker verification methods would be to fuse multiple LLR measures into a unified framework by virtue of the complementary information that each LLR can contribute. Given M different LLR measures, $L_m(U)$, $m = 1, 2,..., M$, a fusion-based LLR measure [3] can be defined as

$$L_{\text{Fusion}}(U) = \mathbf{w}^T \Phi(U) + b, \tag{7}$$

where b is a bias, $\mathbf{w} = [w_1 \, w_2 \ldots w_M]^T$ and $\Phi(U) = [L_1(U) \, L_2(U) \ldots L_M(U)]^T$ are the $M \times 1$ weight vector and LLR-based vector, respectively. The implicit idea of $\Phi(U)$ is that a variable-length input utterance U can be represented by a fixed-dimension characteristic vector via a nonlinear mapping function $\Phi(\cdot)$. Equation (7) forms a nonlinear discriminant classifier, which can be implemented by using the kernel-based discriminant technique, namely the Support Vector Machine (SVM) [10]. The goal of SVM is to find a separating hyperplane that maximizes the margin between classes. Following [10], $\mathbf{w}$ in Eq. (7) can be expressed as $\mathbf{w} = \sum_{j=1}^{J} y_j \alpha_j \Phi(U_j)$, which yields an SVM-based measure:

$$L_{\text{SVM}}(U) = \sum_{j=1}^{J} y_j \alpha_j k(U_j, U) + b, \tag{8}$$

where each training utterance U_j, $j = 1, 2,..., J$, is labeled by either $y_j = 1$ (the positive sample) or $y_j = -1$ (the negative sample), and $k(U_j, U) = \Phi(U_j)^T \Phi(U)$ is the kernel function [10] represented by an inner product of two vectors $\Phi(U_j)$ and $\Phi(U)$. The coefficients α_j and b can be solved by using the quadratic programming techniques [10].

2.1 LLR-Based Multiple Kernel Learning

The effectiveness of SVM depends crucially on how the kernel function $k(\cdot)$ is designed. A kernel function must be symmetric, positive definite, and conform to Mercer's condition [10]. There are a number of kernel functions [10] used in different applications. For example, the sequence kernel [6] can take variable-length speech utterances as inputs. In this paper, we rewrite the kernel function in Eq. (8) as

$$k(U_j, U) = [L_1(U_j) \ldots L_M(U_j)][L_1(U) \ldots L_M(U)]^T = \sum_{m=1}^{M} k_m(U_j, U). \tag{9}$$

Complying with the closure property of Mercer kernels [10], Eq. (9) becomes a composite kernel represented by the sum of M LLR-base sequence kernels [11] defined by

$$k_m(U_j, U) = L_m(U_j) \cdot L_m(U), \tag{10}$$

where $m = 1, 2,..., M$. Since the design of Eq. (9) does not involve any optimization process with respect to the combination of M LLR-base sequence kernels, we further redefine Eq. (9) as a new form, named the LLR-base composite sequence kernel, in accordance with the closure property of Mercer kernels [10]:

$$k_{\text{com}}(U_j, U) = \sum_{m=1}^{M} \beta_m k_m(U_j, U), \tag{11}$$

where β_m is the weight of the m-th kernel function $k_m(\cdot)$ subject to $\sum_{m=1}^{M} \beta_m = 1$ and $\beta_m \geq 0$, $\forall m$. This combination scheme quantifies the unequal nature of M LLR-base sequence kernel functions by a set of weights $\{\beta_1, \beta_2,..., \beta_M\}$. To obtain a reliable set of weights, we apply the MKL [9] algorithm. Since the optimization process is related to the speaker verification accuracy, this new composite kernel defined in Eq. (11) is expected to be more effective and robust than the original composite kernel defined in Eq. (9).

The optimal weights β_m can be jointly trained with the coefficients α_j of the SVM in Eq. (8) via the MKL algorithm [9]. Optimization of the coefficients α_j and the weights β_m can be performed alternately. First we update the coefficients α_j while fixing the weights β_m, and then we update the weights β_m while fixing the coefficients α_j. These two steps can be repeated until convergence. In this work, the MKL algorithm is implemented via the SimpleMKL toolbox developed by Rakotomamonjy et al. [9].

3 Experiments

3.1 Experimental Setup

Our speaker verification experiments were conducted on the speech data extracted from the extended M2VTS database (XM2VTSDB) [12]. In accordance with "Configuration II" described in Table 1 [12], the database was divided into three subsets: "Training", "Evaluation", and "Test". In our experiments, we used "Training" to build each target speaker GMM and background models, and "Evaluation" to estimate the coefficients α_j in Eq. (8) and the weights β_m in Eq. (11). The performance of speaker verification was then evaluated on the "Test" subset.

As shown in Table 1, a total of 293 speakers in the database were divided into 199 clients (target speakers), 25 "evaluation impostors", and 69 "test impostors". Each speaker participated in 4 recording sessions at approximately one-month intervals, and each recording session consisted of 2 shots. In a shot, every speaker was prompted to utter 3 sentences "0 1 2 3 4 5 6 7 8 9", "5 0 6 9 2 8 1 3 7 4", and "Joe took father's green shoe bench out". Each utterance, sampled at 32 kHz, was converted into a stream of 24-order feature vectors, each consisting of 12 mel-

Table 1 Configuration of the speech database

Session	Shot	199 clients	25 impostors	69 impostors
1	1	Training	Evaluation	Test
	2			
2	1			
	2			
3	1	Evaluation		
	2			
4	1	Test		
	2			

frequency cepstral coefficients (MFCCs) [13] and their first time derivatives, by a 32-ms Hamming-windowed frame with 10-ms shifts.

We used 12 ($2 \times 2 \times 3$) utterances/client from sessions 1 and 2 to train the client model, represented by a GMM with 64 mixture components. For each client, the other 198 clients' utterances from sessions 1 and 2 were used to generate the UBM, represented by a GMM with 256 mixture components; 50 closest speakers were chosen from these 198 clients as a cohort. Then, we used 6 utterances/client from session 3, along with 24 ($4 \times 2 \times 3$) utterances/evaluation-impostor, which yielded 1,194 (6×199) client examples and 119,400 ($24 \times 25 \times 199$) impostor examples, to estimate α_j and β_m. However, recognizing the fact that the kernel method can be intractable when a huge amount of training examples involves, we downsized the number of impostor examples from 119,400 to 2,250 using a uniform random selection method. In the performance evaluation, we tested 6 utterances/client in session 4 and 24 utterances/test-impostor, which produced 1,194 (6×199) client trials and 329,544 ($24 \times 69 \times 199$) impostor trials.

3.2 Experimental Results

We implemented two SVM systems, $L_{\text{Fusion}}(U)$ in Eq. (7) ("LLRfusion") and $k_{\text{com}}(U_j, U)$ in Eq. (11) ("MKL_LLRfusion"), both of which are fused by five LLR-based sequence kernel functions defined in Eqs. (1)–(5). For the purpose of performance comparison, we used six baseline systems, $L_{\text{UBM}}(U)$ in Eq. (1) ("GMM-UBM"), $L_{\text{Bengio}}(U)$ in Eq. (6) ("GMM-UBM/SVM"), $L_{\text{Max}}(U)$ in Eq. (2) ("Lmax_50C"), $L_{\text{Ari}}(U)$ in Eq. (3) ("Lari_50C"), $L_{\text{Geo}}(U)$ in Eq. (4) ("Lgeo_50C"), and $L_{\text{Tnorm}}(U)$ in Eq. (5) ("Tnorm_50C"), where 50C represents 50 closest cohort models were used. Figure 1 shows the results of speaker verification evaluated on the "Test" subset in terms of DET curves [14]. We can observe that the curve "MKL_LLRfusion" not only outperforms six baseline systems, but also performs better than the curve "LLRfusion". Further analysis of the results via the minimum Half Total Error Rate (HTER) [14] showed that a 5.76 % relative improvement was achieved by "MKL_LLRfusion" (the minimum HTER = 3.93 %), compared to 4.17 % of "LLRfusion".

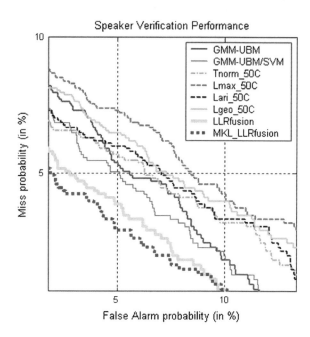

Fig. 1 DET curves for "Test"

4 Conclusion

In this paper, we have presented a new kernel function, named the Log-Likelihood Ratio (LLR)-based composite sequence kernel, for SVM-based speaker verification. This kernel function not only can be jointly optimized with the SVM training via the Multiple Kernel Learning (MKL) algorithm, but also can calculate the speech utterances in the kernel function intuitively by embedding an LLR in the sequence kernel. Our experimental results have shown that the proposed system outperforms the conventional speaker verification approaches.

Acknowledgments This work was funded by the National Science Council, Taiwan, under Grant: NSC101-2221-E-231-026.

References

1. Reynolds DA, Quatieri TF, Dunn RB (2000) Speaker verification using adapted Gaussian mixture models. Digit Signal Proc 10:19–41
2. Rosenberg AE, Delong J, Lee CH, Juang BH, Soong FK (1992) The use of Cohort Normalized scores for speaker verification. Proc, ICSLP
3. Chao YH, Tsai WH, Wang HM, Chang RC (2006) A kernel-based discrimination framework for solving hypothesis testing problems with application to speaker verification. Proceedings of the ICPR
4. Auckenthaler R, Carey M, Lloyd-Thomas H (2000) Score normalization for text-independent speaker verification system. Digit Signal Proc. 10:42–54

5. Bengio S, Mariéthoz J (2001) Learning the decision function for speaker verification. Proceedings of the ICASSP
6. Wan V, Renals S (2005) Speaker verification using sequence discriminant support vector machines. IEEE Trans Speech Audio Proc 13:203–210
7. Campbell WM, Sturim DE, Reynolds DA (2006) Support vector machine using GMM supervectors for speaker verification. IEEE Signal Proc Lett 13
8. Karam ZN, Campbell WM (2008) A multi-class MLLR Kernel for SVM speaker recognition. Proceedings of the ICASSP
9. Rakotomamonjy A, Bach F.R, Canu S, Grandvalet Y (2008) SimpleMKL. J. Mach Learn Res 9:2491–2521
10. Herbrich R (2002) Learning Kernel classifiers: theory and algorithms, MIT Press
11. Chao YH, Tsai WH, Wang HM (2010) Speaker verification using support vector machine with LLR-based sequence Kernels. Proceedings of the ISCSLP
12. Luettin J, Maître G (1998) Evaluation protocol for the extended M2VTS database (XM2VTSDB). IDIAP-COM 98-05, IDIAP
13. Huang X, Acero A, Hon HW (2001) Spoken language processing. Prentics Hall
14. Bengio S, Mariéthoz J (2004) The expected performance curve: a new assessment measure for person authentication. Proceedings ODYSSEY

Edit Distance Comparison Confidence Measure for Speech Recognition

Dawid Skurzok and Bartosz Ziółko

Abstract A new possible confidence measure for automatic speech recognition is presented along with results of tests where they were applied. A classical method based on comparing the strongest hypotheses with an average of a few next hypotheses was used as a ground truth. Details of our own method based on comparison of edit distances are depicted with results of tests. It was found useful for spoken dialogue system as a module asking to repeat a phrase or declaring that it was not recognised. The method was designed for Polish language, which is morphologically rich.

Keywords Speech recognition decisions · Polish

1 Introduction

Research on automatic speech recognition (ASR) started several decades ago. Most of the progress in the field was done for English. It has resulted in many successful designs, however, ASR systems are always below the level of human speech recognition capability, even for English. In case of less popular languages, like Polish (with around 60 million speakers), the situation is much worse. There is no large vocabulary ASR (LVR) commercial software for continuous Polish. Polish speech contains high frequency phones (fricatives and plosives) and the language is highly inflected and non-positional.

D. Skurzok (✉) · B. Ziółko
Department of Electronics, AGH University of Science and Technology,
Al. Mickiewicza 30, 30-059 Kraków, Poland
e-mail: skurzok@agh.edu.pl.
URL: www.dsp.agh.edu.pl

B. Ziółko
e-mail: bziolko@agh.edu.pl.
URL: www.dsp.agh.edu.pl

J. J. (Jong Hyuk) Park et al. (eds.), *Multimedia and Ubiquitous Engineering*,
Lecture Notes in Electrical Engineering 240, DOI: 10.1007/978-94-007-6738-6_19,
© Springer Science+Business Media Dordrecht(Outside the USA) 2013

It is crucial in a spoken dialogue system to not only provide a hypothesis of what was spoken but also to evaluate how likely it is. A simple probability is not always a good measure because its value depends on too many conditions. In case of dialogue systems, additional measure evaluating if the recognition is creditable or not is very useful. A relation to other, non-first hypothesis can provide it. It allows to repeat a question by a spoken dialogue system or choose a default answer for an unknown utterance. The purpose of Confidence Measures (CMs) is to estimate the quality of a result. In speech recognition, confidence measures are applied in various manners.

Existing types and applications of CMs were well summarised [1–3]. CMs can help to decide to keep or reject a hypothesis in keyword spotting applications. They can be also useful in detecting out-of-vocabulary words to not confuse them with some similar vocabulary words. Moreover, for acoustic adaptation, CM can help to select the reliable phonemes, words or even sentences, namely those with a high confidence score. They can be also used for the unsupervised training of acoustic models or to lead a dialogue in an automatic call-centre or information point in order to require a confirmation only for words with a low confidence score. Recently, applying Bayes based CM for reinforced learning was also tested [4]. A CM based on comparison of phonetic substrings was also described [5]. CMs were also applied in a new third-party error detection system [6]. CMs are even more important in speaker recognition. A method based on expected log-likelihood ratio was recently tested in speaker verification [7]. CMs can be classified [2] according to the criteria which they are based on: hypothesis density, likelihood ratio, semantic, language syntax analysis, acoustic stability, duration, lattice-based posterior probability.

2 Literature Review

Results and views on CMs for speech recognition found in the latest papers were analysed while we worked on our method. In some scenarios it is very important to compute CMs without waiting for the end of the audio stream [2]. The frame-synchronous ones can be computed as soon as a frame is processed by the recognition system and are based on a likelihood ratio. They are based on the same computation pattern: a likelihood ratio between the word for which we want to evaluate the confidence and the competing words found within the word graph. A relaxation rate to have a more flexible selection of competing words was introduced.

Introducing a relaxation rate to select competing words implies managing multiple occurrences of the same word with close beginning and ending times. The situation can be solved in two ways. A summation method adds up the likelihood of every occurence of the current word and adds up the likelihood of every occurence of the competing words. A maximisation method keeps only the occurence with the maximal acoustic score.

The frame-synchronous measures were implemented in three ways regarding a context: unigram, bigram and trigram. The trigram one gave the best results on a test corpus.

The local measures estimate a local posterior probability in the vicinity of the word to analyse. They can use data slightly posterior to the current word. However, this data is limited to the local neighbourhood of this word and the confidence estimation does not need the recognition of the whole sentence. Local measures gave better results on a test set.

Two n-gram CMs based evaluations were also recently tested [8] 7-gram based on part-of-speech (POS) tags and 4-gram based on words. The latter was not succesful in detecting wrong recognitions. Applying POS tags in a CM was efficient, probably because it enables analysis on a larger time scale (7-gram instead of 4-gram).

A new CM based on phonetic distance was described [9]. It uses distances between subword units and density comparison (called anti-model by authors). The method employs separate phonetic similarity knowledge for vowels and consonants, resulting in more reliable performance. Phonetic similarities between a particular subword model and the remaining models are identified using training data

$$P\left(X^{\{i\}}|\lambda_{i,1}\right) \geq P\left(X^{\{i\}}|\lambda_{i,2}\right) \geq \cdots \geq P\left(X^{\{i\}}|\lambda_{i,M}\right) \tag{1}$$

where $X^{\{i\}}$ is a collection of training data labeled as model λ_i and $\lambda_{i,m}$ indicates the mth similar model among M subword models compared to the pivotal model λ_i.

Applying of conditional random fields was recently tested [10] for confidence estimation. They allow comparison of features from several sources, namely lattice arc posterior ratio, lattice-based acoustic stability and Levenshtein alignment feature.

3 1-to-3 Comparison

The most widely known CM is based on hypothesis density. It compares the strongest hypothesis with an average of the following n weaker ones (Fig. 1). In our experiments $n = 3$ was empirically found useful and it is a common value for this parameter in other systems as well. Our evaluations were done for sentence error rate. In the first evaluations it worked very well but later on, we found out, that its usefulness is limited in real dialogue applications because it had similar ratio for sentences allowed by a dictionary as for the ones which were not allowed. It was confirmed in later statistical tests with larger dictionaries. This is why we searched for a method based on edit distance comparison and earlier on phonetic substrings [5].

Fig. 1 Algorithm of a standard method of CM by analysis of hypotheses density

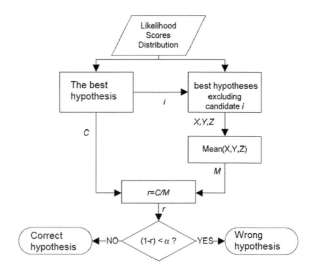

4 Edit Distance Comparison

Edit distance comparison CM was designed and implemented for scenarios where there are several utterances very similar to each other. Such situation is especially common in morphologically rich languages like Polish [11], Czech [12] or Finnish [13]. In this type of scenarios classical CMs frequently fail to help detect wrong recognitions. Our new approach operates by measuring Levenshtein distance [14] in phonetic domain between the strongest hypothesis and the following ones. In this method, the mean of edit distances between the first hypothesis and m following ones is taken as the confidence value. We found that $m = 6$ gives the best results (Fig. 2).

Considering only the mean of distances as the confidence indicator, gives worse results then simple 1-to-3 probability comparison, although both methods can be connected to improve final results. Both methods returns numbers from different range and with different variance. We suggest a following formula as a hybrid approach

$$C = C_{1\,to\,3} + \alpha \bar{D}^\beta, \qquad (2)$$

where C is a final confidence, $C_{1\,to\,3}$ is a confidence calculated using previous method and $\bar{D}$ is a mean of edit distances between the strongest and m following hypothesis. Coefficients α and β are used to scale distance confident and were chosen through optimization. We found that $\alpha = 0.8$ and $\beta = -2$ give the best results.

As it can be concluded, the suggested edit distance comparison method is quite a new approach, which does not fall directly into any of the CM types presented above and listed in literature [2] (Table 1).

Edit Distance Comparison Confidence Measure

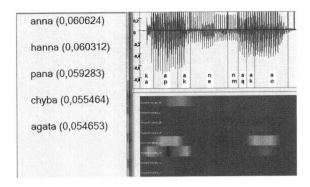

Fig. 2 A screenshot of the developer version of our ASR SARMATA system presents an example of how the described edit distance CM can be applied. The *left* part shows the ranking of top 5 hypotheses and the right one, the time and frequency representation of the analysed audio file. The first three hypothesis have small edit distance between them and the recognition is actually correct

Table 1 Example of calculation of edit distance CM

	Hypothesis	Likelyhood	Distance
1	/anna/	0.120	0
2	/xanna/	0.095	1
3	/panna/	0.080	1
4	/pana/	0.065	2

For this case, let us assume that $m = 3$. The 1-to-3 confidence is C_1 to $_3 = 0.12/((0.095 + 0.08 + 0.065)/3) = 0.12/0.08 = 1.5$. The hybrid confidence (2) is $C = 1.5 + 0.8 ((1 + 1+2)/3)^{-2} = 1.5 + 0.8 \cdot 1.33^{-2} = 1.5 + 0.45 = 1.95$

5 Tests and Results

The standard 1-to-3 method was compared with the edit distance method in a sequence of experiments on as test corpus based on CORPORA [15]. The recordings consists of 4435 audio files, each with one word spoken by various male speakers. The audio files have sampling rate 16 kHz and 16-bit rate. No language model was used in the tests. Some of the words in test corpora were recorded as isolated word, while others were extracted from longer sentences. All tests were made using SARMATA ASR system [11]. The dictionary has 9177 words. 1492 of total 4435 words were recognised correctly (Table 2).

Table 2 Result for different methods ED is an abbreviation of edit distance confidence

	1-to-3	ED	1-to-3 + ED
Precision	0.71	0.38	0.72
Recall	0.65	0.77	0.65
Accuracy	0.79	0.50	0.80
F-measure	0.68	0.51	0.70

6 Conclusions

The suggested CM method based on edit distance enhanced the classical 1-to-3 method in an experiment motivated by real applications and end-user tests. The method was designed for morphologically rich languages, like Polish, as it gives better scores if the strongest hypotheses are phonetically similar. The presented method gives 2 % improvement in F-measure and 1 % improvement in accuracy.

Acknowledgments The project was funded by the National Science Centre allocated on the basis of a decision DEC-2011/03/D/ST6/00914.

References

1. Guo G, Huang C, Jiang H, Wang RH (2004) A comparative study on various confidence measures in large vocabulary speech recognition. Proceedings of international symposium on Chinese spoken language, pp 9–12
2. Razik J, Mella O, Fohr D, Haton J (2011) Frame-synchronous and local confidence measures for automatic speech recognition. Int J Pattern Recognit Artif Intell 25:157–182
3. Wessel F, Schluter R, Macherey K, Ney H (2001) Confidence measures for large vocabulary continuous speech recognition. IEEE Trans Speech Audio Proc 9(3):288–298
4. Molina C, Yoma N, Huenupan F, Garreton C, Wuth J (2010) Maximum entropy-based reinforcement learning using a condense measure in speech recognition for telephone speech. IEEE Trans Audio, Speech Lang Proc 18(5):1041–1052
5. Ziółko B, Jadczyk T, Skurzok D, Ziółko M (2012) Confidence measure by substring comparison for automatic speech recognition. ICALIP, Shanghai
6. Zhou L, Shi Y, Sears A (2010) Third-party error detection support mechanisms for dictation speech recognition. Interact Comput 22:375–388
7. Vogt R, Sridharan S, Mason M (2010) Making confident speaker verification decisions with minimal speech. IEEE Trans Audio Speech Lang Process 18(6):1182–1192
8. Huet S, Gravier G, Sebillot P (2010) Morpho-syntactic post-processing of n-best lists for improved French automatic speech recognition. Comput Speech Lang 24:663–684
9. Kim W, Hansen J (2010) Phonetic distance based condense measure. IEEE Signal Process Lett 17(2):121–124
10. Seigel M, Woodland P (2011) Combining information sources for confidence estimation with crf models. Proceedings of InterSpeech
11. Ziółko M, Gałka J, Ziółko B, Jadczyk T, Skurzok D, Mąsior M (2011) Automatic speech recognition system dedicated for Polish. Proceedings of Interspeech, Florence
12. Nouza J, Zdansky J, David P, Cerva P, Kolorenc J, Nejedlova D (2005) Fully automated system for Czech spoken broadcast transcription with very large (300 k+) lexicon. Proceedings of InterSpeech, pp 1681–1684
13. Hirsimaki T, Pylkkonen J, Kurimo M (2009) Importance of high-order n-gram models in morph-based speech recognition. IEEE Trans Audio Speech Lang Process 17(4):724–732
14. Levenshtein VI (1966) Binary codes capable of correcting deletions, insertions, and reversals. Soviet Phys Doklady 10:707–710
15. Grocholewski S (1998) First database for spoken Polish. Proceedings of international conference on language resources and evaluation, Grenada, pp 1059–1062

Weighted Pooling of Image Code with Saliency Map for Object Recognition

Dong-Hyun Kim, Kwanyong Lee and Hyeyoung Park

Abstract Recently, codebook-based object recognition methods have achieved the state-of-the-art performances for many public object databases. Based on the codebook-based object recognition method, we propose a novel method which uses the saliency information in the stage of pooling code vectors. By controlling each code response using the saliency value that represents the visual importance of each local area in an image, the proposed method can effectively reduce the adverse influence of low visual saliency regions, such as the background. On the basis of experiments on the public Flower102 database and Caltech object database, we confirm that the proposed method can improve the conventional codebook-based methods.

Keywords Object recognition · Visual saliency map · Codebook-based recognition · Code pooling

1 Introduction

Subsequent to the development of the bag-of-features (BoF) method [1] and spatial pyramid matching (SPM) method [2] for object recognition, many studies have been conducted on these types of codebook-based object recognition methods. Some of the studies focused on finding good coding schemes [3], while others

D.-H. Kim
Infraware, Seoul, Korea

K. Lee
Korea Open National University, Seoul, Korea

H. Park (✉)
School of Computer Science and Engineering, Kyungpook National University,
Daegu, Korea
e-mail: hypark@knu.ac.kr

J. J. (Jong Hyuk) Park et al. (eds.), *Multimedia and Ubiquitous Engineering*,
Lecture Notes in Electrical Engineering 240, DOI: 10.1007/978-94-007-6738-6_20,
© Springer Science+Business Media Dordrecht(Outside the USA) 2013

were devoted to pooling of code vectors [4]. However, most of the studies commonly treat all the code vectors from the main object and from the background image with the same importance, and therefore, the codes from the background can have an adverse influence on the recognition performance. To resolve this problem, we propose the use of saliency information, which is often calculated for detecting the reason of interest (ROI), in order to alleviate the effect of the code vectors of the background in the pooling stage. We also propose a generalized pooling method with saliency weight based on the concept of α-mean [5], which is a generalized version of mean operation.

2 Overall Structure of Proposed System

Figure 1 shows the overall process of the proposed object recognition method involving the use of a saliency map; the method is based on a codebook and SPM. We represent an input image I by using the set of scale invariant feature transform (SIFT) descriptors $\{x_m\}_{m=1...M}$, and we then apply the locality-constrained linear coding (LLC) method [3] to obtain a code vector c_m for each descriptor x_m. Once the set of code vectors $\{c_m\}_{m=1...M}$ is obtained, we perform pooling with SPM to get the histogram features of the sub-regions structured in three levels. In the pooling stage, we also use the additional weight value w_m of each c_m, which is obtained by using a saliency map $S(x,y)$. Finally, all the histogram features from the sub-regions are concatenated to obtain a single feature vector for the given image. The feature vector is fed to a linear support vector machine (SVM) classifier to get the recognition result.

In the overall process, the novelty of the present work lies in two steps. First, we calculate the weight value of each code vector by using a saliency map. Second, we propose a generalized pooling method involving the use of the weight values, which we call α-pooling. These two processes are explained in detail in the next section.

3 Pooling of Image Code with Saliency Weight

When we have a saliency map $S(x,y)$ for a given image $I(x,y)$, we first calculate the weight value w_m for each code vector c_m that has been obtained by the LLC method. When a code vector is obtained for a feature descriptor x_m that has been extracted from a local image patch i_m, the corresponding weight value w_m is calculated by simply taking the average of the saliencies in the local image patch, which can be written as

$$w_m = \frac{1}{|i_m|} \sum_{(x,y) \in i_m} S(x, y), \tag{1}$$

where $|i_m|$ denotes the number of pixels in i_m.

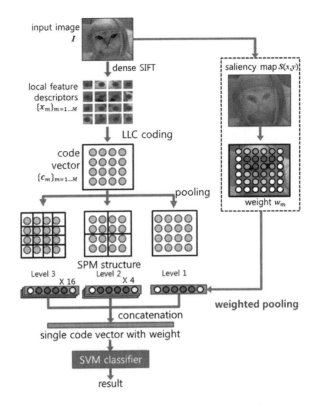

Fig. 1 Overall structure of the proposed method, which combines a codebook-based object recognition method and a saliency map

Once the weight values and code vectors are obtained, we conduct pooling to obtain the histogram features for the given image. In the SPM method, we need to calculate a histogram feature for each sub-region defined in the SPM structure. For example, when we use three-level SPM with grids of dimensions 1×1, 2×2, and 4×4, we can have a set of 21 sub-regions, $\{R_{1,1}^{1\times1}, R_{1,1}^{2\times2}, \ldots, R_{2,2}^{2\times2}, R_{1,1}^{4\times4}, \ldots, R_{4,4}^{4\times4}\}$. When a sub-region R is composed of N code vectors and N weight values, represented as $\{(c_j, w_j)\}_{j=1\ldots N}$, the corresponding histogram feature $\boldsymbol{h}$ is obtained by α-pooling, which is defined as

$$\boldsymbol{h}_\alpha = f_\alpha^{-1}\left(\sum_{j=1}^N f_\alpha(w_j \boldsymbol{c}_j)\right), \quad (2)$$

$$f_\alpha(\boldsymbol{u}) = \begin{cases} \boldsymbol{u}^{\frac{1-\alpha}{2}} & \alpha \neq 1 \\ \log \boldsymbol{u} & \alpha = 1 \end{cases}. \quad (3)$$

This α-pooling is based on a stochastic integration method that is a generalization of various types of mean operations, which is called α-mean [5]. When $\alpha = -1$ and $w_j = 1 (j = 1\ldots N)$ are satisfied, the Eq. (2) becomes equivalent to the well-known sum pooling. Further, when $\alpha = -\infty$ and $w_j = 1 (j = 1\ldots N)$ are satisfied, it gives the formula for the conventional max pooling. By taking arbitrary real values of α and w_j, we can obtain various types of weighted pooling methods. In the experiments, we will show the dependence of the recognition performance on the value of α as well as the weight.

4 Experimental Comparisons

In order to confirm the effect of the saliency weight, we conducted computational experiments with three benchmark datasets: Caltech101 [6], Caltech256 [6], and Flower102 [7]. In each experiment, we extracted a dense set of multi-scale SIFT (PHOW) descriptors with a grid size of 16×16, 24×24, and 32×32 at every 6-pixel steps. In applying LLC, we set its parameter K to be 5. To obtain the saliency value, we adopted the saliency map proposed in [8] and [9].

Figure 2 shows the effect of the weight value and the dependence of the performance of the proposed method on the value of α. From the figure, we can confirm that the use of weights improves the performance for all the databases, regardless of the value of α. We can also see a consistent tendency: a smaller value of α gives better performance. This observation implies that max pooling is superior to sum pooling.

In Table 1, we showed the best accuracies of the proposed method among the results for different values of α shown in Fig. 2. We also showed the results of state-of-the-art methods reported in the literatures. In the case of Flower102, we listed two representative results reported in the original works [7] that has built the database: one was obtained by using only SIFT descriptor and the other was obtained by using the combination of four different descriptors. Though the proposed method cannot achieve the best accuracy obtained by using four descriptors, we can say that our results is promising in the sense that it is superior to the original work under the same condition of single SIFT descriptor. In the case of Caltech databases, we compared the proposed method with a number of recent codebook-based methods. As shown in Table 1, the proposed method gives the best result on Catech101, and the second best results on Catech256. Concerning

Table 1 Experimental results on three benchmark datasets

Method	Flower 102	Caltech 101	Caltech 256
	(20 train)	(15 train)	(30 train)
Weighted α-pooling	$0.633(\alpha = -5)$	$0.691(\alpha = -\infty)$	$0.403(\alpha = -\infty)$
α-pooling	$0.579(\alpha = -5)$	$0.663(\alpha = -\infty)$	$0.392(\alpha = -\infty)$
Nilsback [7] (SIFT)	0.551	–	–
Nilsback [7] (4 descriptors)	0.728	–	–
Yang [2] (sparse code + SIFT)	–	0.670	0.340
Wang [3] (LLC + HOG)	–	0.654	0.412
McCann [10] (SPM variant)	–	0.686	0.395

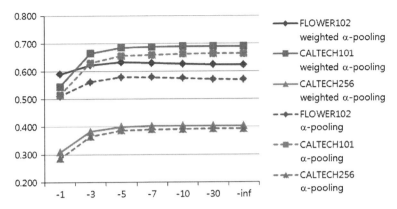

Fig. 2 Dependence of the performance of the proposed method on the value of α and weights

the method suggested in [3], which uses LLC and max pooling, we need to note that it is a special case of the proposed α-pooling method with $\alpha = -\infty$ and $w_j = 1 (j = 1...N)$. Thus, the difference of accuracy between α-pooling and Wang [3] shown in Table 1 is mainly due to the use of different local descriptors (SIFT vs. HOG).

5 Conclusions

On the basis of the LLC and SPM methods, which are state-of-the-art methods of object recognition, we propose a novel weighted α-pooling method in which saliency information and the concept of α-mean are used. By using weighted pooling, we can expect to achieve an efficient representation of a given object image by excluding the background information. By using α-mean, we can have a generalized pooling formula, which can cover sum pooling and max pooling. Experiments with benchmark data show the positive effect of the weight value: it leads to the proposed method showing performance comparable to state-of-the-art methods. The proposed method may be improved further by using more sophisticated local feature descriptors and saliency maps.

Acknowledgments This research was partially supported by the MKE(The Ministry of Knowledge Economy), Korea, under the ITRC(Information Technology Research Center) support program (NIPA-2012- H0301-12-2004) supervised by the NIPA(National IT Industry Promotion Agency); and by the Converging Research Center Program funded by the Ministry of Education, Science and Technology (2012K001342).

References

1. Sivic J, Zisserman A (2003) Video google: a text retrieval approach to object matching in videos. Proceedings of ICCV'03, vol. 2. Los Alamitos, USA, p 1470
2. Yang J, Yu K, Gong Y, Huang T (2009) Linear spatial pyramid matching using sparse coding for image classification. Proceedings of CVPR'09, Miami, USA, pp 1794–1801
3. Wang J, Yang J, Yu K, Lv F, Huang T, Gong Y (2010) Locality-constrained linear coding for image classification. Proceedings of CVPR'10, San Francisco, USA, pp 3360–3367
4. Boureau Y-L, Roux N, Bach F, Ponce J, LeCun Y (2011) Ask the locals: multi-way local pooling for image recognition. ICCV'11, Barcelona, Spain
5. Amari S (2007) Integration of stochastic models by minimizing α-divergence. Neural Comput 19(10):2780–2796
6. http://www.vision.caltech.edu/Image_Datasets. Accessed 20 July 2012
7. Nilsback M-E, Zisserman A (2008) Automated flower classification over a large number of classes. Proceedings of ICVGIP'08, Bhubaneswar, India, pp 722–729
8. Harel J, Koch C, Perona P (2006) Graph-based visual saliency. Proceedings of NIPS'06, Vancouver, Canada
9. Cheng M-M, Zhang G-X, Mitra NJ, Huang X, Hu S-M (2011) Global contrast based salient region detection. Proceedings of CVPR'11, Colorado Springs, USA, pp 409–416
10. McCann S, Lowe DG (2012) Local naïve Bayes nearest neighbor for image classification. Proceedings of CVPR'12, Providence, USA, pp 3650–3656

Calibration of Urine Biomarkers for Ovarian Cancer Diagnosis

Yu-Seop Kim, Eun-Suk Yang, Kyoung-Min Nam, Chan-Young Park, Hye-Jung Song and Jong-Dae Kim

Abstract For the ovarian cancer diagnosis with biomarkers in urine samples, various calibration functions are selected and investigated to compensate the variability of their concentrations. The 15 biomarkers tested in this paper were extracted and measured for the urine samples of 178 patients. Three types of functions were employed to calibrate the biomarkers, including the existing one that divides the biomarker concentration by that of the creatinine. The AUC of the ROC of the calibrated biomarker with each function was chosen to evaluate the performance. Experimental results show that the best performance could be obtained by dividing the concentration of the biomarker by that of the creatinine raised to the power of the optimal exponent that was determined for the maximum AUC of the calibrated biomarker.

Keywords Biomarker · Ovarian cancer · Calibration · Logistic regression · Exponential · AUC

1 Introduction

To prevent the ovarian cancer or increase the possibility of survival, which is one of the most fatal malignant cancer, the development of early diagnosis method or detection of risk factors are paramount [1].

Y.-S. Kim · E.-S. Yang · K.-M. Nam · C.-Y. Park · H.-J. Song · J.-D. Kim (✉)
Department of Ubiquitous Computing, Hallym University, 1 Hallymdaehak-gil,
Chuncheon, Gangwon-do 200-702, Korea
e-mail: kimjd@hallym.ac.kr

Y.-S. Kim · E.-S. Yang · K.-M. Nam · C.-Y. Park · H.-J. Song · J.-D. Kim
Bio-IT Research Center, Hallym University, 1 Hallymdaehak-gil, Chuncheon,
Gangwon-do 200-702, Korea

J. J. (Jong Hyuk) Park et al. (eds.), *Multimedia and Ubiquitous Engineering*,
Lecture Notes in Electrical Engineering 240, DOI: 10.1007/978-94-007-6738-6_21,
© Springer Science+Business Media Dordrecht(Outside the USA) 2013

For the past several decades, considerable investment has been made in the early detection of cancer. However, biopsy is necessary to confirm the cancer, which opposes the goal of early diagnosis that should not be invasive. Biomarkers aim to achieve the early diagnosis, and are defined as markers that can objectively measure whether an organism is in a pathologically normal or abnormal state and the degree of reaction to certain drugs. More specifically, biomarkers can express a pathological state of illness, measure the degree of reaction that an organism shows when treated with certain drugs, and predict a viable treatment to an illness. An ideal tumor biomarker will be the protein fragments detected in the patient's urine or blood that cannot be found in a healthy people [2, 3]. Reference [4] reports the possibility of early detection of ovarian cancer using biomarkers found in urine.

Although urine samples are not useful after 24 h of collection, blood samples are disfavored over urine samples due to the invasiveness of collection and blood-borne diseases [5]. The American Conference of Governmental Hygienists (ACGIH) recommends random urine sampling on basis of the Biological Exposure Indices (BEIs). However random urines have drawbacks due to the variability of urinary output. When measuring the biomarkers in urine samples, the protein quantity in urine can change due to the digested food or the amount of water, and the concentration might also vary according to the time of collection or the sampling method. Much of this variability can be compensated for by adjustment of the concentration of the measured analyte based on the level of creatinine in the urine [5]. Creatinine is the metabolite of the muscle tissue and normally exists in urine. According to ACGIH, approximately 1.2 g of creatinine is produced per day. If the average daily urine volume is 1.2 L (range: 600–2500 ml), the mean creatinine concentration is approximately 1 g/L. Based on this assumption, the creatinine correction will adjust the urine concentration to an average concentration of 1 g/L. Some urines during a day will be above 1 g/L and others will be below 1 g/L, but the analyte concentration will be corrected to a value which would be theoretically equivalent to the value of a urine specimen which has a concentration of 1 g/L [5].

In Ref. [4], the biomarker concentration was calibrated by simply dividing it with the creatinine concentration. However the performance can be much more improved when using another function for calibration as will be shown in this paper. In order to find the best fit function for calibration, this paper uses the area under the curve (AUC) of the receiver operating characteristic curve (ROC) for the calibrated concentrations as the evaluation function to inquire the performance of the several calibration functions. The results show that, for most markers, the best performance could be achieved by dividing the biomarker concentration with the creatinine concentration raised to the power of the exponent smaller than one.

2 Experiments and Results

In this paper, the three functions, m/c^r, e^m/e^c, and $e^m/1 + e^c$ are compared, where m, c, and r are the marker, creatinine concentration, and an exponent to be determined, respectively. When the exponent r in the first function equals to '1', the function will be the same as that used in Ref. [4]. Also note that the marker will not be calibrated with the exponent of '0'. The second and third function utilizes the exponential functions which were selected to increase the value at high concentrations, since the slope generally decreases significantly as the concentration increases in most of the pathological and biological phenomenon. However, these models showed lower performance than the first model for most biomarkers.

In the first model, the AUC of the ROC was calculated for each r from 0 to 15 with an incremental of 0.1, and the optimum exponent was determined when the AUC was the maximum.

In the paper, the proposed method was used to measure the concentration of fifteen biomarkers in 178 urine samples (57 patient samples and 121 healthy samples). In Table 1, the comparison of the AUC when various calibration functions were used for each biomarker is shown. The last two columns show the maximum AUC and the corresponding exponent when the biomarkers were calibrated by dividing with the creatinine value raised to the power of an exponent. The biomarkers in the table are sorted according to the values of this AUC, the highest on the top and the lowest on the bottom. The greatest AUC for each marker are bold-faced. The table is separated by the solid line between the 9 and 10th

Table 1 AUC values after various calibration functions were applied. The greatest AUCs for each marker are bold-faced and the solid line is inserted between the 9 and 10th biomarker rows to distinguish the biomarkers with the AUC < 0.6

	Marker	m	m/c	e^m/e^c	$e^m/1 + e^c$	m/c^r	
						AUC	r
1	**HE4**	0.7791	0.8243	**0.8407**	0.7016	0.8249	0.9
2	**CRP**	0.7667	**0.7809**	0.7123	0.6149	**0.7809**	1.2
3	**TTR**	**0.7705**	0.7319	0.7283	0.6338	**0.7705**	0.0
4	**VCAM**	0.7379	0.7448	0.6910	0.6336	**0.7522**	0.6
5	**NCAM**	0.6654	0.6870	**0.7177**	0.6783	0.7134	0.6
6	**ApocIII**	0.6035	0.6745	**0.6946**	0.6910	0.6842	1.3
7	**MPO**	0.6414	0.6349	0.5939	0.5621	**0.6445**	0.2
8	**PDGF-AA**	0.6000	0.6117	0.5830	0.5782	**0.6194**	0.8
9	**CA 15-3**	0.5493	0.5982	0.5211	0.5205	**0.6021**	0.9
10	**CA 125**	0.5469	0.5866	0.5262	0.5447	**0.5876**	1.1
11	**CA 19-9**	0.5419	0.5801	**0.5817**	0.5750	0.5813	1.2
12	**Apo AI**	0.5434	0.5598	0.5607	0.5450	**0.5803**	1.7
13	**CEA**	0.5249	0.5457	0.5556	**0.5663**	0.5543	1.8
14	*OPN*	0.4891	0.5346	0.5224	0.5320	**0.5536**	1.4
15	**PAI-1**	0.5151	0.5188	**0.5341**	0.5224	0.5240	0.4

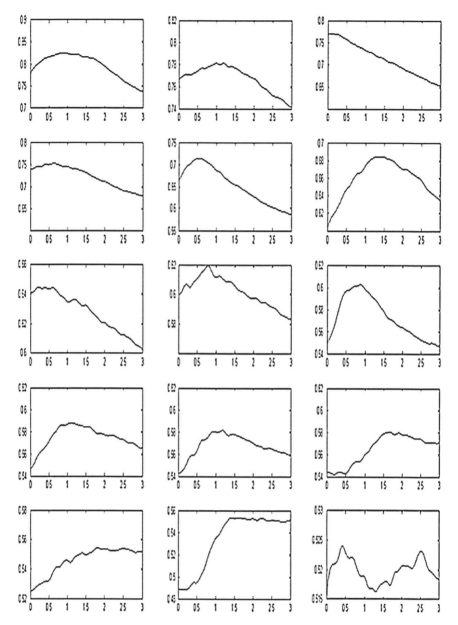

Fig. 1 The AUC variations according to the exponent (From the *top-left*, HE4, CRP, TTR, VCAM, NCAM, ApocIII, MPO, PDGF-AA, CA 15-3, CA 125, CA 19-9, Apo AI, CEA, OPN, PAI-1)

makers, to distinguish the biomarkers with the AUC < 0.6. Note that the biomarkers with the AUC of 0.5 have no information on the disease.

As can be seen in the table, the calibration by the last function in the table shows the best AUC performance for the 6 markers among the 9 top markers. That is, 66 % of the valuable markers showed much better performance when they were calibrated by dividing their concentration with the creatinine concentration raised to the power of the optimal exponent. For 1st, 5th, and 6th markers, this calibration showed only about 2 % less performance than the best calibration method. The optimum exponent was <1 for the 7 markers among the top 9 markers, which opposed the fact that the conventional function is corresponding to the exponent of '1'.

Figure 1 illustrates the AUC variations according to the exponent when the last function was employed for the calibration. All of them showed distinct peaks except for the 3rd and the 15th markers, TTR and PAI-1. This implies that finding the optimum exponent is significant.

3 Conclusion and Discussion

This paper aims to improve the AUC of the ROC for the ovarian cancer diagnosis. The best fit calibration function was explored by comparing several functions that can calibrate the biomarker concentration which was extracted from urine samples. Among them, the calibration that divides the marker concentration with the creatinine concentration raised to the power of the optimum exponent worked best. According to the obtained exponent, this method covers no calibration as well as the existing calibration method where the marker concentration is simply divided by that of the creatinine. The AUC variation according to the exponent in this method was investigated and the distinct peaks were observed, showing that the search of the optimal exponent is more preferable.

Acknowledgments The research was supported by the Research & Business Development Program through the Ministry of Knowledge Economy, Science and Technology (N0000425) and the Ministry of Knowledge Economy (MKE), Korea Institute for Advancement of Technology (KIAT) and Gangwon Leading Industry Office through the Leading Industry Development for Economic Region.

References

1. Choi SK, Cho TI, Lee TJ (2012) Immunohistochemical study of BRCA1, BRCA2, and poly(ADP-ribose) polymerase-1 in ovarian tumors. Korean J Obstet Gynecol 55:8–16
2. Hellstrom I, Heagerty PJ, Swisher EM, Liu P, Jaffar J, Agnew K, Hellstrom KE (2010) Detection of the HE4 protein in urine as a biomarker for ovarian neoplasms. Cancer Lett 296:43–48

3. Hellstrom I, Heagerty PJ, Swisher EM, Liu P, Jaffar J, Agnew K, Hellstrom KE (2010) Detection of the HE4 protein in urine as a biomarker for ovarian neoplasms. Cancer Lett 296:43–48
4. Nolen BM, Lokshin AE (2012) Multianalyte assay systems in the differential diagnosis of ovarian cancer. Expert Opin Med Diagn 6(2):131–138
5. Pacific Toxicology Laboratories http://www.pactox.com/library/article.php?articleID=18

An Iterative Algorithm for Selecting the Parameters in Kernel Methods

Tan Zhiying, She Kun and Song Xiaobo

Abstract Giving a certain training sample set, the learning efficiency almost depends on the kernel function in kernel methods. This inspires us to learn the kernel and the parameters. In the paper, a selecting parameter algorithm is proposed to improve the calculation efficiency. The normalized inner product matrix is the approximation target. And utilize the iterative method to calculate the optimal bandwidth. The defect detection efficiency can be greatly improved adopting the learned bandwidth. We applied the algorithm to detect the defects on tickets' surface. The experimental results indicate that our sampling algorithm not only reduces the mistake rate but also shortens the detection time.

Keywords Kernel methods · Gaussian kernels · Iterative methods · Kernel PCA · Pre-image

1 Introduction

In kernel methods, different kernels have been used to solve a variety of tasks, such as the problems of classification, regression, image de-noising and dimensionality reduction et al. [1]. In support vector machines (SVMs), the three basic kernels polynomial kernels, radial basis functions and sigmoid kernels are successfully

---

T. Zhiying (✉)
1412 Huihong Building, 801 Changwu Mid Road, 213164 Changzhou, Jiangsu, China
e-mail: tanzhiying1010@gmail.com

T. Zhiying · S. Kun
School of Computer Science and Engineering, University of Electronic Science and Technology of China, Chengdu, China

S. Xiaobo
Institute of Advanced Manufacturing Technology, Hefei Institutes of Physical Science, Chinese Academy of Sciences, Changzhou, China

J. J. (Jong Hyuk) Park et al. (eds.), *Multimedia and Ubiquitous Engineering*, Lecture Notes in Electrical Engineering 240, DOI: 10.1007/978-94-007-6738-6_22, © Springer Science+Business Media Dordrecht(Outside the USA) 2013

used [2]. Among the kernels, radial basis functions also named as Gaussian kernels are most widely used for its' stability.

To deal with more complex real problems, some new kernels are proposed. A combination kernel function was obtained by optimizing over a family of data-dependent kernels by Shao et al. [3]. Some regularization techniques are used for learning linear combinations of basic Kernels [4]. Using the learning kernels, the root mean squared error (RMSE) of some classification problems is lower than the single basic kernel. Recently, the non-linear combinations of kernels are also learning by solving the optimization problem. Learning kernels based on a polynomial combination of base kernels has been studied [5]. Multiple kernel learning (MKL) has been recently proposed, which aims at simultaneously learning a kernel and the associated predictor in supervised learning settings [6]. For solving the larger class of problems, the large scale MKL was proposed [7]. And more and more learning methods are constantly being proposed for solving the complex practical problems [8, 9].

Almost all the learning kernels methods are based on the basic kernels and training samples. However, there is little studying on the selection of basic kernels and the parameters. In the paper, we provide an algorithm for learning the parameters of kernels.

2 Kernel Methods

2.1 Kernel PCA

Denote the training set $X = \{x_1, x_2, \ldots, x_N\}$, where the sample $x_i \in R^n$ ($i = 1, 2, \ldots, N$). The nonlinear mapping ϕ maps the samples $x_1, \ldots, x_N$ into the feature space F by [2] $\phi : R^n \to F$, $x \mapsto Y$. Define the kernel matrix $K \in R^{N \times N}$ by $K_{ij} := (\phi(x_i), \phi(x_j))$.

The samples can be centered in feature space by $K_c = (E - I_N)K(E - I_N)$, where the unit matrix $E \in R^{N \times N}, I \in R^{N \times N}(I(i, j) = 1, i, j = 1, 2, \ldots, N)$. The detail calculation of data centering can be found in Schölkopf et al.'s paper [10].

2.2 Pre-Image

To solve the optimization problem of minimizing the reconstruction error, the standard gradient ascent methods were used by the Mika et al [11]. And a modifying iteration method was also proposed to remove outliers from the data vectors by Takahashi and Kurita [12]. In the paper, the local linear property is used to calculate the pre-image by solving [13]

An Iterative Algorithm for Selecting the Parameters

$$\min \rho(t_1, t_2, \ldots, t_p) = ||\phi(\sum_{i=1}^{p} t_i x^i) - P_d \phi(x)||^2$$

$$s.t. \quad \sum_{i=1}^{p} t_i = 1 \quad t_i \geq 0$$

The solution can be obtained by iterative formula

$$t = [[X^p]^T X^p]^{-1} [[X^p]^T X] B \tag{1}$$

where $w_\ell = \sum_{j=1}^{d} \sum_{i=1}^{N} \alpha_i^j k(x, x_i) \alpha_\ell^j$, $\ell = 1, 2, \ldots, N \beta_\ell = w_\ell k(\sum_{i=1}^{p} t_i x^i, x_\ell)$, $(\ell = 1, 2, \ldots, N)$, $B = [\beta_1 / \sum_{\ell=1}^{N} \beta_\ell, \beta_2 / \sum_{\ell=1}^{N} \beta_\ell, \ldots, \beta_N / \sum_{\ell=1}^{N} \beta_\ell]'$. And the initial value of vector t can take $t = [1/p, 1/p, \ldots, 1/p]^T$.

The process of calculating the pre-images can be simply summarized three steps. We firstly calculate the low coordinates $y_i (i = 1, 2, \ldots, N)$ in the feature space using the kernel PCA. Secondly find the p nearest neighbors $x^i (i = 1, 2, \ldots, p)$ by the coordinates $y_i (i = 1, 2, \ldots, N)$. At last, obtain the pre-image by the linear expression of the p nearest neighbors.

3 Theoretical Results and Algorithm

3.1 Theoretical Results

To calculate the optimal bandwidth of Gaussian kernel, establish the following optimization problem

$$\min_{\sigma} F(\sigma) = \sum_{i=1}^{N} \sum_{j=1}^{N} (G(i,j) - K(i,j))^2 \tag{2}$$

where

$$F(\sigma) = \sum_{i=1}^{N} \sum_{j=1}^{N} (\exp(-||x_i - x_j||^2 / 2\sigma^2) - G(i,j)) \tag{3}$$

The necessary conditions of unconstrained optimization problems can be written as $\frac{\partial F(\sigma)}{\partial \sigma} = R(\sigma) = 0$ in the extreme value. the cumulative sum can be seen as the inner product between vector a and vector b, where the vector b is constituted by the coefficients of the function k. The vector a is constituted by items $k(x_i, x_j), k^2(x_i, x_j)(i, j = 1, 2, \ldots, N)$. Then the equation $R(\sigma) = 0$ can be written as $b'a = 0$. In practice, vector b is known, and vector a changes with the variable σ. A plane can be determined by the linear equation $b'x = 0$. Solving the equation $R(\sigma) = 0$ is equivalent to solve equations $b'x = 0$ and $x = a$.

Fig. 1 The curve determined by the parameter

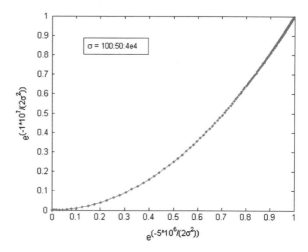

Fig. 2 Solutions' distribution with parameter in two dimension space

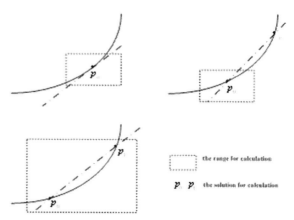

The Fig. 1 shows the changing of two dimensional the vector a following the parameter σ. In Fig. 2 shows three different kinds of solutions' distribution. In Algorithm 1, we will calculate the solution of equation by iterative method [14].

3.2 Algorithm

To calculate the optimal bandwidth σ, we should solve the equation $R(\sigma) = 0$, where

$$R(\sigma) = \sum_{i=1}^{N}\sum_{j=1}^{N} (\exp(-||x_i - x||^2/2\sigma^2) - G(i,j)) \cdot ||x_i - x_j||^2 \cdot \exp(-||x_i - x||^2/2\sigma^2).$$

Based on the above analysis, the equation has at most two roots. We can select the optimal bandwidth σ by the iterative algorithm (Table 1).

An Iterative Algorithm for Selecting the Parameters

Table 1 Iterative algorithm

Algorithm 1
1. Input: $X = [x_1, x_2, \ldots, x_N], \sigma_l(0), \sigma_r(0)$
2. Calculate matrix G, $G(i,j) = \frac{<x_i, x_j>}{\sqrt{<x_i, x_i>}\sqrt{<x_j, x_j>}}$
3. Calculate $R_l(0) = R(\sigma_l(0)), R_r(0) = R(\sigma_r(0))$
4. $nn = 0$
5. while $\left(|R_r(nn) - R_l(nn)| > 1 \times 10^{-6}\right)$
6. $nn = nn+1, \sigma_c = \frac{\sigma_l(nn-1)+\sigma_r(nn-1)}{2}, R_c = R(\sigma_c)$
9.   if $R_l(nn-1) \cdot R_c > 0$
10. $R_l(nn) = R_c, R_r(nn) = R_r(nn-1), \sigma_l(nn) = \sigma_c, \sigma_r(nn) = \sigma_r(nn-1)$
14.   end
15.   if $R_r(nn-1) \cdot R_c > 0$
16. $R_r(nn) = R_c, R_l(nn) = R_l(nn-1), \sigma_r(nn) = \sigma_c, \sigma_l(nn) = \sigma_l(nn-1)$
20.   end
21. end
22. Output: $\sigma = \sigma_l(nn)$

4 Experimental Results

To verify the learning kernel bandwidths' effect, we do some numerical experiments on two data set I and II. The two data sets are from some printed matters. Part samples can be found in Fig. 6. We respectively select 200 samples as the training samples from data sets I and II. We take the initial vaues $\sigma_l(0) = 1 \times 10^2$ and $\sigma_r(0) = 1 \times 10^5$. In Figs. 3, 4 show the iterative results of bandwidths. The bandwidths can fast convergence to the optimal values 14494 and 18375 in the iterative process.

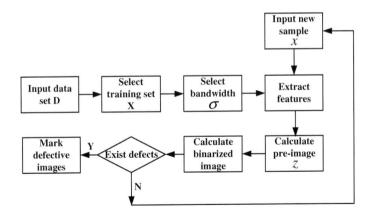

Fig. 3 Solutions' distribution with the parameter

Fig. 4 Solutions' distribution with the parameter

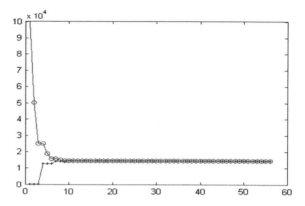

Fig. 5 Solutions' distribution with the parameter

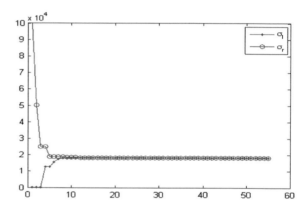

Table 2 Margin specifications

	Test numbers	Bandwidth σ	MRSE
Data 1	1000	14494	1.1834e3
Data 2	724	18375	1.2897e3

Based on the selected 200 training samples, some experiments have been completed on the two data sets. In the experiments, 1000 and 724 test samples are used to verify the rationality of learning bandwidth. We choose 5 nearest neighbors and 80 principal components and the number of iterations is 20 (Figs. 5, 6 and 7), Table 2.

In Fig. 6, shows partial results which include the images with noise, the pre-images and the binary difference images. We can find that the defects can be easily detected by the reconstructed samples using the selected training samples.

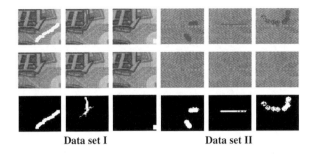

Fig. 6 Solutions' distribution with the parameter

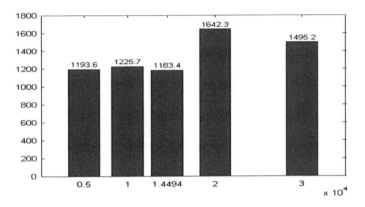

Fig. 7 Solutions' distribution with the parameter

5 Conclusions

We present a method of learning kernel parameters, including an iterative algorithm. The bandwidth of Gaussian kernel has a large influence in calculating the pre-images using the kernel PCA. The method not only can be used to select the parameters, but also can be used to learn the kernel. The sensible of selection theory is also corroborated by some of our empirical results.

References

1. Shawe-Taylor J, Cristianini N (2004) Kernel methods for pattern analysis. Cambridge University, London
2. Schölkopf B, Smola A, Müller KR (1997) Kernel principal component analysis. ICANN: artificial neural networks, pp 583–588
3. Shao JD, Rong G, Lee JM (2009) Learning a data-dependent kernel function for KPCA-based nonlinear process monitoring. Chem Eng Res Design 87:1471–1480

4. Cortes C, Mohri M, Rostamizadeh A (2009) L2 regularization for learning kernels. In: Conference uncertainty in artificial intelligence, pp 109–116
5. Cortes C, Mohri M, Rostamizadeh A (2009) Learning non-linear combinations of kernels. Adv Neural Inf Proc Syst 22:396–404
6. Rakotomamonjy A, Bach FR, Canu S et al (2009) SimpleMKL. J Mach Learn Res 9:2491–2521
7. Bach F (2008) Exploring large feature spaces with hierarchical multiple kernel learning. arXiv preprint 0809:1–30
8. Yi Y, Nan Y, Bingchao D et al (2012) Neural decoding based on Kernel regression. JDCTA Int J Digit Content Technol Appl 6:427–435
9. Shi WY (2012) The algorithm of nonlinear feature extraction for large-scale data set. IJIPM Int J Inf Proc Manage 3:45–52
10. Scholkopf B, Smola A, Muller KR (1998) Nonlinear component analysis as a Kernel eigenvalue problem, vol 10. pp 1299–1319
11. Mika S, Schölkoph B, Smola A et al (2001) Kernel PCA and de-noising in feature spaces. Adv Neural Inf Proc Syst 11:536–542
12. Takahashi T, Kurita T (2002) Robust de-noising by Kernel PCA. Artificial neural networks-ICANN, pp 789–789
13. Tan ZY, Feng Y (2011) A novel improved sampling algorithm. In: Conference communication software and networks, pp 43–46
14. Gerald CF, Wheatley PO (2006) Applied numerical analysis. Pearson Academic, America

A Fast Self-Organizing Map Algorithm for Handwritten Digit Recognition

Yimu Wang, Alexander Peyls, Yun Pan, Luc Claesen and Xiaolang Yan

Abstract This paper presents a fast version of the self-organizing map (SOM) algorithm, which simplifies the weight distance calculation, the learning rate function and the neighborhood function by removing complex computations. Simplification accelerates the training process in software simulation and is applied in the field of handwritten digit recognition. According to the evaluation results of the software prototype, a 15–20 % speed-up in the runtime is obtained compared with the conventional SOM. Furthermore, the fast SOM accelerator can recognize over 81 % of handwritten digit test samples correctly, which is slightly worse than the conventional SOM, but much better than other simplified SOM methods.

Keywords Neural network · Self-organizing map · Handwritten digit recognition · Simplification

1 Introduction

The self-organizing map (SOM) also called Kohonen neural network is a competitive learning artificial neural network proposed by Kohonen in 1982 [1]. It is an unsupervised learning method which has both visualization and clustering properties by discovering the topological structure hidden in the data sets. Essentially the goal of a self-organizing map is to map continuous high-dimensional data onto a discrete low (typically one- or two-) dimensional feature map.

Y. Wang (✉) · Y. Pan · X. Yan
Institute of VLSI Design, Zhejiang University, Hangzhou, People's Republic of China
e-mail: wym85511@gmail.com

A. Peyls · L. Claesen
EDM, Hasselt University, Diepenbeek, Belgium

J. J. (Jong Hyuk) Park et al. (eds.), *Multimedia and Ubiquitous Engineering*,
Lecture Notes in Electrical Engineering 240, DOI: 10.1007/978-94-007-6738-6_23,
© Springer Science+Business Media Dordrecht(Outside the USA) 2013

As a clustering algorithm, the SOM has been applied widely in various fields including pattern recognition, defect inspection and as a data-mining tool to perform classification of high-dimensional data [2, 3]. Research on improving the performance of the SOM has been going on for decades. One of the key issues to overcome is the low speed learning process while obtaining a well trained map. A SOM is well trained if clustering is achieved in a short time and, at the same time, it creates a projection of data into the map strongly related to the distribution of data in the input space. One of the main reasons for this continued research effort is that the amount of data which is to be analyzed can be huge, for instance thousands of high-dimensional image vectors. The simulation of extensive networks with thousands of neurons, each with high-dimensional weights takes relatively much time on state of the art general purpose computers. To solve this problem, this paper presents a fast version of the SOM algorithm and software simulation proves that the SOM has been accelerated to some extent.

The remainder of this paper is organized as follows: Sect. 2 gives a brief overview of related works. Next, Sect. 3 presents the conventional self-organizing map. Section 4 presents our proposed fast self-organizing map algorithm. Section 5 discusses the experimental results on handwritten digit recognition and finally in Sect. 6 the conclusions are drawn.

2 Related Work

To improve both the efficiency and effectiveness of the conventional SOM algorithm, many approaches have been proposed. A first possibility to reduce the runtime of the SOM is to compute initial values for the feature map instead of choosing them randomly in such a way that the training will be accelerated. In [4] the K-means clustering algorithm is used to select initial values for the weight vectors of the neurons, which subsequently reduces the required amount of training steps. Because the SOM offers multiple opportunities to exploit the parallel computing [5], a second way of handling the computational complexity is to transform the SOM algorithm into a distributed algorithm. Lobo et al. developed a distributed SOM in order to speed up the training of the SOM [6]. In order to shorten the processing time the batch version of the SOM has been used by Yu and Alahakoon [7], this version of the SOM is also more suitable for parallel implementation. A third way of accelerating the neural computations is to design simplified SOM algorithm. The weight update step is simplified by removing the non-linear functions in the following papers [8, 9] and therefore results in a more hardware-friendly version of the SOM algorithm. Nevertheless, these simplified methods suffer from a low recognition accuracy and are hardly effective in complex applications. In this context, a fast SOM algorithm is proposed in this paper which not only speeds up the training process but also promises a similar recognition accuracy with the conventional SOM.

3 Self-Organizing Map Algorithm

1. Initialization step: At the start of the SOM algorithm, typically all the weights w_j of the neurons are initialized with random values.
2. Compute the distance between the training vector $X = \{x_1, \ldots, x_M\}$ and each neuron N_j with weight w_j, using the Euclidean distance function:

$$D_j = \sqrt{\sum_{i=1}^{M} (x_i - w_{ji})^2} \tag{1}$$

3. Define the winning neuron as the neuron with the minimum distance.
4. Update each neuron according to the following update function:

$$w_j(t+1) = w_j(t) + \alpha(t) \cdot N_{j,I(X)}(t) \cdot (X - w_j(t)) \tag{2}$$

$w_j(t+1)$ is the updated weight vector, $\alpha(t)$ the learning rate and $N_{j,I(X)}(t)$ the topological neighbourhood value at training step t. $I(X)$ is the winning neuron.

5. Update the neighborhood function and the learning rate.
6. Repeat steps 2–5 for the next training vector.

4 Fast Self-Organizing Map Algorithm

Distance calculation The conventional self-organizing map uses the Euclidean norm as the distance calculating function (see Eq. 1), however because it involves the squaring of values and a square root, the Euclidean distance computation is time-consuming for software prototype and also resource-intensive for hardware implementation. Following [8, 10], we use the Manhattan distance which is computationally simpler for calculating the distance between vectors.

$$D_j = \sum_{i=1}^{M} |(x_i - w_{ji})| \tag{3}$$

Learning Rate Function The learning rate is typically defined as the following exponential function.

$$\alpha(t) = \alpha_0 \cdot exp\left(\frac{-t}{\tau_\alpha}\right) \tag{4}$$

Note that because of the multiplication, division and exponential function, this function will cost too much learning time. To reduce the computational complexity imposed by the exponential calculation of the conventional SOM, an alternative formula is selected to substitute the conventional Eq. 4. Actually, this alternative is the first term of the Taylor series expansion of Eq. 4.

$$\alpha = \alpha_0 \left(1 - \frac{t}{T}\right) \tag{5}$$

Neighborhood Function When using Kohonen's self-organizing map, the distance in the feature map between the neurons influences the learning process. A typical neighborhood function is shown in Eq. 6, which decreases not just over time, but also depends on the topological distance of the two neurons in the net.

$$N_{x,y}(t) = exp\left(\frac{-d_{x,y}^2}{2\sigma^2(t)}\right) \tag{6}$$

Here $d_{x,y}$ is the distance between node x and node y, more specifically it is the physical distance between the nodes in the feature map and $\sigma(t)$ the time dependent value responsible for decreasing the neighborhood size over time.

However our proposed neighborhood function ignores any influence of time and only depends on the topological distance between two neurons, which is computed by the Euclidean norm. For each neuron within the neighborhood size ns, the neighborhood parameter is calculated as shown in Eq. 7. Neurons outside this neighborhood area will not be updated.

$$N_{x,y} = \begin{cases} e^{-2d_{x,y}^2} & \text{if } d(x,y) \le ns \\ 0 & \text{if } d(x,y) > ns \end{cases} \tag{7}$$

The weight update function depends on both the learning rate function and the neighborhood function. In Fig. 1, respectively the results of the conventional weight update function and our proposed weight update function are shown in the condition of $\alpha_0 = 1, d_{x,y} = [0,4], t = [0,10000]$. Note that the shapes of the 3D charts based on these functions are similar, which motivates the similarity in performance between both versions. The performance results will be given in the last section.

5 Case Study: Handwritten Digit Recognition

The performance of the SOM was tested in the field of handwritten digit recognition and the MNIST database was chosen to train and test the feature map [11]. We evaluated the proposed fast SOM by a software simulation on a PC with a general purpose processor clocked at 2.1 GHz and 2 GB of SDRAM. In Fig. 2, the runtime with varying amounts of iterations and varying amounts of neurons is

A Fast Self-Organizing Map Algorithm for Handwritten Digit Recognition

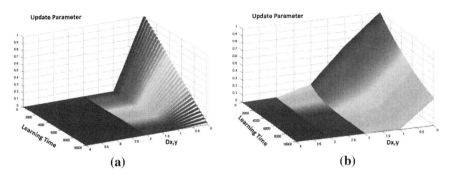

Fig. 1 Comparison of weight update functions. **a** Proposed weight update function. **b** Conventional weight update function

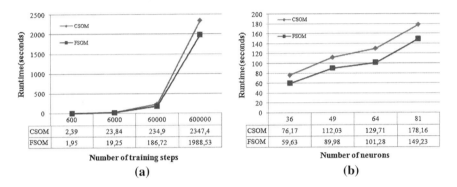

Fig. 2 Runtime comparison between conventional and fast SOM. **a** Runtime with different training steps. **b** Runtime with varying numbers of neurons

compared between the proposed fast SOM and the conventional SOM respectively. By using the Manhattan distance metric, a simplified learning rate and neighborhood function, no multiplications and exponential computations is required, due to this a reduction in time ranging from 15 to 20 % can be achieved.

Furthermore, it is also important to note that the proposed fast SOM algorithm is able to obtain this speed-up while maintaining an accuracy which is similar compared to the conventional SOM. Our fast SOM algorithm outperforms other simplified SOM algorithms such as [8, 9].

In Table 1 the recognition accuracy of the conventional, our fast version and also Pena's [10] SOMs are shown. These were obtained by various numbers of iterations and each iteration equals training the feature map with 60,000 input vectors. Afterwards, the SOM is tested with 10,000 test samples. Finally the feature maps of the conventional and proposed SOMs are shown in Fig. 3. The neurons of both maps clearly organized themselves and clusters can be distinguished.

Table 1 Recognition accuracy on MNIST database

Iterations	Conventional SOM (%)	Proposed fast SOM (%)	Pena's SOM (%)
1	82.34	81.83	64.96
10	84.93	83.24	66.03
100	85.01	83.79	66.81
200	85.74	84.13	67.24

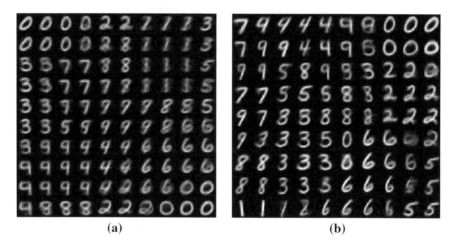

Fig. 3 Feature map after training with MNIST dataset. **a** Feature map of proposed SOM. **b** Feature map of conventional SOM

6 Conclusion

This paper proposes a fast SOM algorithm for handwritten digit recognition which simplifies the conventional SOM by removing complex computations in the weight distance calculation, the learning rate function and the neighborhood function. After evaluating the performance in software simulation, we conclude that the proposed fast SOM algorithm can reach the goal of accelerating to some extent, maintain similar recognition accuracy compared to the conventional SOM and performs much better than other simplified SOM methods.

References

1. Kohonen T (1990) The self-organizing map. Proc IEEE 78(1):1464–1480
2. Kohonen T, Kaski S, Lagus K et al (2000) Self organization of a massive document collection. IEEE Trans Neural Netw 11(3):574–585
3. Silven O, Niskanen M, Kauppinen H (2003) Wood inspection with non-supervised clustering. Mach Vis Appl 3:275–285

4. Mu-Chun S, Hsiao-Te C (2000) Fast self-organizing feature map algorithm. IEEE Trans Neural Netw 11(3):721–732
5. Nordström T (1992) Designing parallel computers for self organizing maps. In: Fourth Swedish workshop on computer system architecture
6. Lobo VJ, Bandeira N, Moura-Pires F (1998) Distributed Kohonen networks for passive sonar based classification. In: International conference on multisource-multisensor information fusion, Las Vegas
7. Yaohua Y, Damminda A (2006) Batch implementation of growing self-organizing map. In: International conference on computational intelligence for modelling control and automation, and international conference on intelligent agents, web technologies and internet commerce
8. Pena J, Vanegas M (2006) Digital hardware architecture of Kohonen's self organizing feature maps with exponential neighboring function. In: IEEE international conference on reconfigurable computing and FPGA
9. Agundis R, Girones G, Palero C, Carmona D (2008) A mixed hardware/software SOFM training system. Computaciny Sistemas 4:349–356
10. Porrmann M, Witkowski U, Ruckert U (2006) Implementation of self-organizing feature maps in reconfigurable hardware. In: FPGA implementations of neural networks. Springer, Heidelberg, pp 247–269
11. LeCun Y, Cortes C, The MNIST database of handwritten digits. http://yann.lecun.com/exdb/mnist/

Frequent Graph Pattern Mining with Length-Decreasing Support Constraints

Gangin Lee and Unil Yun

Abstract To process data which increasingly become larger and more complicated, frequent graph mining was proposed, and numerous methods for this has been suggested with various approaches and applications. However, these methods do not consider characteristics of sub-graphs for each length in detail since they generally use a constant minimum support threshold for mining frequent sub-graphs. Small sub-graphs with a few vertices and edges tend to be interesting if their supports are high, while large ones with lots of the elements can be interesting even if their support are low. Motivated by this issue, we propose a novel frequent graph mining algorithm, Frequent Graph Mining with Length-Decreasing Support Constraints (FGM-LDSC). The algorithm applies various support constraints depending on lengths of sub-graphs, and thereby we can obtain more useful results.

Keywords Frequent graph mining · Length-decreasing support constraint · Sub-graph

1 Introduction

As data generated from the real world have been complicated and large increasingly, previous frequent pattern mining methods, which find frequent patterns from simple database composed of items, have been faced with limitations that cannot deal with these large and complex data. Thereafter, to overcome this problem, frequent graph mining has been proposed [2–5]. However, existing frequent graph mining methods

G. Lee · U. Yun (✉)
Department of Computer Science, Chungbuk National University,
Cheongju-si, South Korea
e-mail: yunei@chungbuk.ac.kr

G. Lee
e-mail: abcnarak@chungbuk.ac.kr

J. J. (Jong Hyuk) Park et al. (eds.), *Multimedia and Ubiquitous Engineering*,
Lecture Notes in Electrical Engineering 240, DOI: 10.1007/978-94-007-6738-6_24,
© Springer Science+Business Media Dordrecht(Outside the USA) 2013

extract frequent sub-graphs with only one minimum support constraint which is set in the early mining procedure regardless of sub-graphs' lengths. Therefore, they have the following problem. Small sub-graphs having a few vertices and edges tend to be interesting if they have high support values. In contrast, large sub-graphs having many vertices and edges can be interesting even though they have low supports. However, the previous methods cannot find interesting large sub-graph patterns with lower supports than a given minimum support threshold since the threshold is fixed regardless of patterns' lengths. To solve the problems, we propose a novel frequent graph mining algorithm, called Frequent Graph Mining with Length-Decreasing Support Constraints (FGM-LDSC).

2 Related Work

Frequent graph mining began from Broad First Search (BFS)-based methods, and thereafter, Depth First Search (DFS)-based mining methods have been studied actively. In addition, graph mining can be applied in other data mining area such as classification [7], and regression analysis [7], and so on. As fundamental graph mining algorithms, there are famous algorithms such as FFSM, gSpan, Gaston [4, 5], etc. Especially, Gaston is a state-of-the-art algorithm which has the fastest runtime performance among them. In addition, there are numerous graph mining algorithms such as applying weight conditions [1, 2], using abbreviated notations called maximal and closed sub-graphs [6], finding frequent sub-graphs with a strong correlation [3], and so on.

LPMiner/SLPMiner [8] is a fundamental frequent pattern mining algorithm applying length-decreasing support constraints. Thereafter, WSLPMiner [9] was proposed, which can mine weighted sequential frequent patterns in the same environment, where the length means the number of items belonging to any pattern (or a set of items). However, these algorithms deal with only itemset databases, and therefore, they are not suitable for mining frequent sub-graphs from databases consisting of graphs.

3 Frequent Graph Mining with Length-Decreasing Support Constraints

3.1 Length-Decreasing Support Constraints on Frequent Graph Mining

Previous frequent graph mining methods generally consider only one standard, a single minimum support threshold when they extract frequent sub-graphs. However, this is unsuited for determining whether all of the sub-graphs are actually

Frequent Graph Pattern Mining with Length-Decreasing Support Constraints

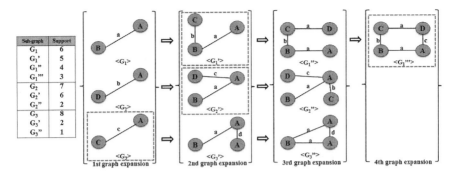

Fig. 1 An example of graph expansions

valid or not. Recall that certain sub-graphs with a small number of vertices and edges tend to be interesting if they have high supports. In contrast, other sub-graphs having lots of the elements can be interesting even though they have relatively low supports.

Example 1 Figure 1 represents an example of graph expansions for mining frequent sub-graphs, where we assume that sub-graphs within the red dotted rectangles are interesting patterns and have to be extracted. In general frequent graph mining, a minimum support threshold has to be set as 3 to mine all of the interesting sub-graph patterns. However, since the method extracts not only interesting patterns but also all of the frequent but uninteresting ones with 3 or more supports such as $\{G_1, G_1'', G_2\}$, it eventually repeats mining operations many times to generate these meaningless sub-graphs.

To reduce these inefficient operations and find only interesting sub-graphs, we propose length-decreasing support constraints suitable for graph structures.

Definition 1 A length of any sub-graph is determined by vertices and edges composing the sub-graph. If a certain sub-graph, G has a path or free tree structure, its length, $length(G)$ is denoted as $length(G)$ = # of vertices included in G. On the other hand, consider that G has a cyclic graph form. Let G_{prev} be a graph such as a path or a free tree just before G is expanded as a cyclic graph. Then, $length(G)$ is calculated by the equation, $length(G)$ = # of vertices in G_{prev} + # of cyclic edges added to G.

FGM-LDSC assigns minimum support thresholds with respect to each sub-graph's length depending on Definition 1, where these threshold values are gradually decreased from high to low values.

3.2 Pruning Strategy Retaining Anti-Monotone Property

As lengths of generated sub-graphs increase in FGM-LDSC, minimum support thresholds corresponding to each length are inversely decreased. In this environment, if we conduct mining process as a general method, fatal pattern losses can be caused since it does not satisfy Anti-monotone property (or downward closure property). To consider and overcome this problem, we propose measures and strategies for effectively removing unneeded sub-graphs as well as satisfying Anti-monotone property.

Definition 2 Let *GDB* be a certain graph database and *GL* be a set of lengths for sub-graphs generated from *GDB*, denoted as $GL = \{gl_1, gl_2, gl_3, ..., gl_n\}$. Then, a set of minimum supports for *GL*, *MS* is denoted as $MS = \{ms_1, ms_2, ms_3, ..., ms_n\}$, where subscripts means sub-graphs' lengths and the relation, $ms_i \geq ms_j$ is satisfied for $1 \leq i < j \leq n$ depending on the length-decreasing support constraint technique. Let *min(MS)* be the lowest support among the values of *MS*, and then, we use the *min(MS)* as a minimum support threshold since the value guarantee Anti-monotone property.

$$min(MS) = min_{1 \leq k \leq n}(ms_k) \tag{1}$$

Depending on the Definition 2, the minimum support for the sub-graph with the longest length is assigned as *min(MS)*. Thus, if any sub-graph, *G* has a support less than *min(MS)*, it means that *G* and all of the possible supper patterns of G also have lower supports than *min(MS)* since their supports become smaller as *G* is gradually expanded according to Definition 2, which satisfies Anti-monotone property. Consequently, it is certain that pruning *G* and *G*'s super patterns does not cause any problem.

Example 2 Consider Fig. 1 again, where *MS* is set as $MS = \{ms_1, ms_2, ms_3, ms_4\} = \{8, 5, 5, 3\}$ and we assume that $G3$, G_1', G_2', and G_1''' are interesting sub-graphs which FGM-LDSC has to extract. Then, $min(MS) = 3$ according to Eq. (1). In the first graph expansion ($ms_1 = 8$), G_3 is only a valid pattern since its support is larger than 8, while the others, G_1 and G_2 become invalid ones. However, FGM-LDSC does not prune them since their supports are higher than 3(= *min(MS)*). In the next expansion ($ms_2 = 5$), G_1' and G_2' are interesting sub-graphs while G_3' is meaningless graph and also pruned permanently since its support is lower than not only 5(= ms_2) but also 3(= *min(MS)*). Especially in here, we can show that G_1' and G_2', which are infrequent sub-graphs at the first expansion step, are changed as frequent patterns in the current step due to the *min(MS)*. In the third ($ms_3 = 5$), G_1'' is not pruned since its support is higher than 3 although it is lower than 5, while G_2'' is permanently pruned since its support < *min(MS)*. G_3'' is not even considered due to pruning G_3'. In the last expansion ($ms_4 = 5$), G_1''', which grow

Frequent Graph Pattern Mining with Length-Decreasing Support Constraints

```
input:    a graph database, GDB, a set of supports for each length, MS
output:   a set of frequent sub-graph patterns, S

Mining_graph_patterns(GDB, δ, λ, ω)
1. calculate min(MS) // according to equation (1)
2. find all vertices and edges such that their support ≥ min(MS) in GDB
3. for each vertex, v in a set of the found frequent vertices, V do
4.     a sub-graph, SG ← v
5.     a set of valid edge, E' ← edges which can be attached to v among the found frequent edges, E
6.     current graph state, GS ← "path"
7.     S = S ∪ Expanding_graph(SG, E', GS)
─────────────────────────────────────────────────────────────────────────────
Expanding_graph(a sub-graph SG, a set of edges E, current graph state GS)
1.  for each edge, e in E do
2.     if GS is "path" or "free tree" do
3.         generate an expanded path or free tree, SG' of SG adding e and a corresponding vertex, v
4.         calculate length(SG') // depending on Definition 1
5.     else generate an expanded cyclic graph, SG' of SG adding only e
6.         calculate length(SG') // depending on Definition 1
7.     select current minimum support for length(SG'), ms from MS
8.     if support of SG' ≥ min(MS) δ do
9.         if support of SG' ≥ ms do
10.             S = S ∪ SG'
11.         else discard SG' // however, SG' is not pruned
12.     else e ← the next edge in E and goto line 1 // SG' is pruned
13.     E' ← a set of valid edges that can be attached to SG'
14.     GS ← current graph state of SG'
15.     S = S ∪ Expanding_graph(SG', E', GS)
```

Fig. 2 FGM-LDSC algorithm

from the infrequent G_1'', becomes a frequent and interesting sub-graph again since it satisfies the $min(MS)$ condition.

3.3 FGM-LDSC Algorithm

Figure 2 presents frequent graph mining procedure performed by FGM-DLSC. In the function: *Mining_graph_patterns*, FGM-DLSC calculates $min(MS)$ with respect to the inputted MS. After that, it finds frequent vertices and edges satisfying the $min(MS)$ condition and extracts frequent sub-graph patterns by expanding graphs regarding the found elements. After the *mining_graph_patterns* function calls its sub-function, *Expanding_graph*, FGM-LDSC conducts the graph expansion step and computes the length of the resulting graph for each edge, where it performs appropriate operations depending on whether the current graph state is a path, a free tree, or a cyclic graph. Thereafter, it finds the *ms* value corresponding to the calculated length from MS, and determines whether the currently expanded graph is frequent or has to be pruned. Then, it conducts a series of processes for SG' satisfying the condition of line 8, and continues the graph expansion steps with respect to SG' through recursive calls of this routine itself.

Fig. 3 Support constraints of DTP and PTE datasets

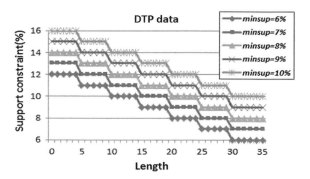

4 Performance Analysis

4.1 Experimental Environment

In this section, performance evaluation results for the proposed algorithm, FGM-FDSC are presented. A target algorithm compared to FGM-FDSC is Gaston [4, 5], which is a state-of-the-art frequent graph mining algorithm. The two algorithms were written as the C++ language and ran with 3.33 GHz CPU, 3 GB RAM, and WINDOWS 7 OS environment. For these experiments, we used a real graph dataset, named DTP. Details of the dataset are available at [4, 5]. Figure 3 represents a distribution of length-decreasing support constraints for the dataset, DTP.

4.2 Experimental Results

Figure 4 shows the results for the number of frequent sub-graph patterns and runtime performance for the DTP dataset. As shown in the left figure, FGM-LDSC dramatically reduces sub-graph patterns, which are unnecessarily generated in mining process, by applying the proposed strategies and techniques. In contrast, Gaston extracts the enormous number of sub-graphs since it mines all of the patterns with higher supports than the single and fixed minimum support threshold. Especially, pattern results generated by Gaston are sharply increased as the threshold becomes low while the results by FGM-LDSC increase slightly and consistently since our algorithm selectively extracts actually interesting sub-graphs for each length. In the right part of Fig. 4, Gaston requires more time resources compared to FGM-LDSC in all of the cases since the Gaston finds and extracts all of the possible frequent sub-graphs not considering whether generated sub-graphs are really interesting or not. Especially when the minimum support threshold is lowered from 7 to 6 % in DTP, we can observe that corresponding runtimes of the two algorithms are greatly increased. However, our FGM-LDSC has an increasing

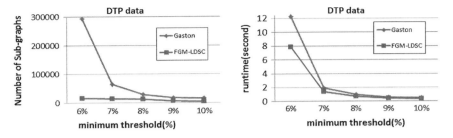

Fig. 4 The number of frequent sub-graphs and runtime results in DTP dataset

rate smaller than that of Gaston, and a gap between them becomes bigger whenever the threshold is lowered. This runtime interval occurs due to the meaningless sub-graphs pruned by Definition 2.

5 Conclusion

In this paper, we proposed a frequent graph mining algorithm with length-decreasing support constraints. Through the proposed algorithm, FGM-LDSC, we could obtain interesting sub-graphs having not only high supports and a few vertices and edges but also relatively low supports and a lot of the elements. Moreover, through the suggested pruning strategies and techniques, we demonstrated that our algorithm outperforms the previous method in terms of mining efficiency, as shown in the experimental results in this paper. Our algorithm can be applied to the other fields such as maximal/closed frequent graph mining, weighted frequent graph mining, and so on, and we expect that the strategies and techniques of our FGM-LDSC will contribute to advancing their mining performance in common with this paper.

Acknowledgments This research was supported by the National Research Foundation of Korea (NRF) funded by the Ministry of Education, Science and Technology (NRF No. 2012-0003740 and 2012-0000478).

References

1. Günnemann S, Seidl T (2010) Subgraph mining on directed and weighted graphs. In: Proceedings of the 14th Pacific-Asia conference on knowledge discovery and data mining, pp 133–146
2. Hintsanen P, Toivonen H (2008) Finding reliable subgraphs from large probabilistic graphs. Data Mining Knowl Discov 17(1):3–23
3. Lee G, Yun U (2012) An efficient approach for mining frequent sub-graphs with support affinities. In: Proceedings of the 6th international conference on convergence and hybrid information technology, Korea, pp 525–532

4. Nijssen S, Kok JN (2004) A quickstart in frequent structure mining can make a difference. In: Proceedings of the tenth ACM SIGKDD international conference on knowledge discovery and data mining, pp 647–652
5. Nijssen S, Kok JN (2005) The Gaston tool for frequent subgraph mining. Electron Notes Theor Comput Sci 127(1):77–87
6. Ozaki T, Etoh M (2011) Closed and maximal subgraph mining in internally and externally weighted graph databases. In: 25th IEEE international conference on advanced information networking and applications workshops, pp 626–631
7. Saigo H, Nowozin S, Kadowaki T, Kudo T, Tsuda K (2009) gBoost: a mathematical programming approach to graph classification and regression. Mach Learn 75(1):69–89
8. Seno M, Karypis G (2005) Finding frequent patterns using length-decreasing support constraints. Data Min Knowl Disc 10(3):197–228
9. Yun U, Ryu KH (2010) Discovering important sequential patterns with length-decreasing weighted support constraints. Int J Inf Technol Decis Making 9(4):575–599

An Improved Ranking Aggregation Method for Meta-Search Engine

Junliang Feng, Junzhong Gu and Zili Zhou

Abstract A meta-search engine transmits the user's query simultaneously to several individual search engines and aggregate results into a single list. In this paper we conduct comparisons on several existing rank aggregation methods. Then based on those comparisons, an improved ranking aggregation method is proposed for meta-search engine. This method combines merits of the Borda's method and scaled footrule method. Extensive experiments show that this improved method outperforms the alternatives in most cases.

Keywords Search engine · Meta-search engines · Rank aggregation · Borda · Scaled footrule

1 Introduction

With the explosive growth of internet information, an effective search engine becomes more and more important for users to find their desired information from billions of web pages. Although the ranking algorithms (such as PageRank [1]) in search engines have been upgraded fast, but it's still impossible for one single search engine to cover all the web pages even for some famous general search engines, like Google and Bing. For specific queries, different search engines may

J. Feng (✉) · J. Gu · Z. Zhou
Department of Computer Science and Technology,
East China Normal University, Shanghai, China
e-mail: jlfeng@ica.stc.sh.cn

J. Gu
e-mail: jzgu@ica.stc.sh.cn

Z. Zhou
e-mail: zlzhou@ica.stc.sh.cn

J. J. (Jong Hyuk) Park et al. (eds.), *Multimedia and Ubiquitous Engineering*,
Lecture Notes in Electrical Engineering 240, DOI: 10.1007/978-94-007-6738-6_25,
© Springer Science+Business Media Dordrecht(Outside the USA) 2013

only search a subset of the internet. Meta-search engine [2] may help to improve this problem. Constructing a meta-search engine is quite desirable, while the most challenging problem for meta-search is the ranking aggregation method [3].

In this paper, a meta-search engine named ICASearch is implemented. Besides this, we make comparison among several ranking aggregation methods [4], and propose an improved ranking aggregation method. The proposed ranking method is evaluated on our meta-search engine system. The experiment results show that this optimal method has more precise results than the general methods.

2 Meta-Search Engine

A meta-search engine is a system, which fuses the search results from several individual search engines into a single result list. So it enables users to provide search criteria only once and access several search engines simultaneously [5]. When a query arrives, the meta-search engine forwards the query to several constituent components. Then the constituent components process the query and dispatch the query to several general search engines. Each engine responds to the query with a ranked result lists. Finally, the meta-search engine merges all the results lists, and returns the merged list to the user. Now meta-search systems have drawn attentions from both academic and commercial areas.

For web search engines, we only focus on the results in first 1 or 2 pages, that's to say, we only need to consider the partial list, the top 20–50 results from each engine, and merge these lists into a final result list for our user. This is different from merging the full list of the results. It is a more challenging task for aggregation.

Meta-search engine has some advantages over general search engines. Firstly, a more improved precision by merging multiple results, particularly for the web search engine. Secondly, it can provide a more consistent and reliable performance than individual search engines [6]. Thirdly, the architecture [7] of meta-search engine has a better modularity. It allows a single search engine system to be divided into smaller, special components.

3 Architecture of ICASearch

ICASearch, is a meta-search engine aiming to provide better search results for its users. The architecture of ICASearch is depicted in Fig. 1. There are three modules in the system: (1) search engine module; (2) controller module; (3) third party module. We will present the internal details in the following sections.

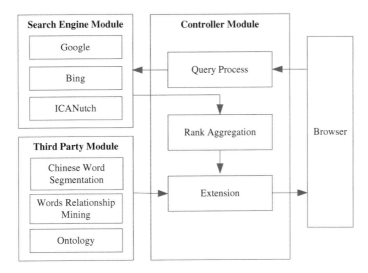

Fig. 1 The Architecture of ICASearch

3.1 Search Engine Module

The search engine module contains several individual search engines. In this system, we use three search engines: Google, Bing and ICANutch. ICANutch is a local search engine. It crawls web pages mainly on internet and mobile applications news. Our experiments are also conducted on these topics.

3.2 Controller Module

This module contains three parts: (1) Query processing. It performs word segmentation when user submits a query, and then dispatches it to the search engine module. (2) Rank aggregation. When relevant results are returned from Search Engine Module, it will merge the results into the final result. (3) Extension. After rank aggregation, we will use the third party module to get relationships and relevance words for the query. This information is shown on the web page to facilitate our user in further search.

3.3 Third Party Module

Three-third party APIs are invoked in this system. (1) Chinese Word Segmentation plugin. As our queries are mainly in Chinese currently, so it's appropriate for us to use Chinese word segmentation plugin to get a more precise understanding of user's query. (2) Words Relationship Mining API. We use this API to extend our search results. (3) Ontology library. There is an ontology system which provides us with some ontologies to do query extensions on our search results.

4 Rank Aggregation Methods

There exist various methods for aggregate result from difference rank-ordered lists [3, 4]. In most case the methods can be classified with the following rule: (1) based on the score; (2) based on the rank; (3) required the training data or not [8]. In this paper, we will mainly discuss the methods based on the rank.

4.1 Preliminaries

Let U presents the universe, a set of items. An ordered list τ with respect to U is an ordering of a subset $S \subseteq U$, i.e., $\tau = [x_1 \geq x_2 \geq \ldots \geq x_k]$, with each $x_i \in S$, and $\geq$ is some ordering relation on S. If $i \in \tau$, let $\tau(i)$ denotes the position or rank of i (a high ranked or preferred element has a low-numbered position in the list). For a list τ, $|\tau|$ denotes the number of elements. With $w^\tau(i)$ we will denote the normalized weight of item $i \in \tau$ in ranked list τ [5].

4.2 Borda's Method

In Borda's method, we use the Borda rank normalization [9] to calculate the $w^\tau(i)$, for an item $c \in U$

$$w^\tau(c) = \begin{cases} 1 - \frac{\tau(c)-1}{|U|}, & \text{if } c \in \tau \\ \frac{1}{2} - \frac{|\tau|-1}{2 \cdot |U|}, & \text{if } c \notin \tau \end{cases} \tag{1}$$

For given ordered lists $\tau_1, \tau_2, \ldots, \tau_k$, then for each element $c \in S$ and list τ_i, we assign the $w^\tau(c)$ to each c in τ_i as $B_i(c)$, so the total Borda score $B(c)$ is defined as $\sum_{i=1}^{k} B_i(c)$. After calculated all the Borda score, we could sort the result in decreasing order by the total Borda score. The computation complexity of this method is $O(n^2)$, n denotes the total size of the partial list results.

4.3 Scaled Footrule Optimization Method

The scaled footrule method use the footrule distances to rank the various results. In the full list scenario, the footrule optimal aggregation can be solve by construct a bipartite graph from the lists and compute the minimum cost perfect matching [6].

For partial lists $\tau_1, \tau_2, \ldots, \tau_k$, we defined a weighted bipartite graph (C, P, W). C denotes the set of nodes to be ranked. $P = \{1, 2, 3, \ldots, n\}$ denotes the set of available positions. The weight $W(c, p)$ is the total footrule distance of ranking the element c in position p, given by

$$W(c, p) = \sum_{i=1}^{k} |\tau_i(c)/|\tau_i| - p/n| \tag{2}$$

So this problem has been converted to calculate the minimum cost perfect matching problem in a bipartite graph. In this paper, we use the Kuhn–Munkres algorithm to solve this matching problem. The computation complexity of the algorithm is $O(n^3)$, n denotes the total size of the partial list results.

4.4 B-F-Rank Method

The Borda method focus the position on the initial return lists, and the scaled footrule optimization method will consider not only the original positions, but also consider the final rank positions (as the bipartite graph defined in Sect. 4.3). After research, we found that, in the first few results the Borda method are more accurate, but the precision declined quickly while the result size increase increasing. The precision change in scale footrule method is relatively stable. So we propose a method named B-F-Rank, which combines the two methods to rank the final result, and suppose it will get more accurate results. There are three steps for the method.

1. Get two aggregated result list L_B and L_F. L_B is the result list ordered by Borda's method, while L_F is ordered by scaled footrule method. As we know, the elements in the L_B and L_F is the same, the only difference is the ranking position.
2. Use the Eq. 1, to normalize the L_B and L_F, so for each element c, we get two weights, $w_B(c)$ and $w_F(c)$, that are been normalized by L_B and L_F respectively.
3. The new weight of element c is given by

$$w_{B-F-Rank}(c) = \alpha \cdot w_B(c) + \beta \cdot w_F(c) \tag{3}$$

α and β is real value and $\alpha + \beta = 1.0$. Then rank the element list by $w_{B-F-Rank}$ in decreasing order and we will get the final result ordered by B-F-Rank method.

5 Experimental Evaluation and Results

In our system, three search engines, Google, Bing, and ICANutch are used. The Borda's method and scaled footrule method are taken as benchmark methods. We prepared 10 queries, all those queries focus on the internet and mobile applications news. This experiment setting decreases the impact of the diverse result set and lets us focus on the rank aggregation method optimization. In each query round, we select the top 50 results from each engine, and after rank aggregation use specific method, our measurement is based on the precision of the top 50 of the final results. The precision is assessed by human judges. Average precision and precision in top-N results (P@N) are chosen as evaluation criteria. Figure 2 is the precision for the methods run in each query round, it shows the B-F-Rank result is more precise.

Figure 3 shows, at the rank N-th position, the average precision of 10 queries' results that use specific method. The B-F-Rank method shows a more stable and better precision curve than the other two methods.

Fig. 2 Precision of results

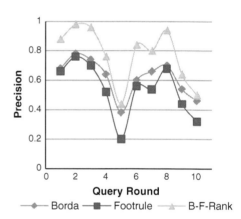

Fig. 3 P@N

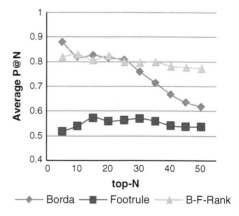

An Improved Ranking Aggregation Method for Meta-Search Engine

Table 1 Average precision of the three methods

Methods	Borda	Scaled footrule	B-F-Rank
Average precision	0.618	0.538	0.774

Table 1 presents the average precision of the three methods in top 50 results. The result shows that the B-F-Rank method outperforms the alternatives in most cases.

6 Conclusions and Future Work

In this paper, several rank aggregation methods are discussed and evaluated. An improved ranking aggregation method named B-F-Rank is proposed. The evaluation result shows that the proposed method outperforms classical methods, Borda's method and scale footrule optimization method, in most cases.

The future work involves, incorporating more search engines in our study and adding semantic and ontology extension in the queries. Furthermore we could also incorporate term similarities and correlation in our aggregation method.

Acknowledgments The work is supported by the Shanghai Scientific Development Foundation with Grant No. 11530700300, and Shandong Province Young and Middle-Aged Scientists Research Awards Fund with No. BS2010DX012.

References

1. Brin S, Page L (1998) The anatomy of a large-scale hypertextual web search engine. Comput Netw 30(1–7):107–117
2. Aslam JA, Montague M (2001) Models for metasearch. In: Proceedings of the 24th annual international ACM SIGIR conference on research and development in information retrieval. ACM Press, New York, pp 276–284
3. Dwork C, Ravi K, Moni N, Siva Kumar D: Rank aggregation methods for the web. In: Proceedings of the tenth international conference on World Wide Web. ACM Press, New York, pp 613–622
4. Farah M, Vanderpooten D (2007) An outranking approach for rank aggregation in information retrieval. In: Proceedings of the 30th annual international ACM SIGIR conference. ACM Press, New York, pp 591–598
5. Woodley Alan P, Geva S (2005) ComRank: metasearch and automatic ranking of XML retrieval systems. In: International conference on cyberworlds. IEEE Press, Singapore, pp 146–154
6. Montague M, Aslam JA (2001) Metasearch consistency. In: Proceedings of the 24th annual international ACM SIGIR conference on research and development in information retrieval. ACM Press, New York, pp 386–387

7. Gulli A, Signorini A (2005) Building an open source meta-search engine. In: Special interest tracks and posters of the 14th international conference on World Wide Web. ACM Press, New York, pp 1004–1005
8. Aslam JA, Montague M (2001) Models for metasearch. In: Proceedings of the 24th annual international ACM SIGIR conference on research and development in information retrieval. ACM Press, New York, pp 276–284
9. Renda ME, Straccia U (2003) Web metasearch: rank vs. score based rank aggregation methods. In: Proceedings of the 2003 ACM symposium on applied computing. ACM Press, New York, pp 841–846

Part V
Multimedia and Ubiquitous Computing Security

Identity-Based Privacy Preservation Framework over u-Healthcare System

Kambombo Mtonga, Haomiao Yang, Eun-Jun Yoon and Hyunsung Kim

Abstract The digitization of patient health information has brought many benefits and challenges for both the patient and doctor. But security and privacy preservation have remained important challenges for wireless health monitoring systems. Such concerns may result in reluctance and skepticism towards health systems by patients. The reason for this skepticism is mainly attributed to the lack of assurances about the way patient health information is handled and the implications that may result from it on patients' privacy. This paper proposes an identity-based privacy preservation framework over u-healthcare systems. Our framework is based on the concepts of identity-based cryptography and non-interactive key agreement scheme using bilinear pairing. The proposed framework achieves authentication, patient anonymity, un-traceability, patient data privacy and session key secrecy, and resistance against impersonation and replay attacks.

Keywords u-healthcare · Privacy preservation · Data privacy · Identity-base encryption · Non-interactive key agreement

---

K. Mtonga
Department of IT Convergence, Kyungil University, 712-701, Kyungsansi, Kyungpook Province, Korea

H. Yang
College of Computer Science and Engineering, UEST of China, Chengdu 610054, China

E.-J. Yoon · H. Kim (✉)
Department of Cyber Security, Kyungil University, 712-701, Kyungsansi, Kyungpook Province, Korea
e-mail: kim@kiu.ac.kr

E.-J. Yoon
e-mail: ejyoon@kiu.ac.kr

J. J. (Jong Hyuk) Park et al. (eds.), *Multimedia and Ubiquitous Engineering*, Lecture Notes in Electrical Engineering 240, DOI: 10.1007/978-94-007-6738-6_26, © Springer Science+Business Media Dordrecht(Outside the USA) 2013

1 Introduction

Advances in telecommunication technology have made possible data transmission over the wireless system. This has enabled remote patient monitoring systems which collects disease-specific metrics via wireless biomedical devices used by patients. The collected health data is transmitted to a remote server for storage and later examination by the healthcare professionals. However, the different usage scenarios of remote monitoring systems e.g. in-hospital and home monitoring have resulted in diverse security and privacy concerns [1, 2]. Ensuring privacy and security of health information, including information in the electronic health record (EHR), is the key component to build the trust required to realize the potential benefits of electronic health information exchange [3]. Many protocols to enhance privacy and security of remotely collected patient health information have been put forward by researchers [4–7].

In this paper, we propose an identity-based privacy preserving framework over u-healthcare systems. In our framework; (1) Identity-based cryptography (IBC) is adopted to ensure the secure transmission, receiving, storing and access of patient data. (2) The doctor can give feedback directly to the patient on his/her health condition. (3) The patient and doctor can establish a secure channel directly by establishing session key with non-interactive manner.

2 Preliminaries

In this section we briefly review our threat model and present notations used throughout the paper. For details on Bilinear pairing, Bilinear Diffie-Hellman problem and non-interactive identity-based key agreement please refer to [8].

2.1 Threat Model

There are many threats to patient privacy and security in remote health monitoring system. Some of these threats include: data breach by insiders, insider curiosity, accidental disclosure and unauthorized intrusion of network system by outsiders [9]. In our framework, we aim to enhance patient data and identity privacy against both insider and outsider attacks i.e. attacks being provoked by an entity that is part of the network or by an outside entity who has, somehow, gained access to the network. For outside attackers, authentication and IBC-based data encryption have been adopted in order to prevent the attacker from gaining access to patient's data. In addition, a patient uses pseudo-ID when sending his/her health data and the data is stored encrypted, hence even if an insider accesses the patient's records, patient identity remains protected. Figure 1 illustrates our threat model.

Identity-Based Privacy Preservation Framework over u-Healthcare System

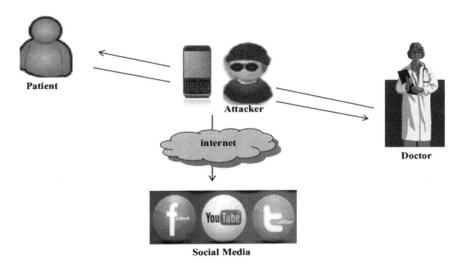

Fig. 1 Threat model to patient privacy

2.2 Notations

Table 1 introduces the notations used throughout the remainder of this paper.

Table 1 Notations

Notation	Meaning
TA	Trusted authority
U_i	Patient i
D_l	Doctor l
S_x	Private key for entity x
PR_{Ui}	Set of private keys for patient i
Q_x	Public key for entity x
ID_x	Identity of entity x
$SK_{x\text{-}y}$	Shared key between entity x and y
pid_j	jth pseudo-ID for patient i
$H_1(.)$	Hash function; $H_1: \{0, 1\}^* \rightarrow G_1$
$H_2(.)$	Hash function; $H_2: \{0, 1\}^* \rightarrow Z_q^*$
M	Patient health information
M'	Medical advice from doctor
T_x	Time stamp generated by entity x
$\hat{e}$	Bilinear map
A	Master secret key for health monitoring server
$\|$	Concatenation

3 Privacy Preservation Framework

In this section, we propose an identity-based privacy preservation framework over u-healthcare systems.

3.1 System Initialization

In our framework, health monitoring server performs the role of trusted authority. To initialize the system, the health monitoring server first runs the set up for bilinearity as mentioned in Sect. 2.1A. The health monitoring server then chooses a random number $a \in Z_q^*$ as the master key and computes the corresponding public key $Q_{TA} = aP$. It also chooses two secure collision free hash functions $H_1(.)$ and $H_2(.)$, where $H_1(.) : \{0, 1\}^* \rightarrow G_1$ and $H_2(.) : \{0, 1\}^* \rightarrow G_2$. The server then publishes the public system parameters as $\{G_1, G_2, q, P, Q_{TA}, H_1(.), H_2(.)\}$ and keeps the master key a, secret (Fig. 2).

3.2 Registration

Let U_i be a patient seeking medical help from doctor D_l. Since each doctor of the hospital is registered with the health monitoring server and the server keeps a profile of the doctor, U_i registers directly with the health monitoring server and he/she is assigned a doctor depending on her/his health problem. To register, U_i

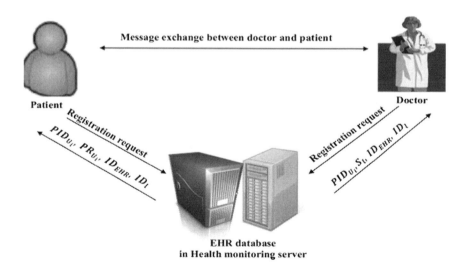

Fig. 2 Privacy preserving framework configuration

submits identity ID_i e.g. an email address or social security number to server. The server first validates ID_i. If the validation is successful, the server then chooses a family of n un-linkable pseudo-IDs given by

$$PID_{Ui} = \{pid_0, pid_1 \ldots, pid_{j-1}, pid_j, pid_{j+1} \ldots, pid_{n-1}\} \quad (1)$$

For each pseudo-ID pid_j in PID_{Ui}, the server computes the public key $Q_j = H_1(pid_j)$ and the corresponding private key $S_j = aH_1(pid_j)$, such that the families of public and private keys are

$$PB_{Ui} = \{Q_0, Q_1, Q_2 \ldots, Q_{j-1}, Q_i, Q_{j+1} \ldots, Q_{n-1}\} \quad (2)$$

$$PR_{Ui} = \{S_0, S_1, S_2 \ldots, S_{j-1}, S_i, S_{j+1} \ldots, S_{n-1}\} \quad (3)$$

To allow revocation, the server adds an expiry date into pid_j such that each of the public keys $Q_j = H_1(pid_j)$ is valid only before the specified expiry time t_j. After the specified time, the corresponding private key $S_j = aH_1(pid_j)$ is revoked automatically. We also assume that the patient can only use the pseudo-IDs pid_j, $0 \leq j \leq n - 1$ sequentially. Finally, the server sends the whole tuples $\{PID_{Ui}, aH_1(pid_j)\}$ for $0 \leq j \leq n - 1$ to U_i via a secure channel. With these pseudo-IDs, the patient can constantly change his/her pseudo-ID to achieve anonymity and un-traceability during communication process in the u-healthcare environment.

Doctor D_l registers with the health monitoring server by providing his/her true identity ID_l. The server then computes $S_l = aH_1(ID_l)$ the private key and $Q_l = H_1(ID_l)$ the public key for doctor. It also computes $S_{EHR} = aH_1(ID_{EHR})$ and $Q_{EHR} = H_1(ID_{EHR})$, the key pair for EHR.

3.3 Patient Health Information Transfer to EHR

To send her health data, U_i carries out the following steps:

- Pick a valid pseudo-ID pid_j in the sequence and corresponding private key $S_j = aH_1(pid_j)$.
- Using this private key, the patient computes a session key shared with doctor D_l as $SK_{i-l} = \hat{e}(aH_1(pid_j), H_1(ID_l)) = \hat{e}(Q_j, Q_l)^a$, and another session key shared with EHR as $SK_{i-EHR} = \hat{e}(aH_1(pid_j), H_1(ID_{EHR})) = \hat{e}(Q_j, Q_{EHR})^a$.
- Using SK_{i-l}, patient applies *IBC*-based encryption on her health data M as $C = E_{SKi-l}(M)$.
- Use EHR's public key Q_{EHR} to encrypt message $Y = E_{Q_{EHR}}(T_i \| ID_l \| pid_j \| SK_{i-HER} \| C)$ and send Y to EHR.

Table 2 Patient health information table by EHR

Doctor ID	Patient ID	PHI
ID_l	pid_j	C
:	:	:

3.4 Patient Health Information Verification and Storing by EHR

When EHR receives patient health information, it carries out the following authentication steps:

- Use S_{EHR} to decrypt Y as $\{T_i \| ID_l \| pid_j \| SK_{i\text{-}EHR} \| C\} = D_{SEHR}(Y)$.
- Check if time stamp T_i is valid by verifying if $T_i\text{-}T_{EHR} < \Delta T$ is satisfied, where T_{EHR} is time the message is received by EHR and ΔT is predefined transmission delay. If the verification is not success, the message is rejected and HER generates a message asking patient to resend the message. This also helps to overcome replay attacks. Otherwise, EHR proceeds to verify the identity of doctor, ID_l.
- Once the verification of doctor is successful, EHR proceeds to compute $SK_{EHR\text{-}i} = \hat{e}(aH_1(ID_{EHR}), H_1(pid_j))$ using the received pid_j.
- EHR then checks if $SK_{EHR\text{-}i} = SK_{i\text{-}EHR}$. Note that $SK_{EHR\text{-}i}$ is equal to $SK_{i\text{-}EHR}$ since; $SK_{i\text{-}EHR} = \hat{e}(aH_1(pid_j), H_1(ID_{EHR})) = \hat{e}(Q_j, Q_{EHR})^a = \hat{e}(Q_{EHR}, Q_j)^a = SK_{i\text{-}EHR}$.

If the above holds, U_i is authenticated by EHR, and EHR stores $\{ID_l, pid_j, C\}$ in the database as shown in Table 2.

3.5 Patient Health Information Recovery by Doctor

To access patient's health data M, the doctor first gets her/himself authenticated to EHR by carrying out the following steps:

- Computes $SK_{l\text{-}EHR} = \hat{e}(aH_1(ID_l), H_1(ID_{EHR})) = \hat{e}(Q_l, Q_{EHR})^a$.
- Sends $V = E_{QEHR}(T_l \| pid_j \| ID_l \| SK_{l\text{-}HER})$ encrypted with the public key of EHR, Q_{EHR}, as a request for the patient's health information. Since the doctor is aware that each of the pseudo-IDs has an expiry date and that they are used sequentially, when pid_j is chosen, the doctor chooses the one that is valid and current. Hence D_l can request for specific patient health information from EHR depending on the specified pid_j.
- When EHR receives the request V, it uses its private key S_{EHR} to decrypt the request, i.e. $D_{SEHR}(V) = \{T_l \| pid_j \| ID_l \| SK_{l\text{-}EHR}\}$, and then checks if the received timestamp T_l satisfies the condition $T_l - T_{EHR} < \Delta T$. If the verification is

Identity-Based Privacy Preservation Framework over u-Healthcare System 209

success, EHR proceeds to check if the received pseudo-ID pid_j for U_i and the identity ID_l for D_l match the ones received from the patient.

- EHR then uses ID_l to compute a session key shared with D_l, $SK_{EHR-l} = \hat{e}(aH_1(ID_{EHR}), H_1(ID_l))$ and check if $SK_{EHR-l} = SK_{l-EHR}$ holds. If the equation holds, EHR authenticates the doctor and sends $\{C, pid_j\}$ to the doctor. Note that $SK_{EHR-l} = \hat{e}(aH_1(ID_{EHR}), H_1(ID_l)) = \hat{e}(Q_{EHR}, Q_l)^a = \hat{e}(Q_l, Q_{EHR})^a = SK_{l-EHR}$.

When D_l receives $\{C, pid_j\}$, he/she verifies pid_j. Since D_l already has pid_j, she/he can pre-compute $SK_{l-i} = \hat{e}(aH_1(ID_l), H_1(pid_j))$ in advance or could compute after receiving the message from EHR. With this key, D_l successfully decrypts C, i.e. $M = D_{SKl-i}(C)$.

The doctor can now analyze the patient's health information such that if there is need for immediate medical advice for the patient, the doctor generates medical advice M' and encrypts it with the session key SK_{l-i}. This session key is used to establish a secure channel for the subsequent communication sessions between the doctor and the patient till t_j the expiry date of pid_j.

For mutual authentication, after computing SK_{l-i}, the doctor can compute an authentication code, $Auth = H_2(SK_{l-i}\|pid_j\|ID_l)$ and send it together with the response M' encrypted with SK_{l-i} as $\{E_{SKl-i}(M'), Auth\}$. Upon receiving the doctor's response, U_i also generates a verification code $Veri = H_2(SK_{l-i}\|pid_j\|ID_l)$ and checks if $Auth = Veri$. If the equation holds, then U_i believes that the medical advice M' came from legitimate doctor and that he/she has established a secure channel using the key SK_{l-i}. Otherwise the patient rejects the session. This protects the patient from bogus health advices from adversaries which could result in drug overdose or worse still unnecessary death.

4 Conclusion

In this article, we have presented a privacy preserving security framework over u-healthcare system. In our framework, patients are only pseudonymously identified hence protecting the patient from negative effects of identity theft such as fraudulent insurance claims by adversaries. However, since health monitoring server knows the patients' real identity, in case of apparent abuse via judicial procedure, this real identity can be revealed.

Acknowledgments This work was supported by the National Research Foundation of Korea Grant funded by the Korean Government (MEST) (NRF-2010-0021575) and was partially supported by the MKE (The Ministry of Knowledge Economy), Korea, under the ITRC (Information Technology Research Center) support program (NIPA-2012- H0301-12-2004) supervised by the NIPA (National IT Industry Promotion Agency).

References

1. Varshney U (2003) Pervasive healthcare. IEEE Comput 36(12):138–140
2. Ng HS, Sim ML, Tan CM (2006) Security issues of wireless sensor networks in healthcare applications. BT Technol J 24(2):138–144
3. Appari A, Johnson ME (2010) Information security and privacy in healthcare: current state of research. Int J Internet Enterp Manag 6(4):279–314
4. Huang Q, Yang X, Li S (2011) Identity authentication and context privacy preservation in wireless health monitoring system. Int J Comput Netw Inf, Security, pp 53–60
5. Layouni M, Verslype K, Sandikkaya MT (2009) Privacy-preserving telemonitoring for eHealth data and applications security. LNCS 5645:95–110
6. Hasque MM, Pathan AK, Hong CS (2008) Securing u-Healthcare sensor networks using public key based scheme. In: ICACT, pp 17–20
7. Sax U, Kohane I, Mandl KD (2005) Wireless technology infrastructures for authentication of patients: PKI that rings. J Am Med Informatics Assoc 12(3):263–268
8. Kim H (2012) Non-interactive hierarchical key agreement protocol over hierarchical wireless sensor networks. Commun Comput Inf Sci 339(5):86–93
9. Dixon P (2006) Medical identity theft: the information crime that can kill you. The world privacy forum

A Webmail Reconstructing Method from Windows XP Memory Dumps

Fei Kong, Ming Xu, Yizhi Ren, Jian Xu, Haiping Zhang and Ning Zheng

Abstract Retrieving the content of webmail from physical memory is one key issue for investigators because it may provide with useful information. This paper proposes a webmail evidence reconstructing method from memory dumps on Windows XP platform. The proposed method uses mail header format defined in RFC2822 and HTML frame based on specific webmail server to locate header and body respectively. Then webmail is reconstructed based on matching degree between FROM, TO(CC/BCC), DATE and SUBJECT fields of header and corresponding content extracted from body. The experiment results show that this method could reconstruct the webmail from memory dumps.

Keywords Digital forensics · Webmail · Memory dumps

F. Kong · M. Xu (✉) · Y. Ren · J. Xu · H. Zhang · N. Zheng
College of Computer, Hanzhou Dianzi University, Hangzhou, China
e-mail: 100061040@hdu.edu.cn

F. Kong
e-mail: mxu@hdu.edu.cn

Y. Ren
e-mail: renyz@hdu.edu.cn

J. Xu
e-mail: jian.xu@hdu.edu.cn

H. Zhang
e-mail: zhanghp@hdu.edu.cn

N. Zheng
e-mail: nzheng@hdu.edu.cn

J. J. (Jong Hyuk) Park et al. (eds.), *Multimedia and Ubiquitous Engineering*,
Lecture Notes in Electrical Engineering 240, DOI: 10.1007/978-94-007-6738-6_27,
© Springer Science+Business Media Dordrecht(Outside the USA) 2013

1 Introduction

Traditional digital forensic is mainly on permanent storage devices, but it could not work well to obtain volatile information. Webmail data in memory can provide wealthy of information such as crime plan of suspect and relationship of crime organization. And webmail reconstruction, for clues to solve the case and provide evidence in court, will play an increasingly important role.

To the best of our knowledge, there is no work focusing on webmail reconstructing from memory dumps. However, some past forensic research about web browser and email client has been done and achieved positive results. Rachid Hadjidj developed one e-mail forensic analysis software tool [1]. Pereira presented a method to recover Firefox remnants [2]. Oh et al., proposed a method [3] to collect information relevant to the case. But their methods cannot be directly used in webmail forensic from memory.

Here we study the reconstruction of webmail resided in memory, using matching degree between related mail fields. Our method scans memory dumps and locates web mail header and body using string match method without knowing process or kernel data structure.

2 Our Method

2.1 Basic Steps

The proposed method works in three steps: getting memory data, preprocessing dump file and reconstructing webmail.

We dump memory data and save them as a file following the rules [4]. Webmail accesses to mail server with HTTP, and there are three methods for HTTP data compressing. Specifically, we write a program to decompress dumps. If the traversed data is compressed data, it will be decompressed and then outputted, otherwise, the data will be directly outputted. Program read 4 KB data every time. If compressed data is separated in two blocks, program will read next block and decompress. The basic idea of reconstructing webmail is to locate and recover all headers and bodies of webmail, and then match them. According to RFC2822 [5], a complete webmail header must have FROM, TO(CC/BCC) and DATE fields. Every field has a field name, followed by a colon, a field body, and terminated by a CRLF. Since webmail content is processed in HTML form, so content of webmail are inserted with some HTML tags. Using these tags as fingerprint, we can find the body content.

2.2 Matching Degree Metrics

It is supposed that a full webmail includes a subject, which correlates with webmail body. Also, it contains a receiver's name, a sender's name and date at the greeting part and end part of message. Based on this assumption, the possibility of a header and body matched as one mail can be estimated based on the relation between the text of webmail body and DATE, FROM, TO(CC/BCC), SUBJECT field of the header.

Let D_H = $\{d|\ d$ is a normalized date string from the DATE field in header $H\}$;

E_H = $\{e|e$ is an email address in FROM, TO(CC/BCC) field from header $H\}$;

W_H = $\{w|\ w$ is w is a word in SUBJECT field from header $H\}$;

D_B = $\{d|\ d$ is a normalized date string from text in mail body $B\}$;

E_B = $\{e|\ e$ is an email address from text in mail body $B\}$;

W_B = $\{w|\ w$ is w is a word from text in mail body $B\}$

Definition 1 Matching degree (*MD*) between a mail header H and a body B denotes the probability of them belonged to a same webmail. It can be measured as:

$$MD(H,B) = \frac{|D_H \cap D_B| + |E_H \cap E_B| + |W_H \cap W_B|}{|D_H| + |E_H| + |W_H|} \times 100\ \%$$

2.3 Related Algorithms

Idea of Go-Back-N in TCP protocol is taken to process length-fixed string in this paper. When reading data in dump file, it is not reading from the end but from the last N-1 bytes of last block. N is the length of longest string that would be located. This strategy could locate strings that across 2 blocks.

(1) Webmail header recovering algorithms

MailHeaderRecovering is used to locate three required fields and recover webmail header. HeaderCompleteness is called to judge whether these fields are complete. If all three fields are complete, headers are saved to array *mh*[]. If one field terminates with non-ascii symbol or '\0', it is incomplete; otherwise it is separated.

```
Algorithm 1: MailHeaderRecovering(F)
Input: the decoded memory dump files F
Output: Mail Header Array mh[]
  read one block data b from dump file F;
  While(!EOF(F))
    if(b~="From:"&&b~="To{cc|bcc}:"&&b~="Date:"
        &&HeaderCompleteness(b)==CompleteHeader)
        mh[i++]←mail header in b;
    read next block c from dump file F;
    if((b~="From:"||b~="To{cc|bcc}:"||b~="Date:")
        &&HeaderCompleteness(b)==SeparateHeader)
        b←b,c;
    else b←c;
Algorithm 2: HeaderCompleteness(b)
Input: a block data b from file F,
Output: according the status of required fields, it re-
turns CompleteHeader, SeparateHeader or IncompleteHeader.
  Flag{Date,From,To}←SeperateField;
  foreach field in {Date,From,To}{
    if(b~=field)
        if(field is terminated by CRLF)
            Flag{field}←CompleteField;
        else if(field is terminated by Non-ASCCII or '\0')
            Flag{field}←IncompleteField;
        else Flag{field}←SeperateField;
    if(all of Flag{Date},Flag{From} and Flag{To} are Com-
        pleteField)
        return CompleteHeader;
    else if(Flag{Date},Flag{From} or Flag{To} is
        IncompleteField)
        return IncompleteHeader;
    else return SeparateHeader;
```

(2) Webmail body recovering algorithms

There are some HTML tags before and after body message, and they will be called starting tags and ending tags for short. Variables of *startTagOffset* and *endTagOffset* are used to record offset of starting and ending tag respectively. Complete webmail body will be saved to array *mb*[]. Details are listed in algorithm 3.

```
Algorithm 3: MailBodyRecovering(F)
Input: the decoded memory dump files F
Output: Mail Body Array mb[]
  startTag,endTag←starting tag,ending tag;
  startTagOffset,endTagOffset,startTagFlag←0;
  read one block data b from dump file F;
  while(!EOF(F)
    if(b~=startTag)
        startTagOffset←offset of startTag;
        startTagFlag←1;
    else if(b~=endTag)
        endTagOffset←offset of endTag;
        if(startTagFlag==1)
            startTagFlag←0;
            ma[j++]←Copy(startTagOffset,endTagOffset);
    read next block c;
    b←c;
```

(3) Matching algorithms

MatchHeaderAndBody is used to match the recovered webmail headers and bodies. It saves results to *ma*[] according match degree.

```
Algorithm 4: MatchHeaderAndBody(mh[],mb[])
Input: the recovered mail header array mh[] and mail body
array mb[];
Output: mail array ma[];
  foreach B in mb[]
    DB,EB,WB←GetNormalizedDateSet(B),
            GetEmailAddressSet(B),GetWordSet(B);
    MaxMD←0;
    foreach H in mh[]
        D_H,E_H,W_H←GetNormalizedDateSet(H),
                GetEmailAddressSet(H),GetWordSet(H);
```

$$MD \leftarrow (|D_H \cap D_B| + |E_H \cap E_B| + |W_H \cap W_B|)/(|D_H| + |E_H| + |W_H|) \times 100\%$$

```
        if(MaxMD<MD) MatchingHeader=H; MaxMD←MD;
    if(MaxMD>Threshold) ma[i++]←MatchingHeader,B;
```

3 Experiments

3.1 Experiment Preparations

VMWare is used in the experiment because its function of taking snapshot could minimize related interferences when imaging the system. The operating system of host is Windows XP, and version of VMWare is 7.1.5 build-491717 with 128 RAM. IE8 and chrome (ver. 20) are chosen according to StatCounter's statistical data of Oct 2012.

The same recipient and sender do not affect results once the webmail server is chosen. For simplicity, mem_exp@126.com is chosen as test account. "mail.126.com" is a popular website of NTES in China, which provides free webmail service.

One message called "angel" is chosen in experiments, which has 254 words. Twenty copies of angel named angel01, angel02, angel03 … angel20 and every word in the message will be added the same number as the title. Twenty messages are chosen for there are only latest twenty mails listed in first page of inbox.

Then we power on the virtual machine, open the web browser, IE or chrome only one is chosen in one snapshot, and IE image or chrome image is called for short. We sign in, select the seed emails "Angel" series and take snapshot as the ground truth image. Then we close web browser and take another snapshots without other system activities except screensaver.

3.2 Analyses

During preprocess step open source library zlib is used (NTES data takes gzip format) and KMP algorithm is used to locate gzip data header 0X1F8B08. Then we manually analyze the data and found that FROM field is like 'from':['xxx <mem_exp@126.com>']CRLF. The strings of TO and DATE field are 'to': ['"some Chinese characters"<mem_exp@126.com>']CRLF and 'sentDate':new Date (yyyy, MM, dd, hh, mm, ss)CRLF respectively. In experiment, the starting tag string is adjust to "<style>HTML{word-wrap:break-word;}" and the ending tag string is "<script language="javascript">try{parent.JS.modules[window.name]. content.setHeight();}".

Memory data will expire within definite time period [6]. We take one extra experiment to test time span and time limitation of this experiment gave was 5 min.

After preprocessing ground truth data, we manually analyze these data to find all headers and bodies, number is listed in total line (Tables 1 and 2) with pair H–B, which stand for number of copies of header's and body's respectively. Number of headers and bodies found by program is listed in located line.

This table indicates (1) copies of header are more than body, and (2) copies in memory vanished quickly at the first few minutes. The possible reasons are (1) another body format without HTML tag can not be found by program; browser has already requested the latest twenty mail headers' information when users log in. (2) Some data is from network packets and vanished when they were flushed.

Table 1 Copies of mail header and mail body located in chrome image over time

Result	0 min H–B	1 min H–B	3 min H–B	5 min H–B
Located	262-37	124-22	105-20	67-14
Total	268-37	126-23	106-21	70-15
Correctness	0.977-1	0.984-0.957	0.991-0.952	0.957-0.933

Table 2 Copies of mail header and mail body located in IE image over time

Result	0 min H–B	1 min H–B	3 min H–B	5 min H–B
Located	248-34	111-22	87-20	49-13
Total	252-34	113-23	92-21	54-14
Correctness	0.984-1	0.982-0.957	0.946-0.952	0.907-0.929

In matching step, we extracted receiver's name from body before the first comma, sender's name and date from the last two lines. We extract names from three fields, which before symbol "@". Year, month and day is extracted from DATE field.

The highest matching degree in these twenty mails is 1/3, so threshold is set to 0.3. The matching result of twenty mails is complete matched. Given these twenty mails have high matching degree between subject and bodies, another 96 common English text mails are chosen to test the algorithm 4 and 78 mails are matched correctly.

4 Conclusion and Future Work

This paper solve the problem of webmail reconstruction in Windows XP memory dumps, there are still some future work and research needed to do. (1) Body content of webmail that without HTML tags. (2) Designed webmail that has irrelevant fields in header with body content. (3) Webmail with multi-media type files or attachment.

Acknowledgments This work is supported by NSFC (No. 61070212 and 61003195), Zhejiang Province NSF of China (No. Y1090114 and LY12F02006), Zhejiang Province key industrial projects in the priority themes of China (2010C11050), and the science and technology search planned projects of Zhejiang Province (No. 2012C21040).

References

1. Hadjidj R, Debbabi M, Lounis H et al (2009) Towards an integrated e-mail forensic analysis framework. Proc Digital Invest 5:124–137
2. Pereira MT (2009) Forensic analysis of the Firefox 3 Internet history and recovery of deleted SQLite records. Proc Digital Invest 5:93–103
3. Oh J, Lee S, Lee S (2011) Advanced evidence collection and analysis of web browser activity. Proc Digital Invest 8:62–70
4. Vömel S, Freiling FC (2012) Correctness, atomicity, and integrity: defining criteria for forensically-sound memory acquisition. Proc Digital Invest 9:125–137
5. http://www.ietf.org/rfc/rfc2822.txt
6. Solomon J, Huebner E, Bem D, Szezynska M (2007) User data persistence in physical memory. Proc Digital Invest 4:68–72

On Privacy Preserving Encrypted Data Stores

Tracey Raybourn, Jong Kwan Lee and Ray Kresman

Abstract Bucketization techniques allow for effective organization of encrypted data at untrusted servers and for querying by clients. This paper presents a new metric for estimating the risk of data exposure over a set of bucketized data. The metric accounts for the importance of bucket distinctness relative to bucket access. Additionally, we review a method of controlled diffusion which improves bucket security by maximizing entropy and variance. In conjunction with our metric we use this method to show that the advantages of bucketization may be offset due to a loss of bucket security.

Keywords Privacy · Trust · Bucketization · Encryption · Multimedia databases

1 Introduction

Data is a valuable asset in modern enterprise, and the need to facilitate a variety of multimedia types such as voice, video, text, and images is ever more imperative. The Database As a Service (DAS) model is one system promoted to minimize the overall costs of asset ownership [7]. Private clouds have used DAS for very large, non-relational and multimedia databases, such as search engines [13]. Medical institutes have also used the service to store a variety of digitized patient images. Despite its advantages, outsourcing raises concerns over data confidentiality when

---

T. Raybourn · J. K. Lee (✉) · R. Kresman
Department of Computer Science, Bowling Green State University, Bowling Green, OH 43403, USA
e-mail: leej@bgsu.edu

T. Raybourn
e-mail: traybou@bgsu.edu

R. Kresman
e-mail: kresman@bgsu.edu

J. J. (Jong Hyuk) Park et al. (eds.), *Multimedia and Ubiquitous Engineering*, Lecture Notes in Electrical Engineering 240, DOI: 10.1007/978-94-007-6738-6_28, © Springer Science+Business Media Dordrecht(Outside the USA) 2013

the service provider is untrusted [3]. Database encryption is a typical solution, where the client downloads and decrypts records from the server for further processing [4]. Most encryption ciphers, however, do not support SQL queries, resulting in query methods that return unwanted records and perform unnecessary decryption [3]. Balancing privacy and efficiency is the focus of much research on querying encrypted databases [1, 2, 4, 8, 12, 14, 15].

Bucketization is one technique for executing range queries over encrypted data on a DAS server [3, 4]. Encrypted records are divided into buckets, each of which has an ID and a range defined by its minimum and maximum values. In a multimedia environment, the DAS may maintain a grid of nodes, with each node housing a particular type of multimedia data (voice, video, text, images, etc.) [10, 13]. The client holds indexing information about the range of each bucket on the server. Client queries are mapped to the set of buckets that contain any value satisfying the query. The relevant buckets are then requested from the server.

To illustrate, suppose a film production company outsources its video clip database to Amazon's Relational Database Service (RDS), and Amazon uses two buckets: bucket B1 holds a total of 500 clips for every year between 1990 and 2001; and B2 holds a total of 400 clips for 2002 through 2012 of which 100 are made after 2007. A client query for clips shot after 2007 consults the local bucket index, and sends a request for B2 to the server. All records from B2 are downloaded, decrypted, and processed as needed. A client query for clips shot before 2008, will return and decrypt both B1 and B2. In either case, the returned records include clips beyond the desired range, known as false positives, which must be filtered out at the client. False positives are especially problematic for multimedia databases, where retrieval and decryption of non-text data (video, audio, images, etc.) means high computational overhead for the client [13].

While false positives are considered an acceptable cost of bucketization, there are strategies to minimize them [1, 4, 11, 15]. We briefly discuss three of these bucketization methods: Query Optimal Bucketization, Controlled Diffusion, and Deviation Bucketization. Some detail is given to facilitate the reader's understanding of our experiments, for which these algorithms are implemented.

Hore et al. [4] presented a Query Optimal Bucketization (QOB) algorithm that minimizes bucket cost, where cost is a function of the value range and value frequencies in the bucket (Eq. 1). QOB generates an optimal solution to the problem of bucketizing a set of values, $V = v_i, \ldots, v_n$, using at most M buckets, where each value, $v_1 < \ldots < v_n$, occurs at least once in V. Each bucket covers the values $(v_i, v_j]$ and the bucket cost BC is given by,

$$BC(i,j) = (v_j - v_i + 1) * \sum_{i \leq t \leq j} f_t, \tag{1}$$

where f_t is the frequency of each distinct bucket value. The algorithm computes the summed cost of every two bucket combination over the data set, partitioning the bucket pair that returns the minimum cost. QOB reduces false positives by minimizing the total cost of all buckets.

Hore et al. [4] introduced a method of controlled diffusion, which allows a bounded performance degradation (K) in order to improve optimal (QOB) bucket privacy, as measured by entropy and the variance. Controlled diffusion creates a new set of M approximately equi-depth buckets, called composite buckets (CB), and redistributes (*diffuses*) elements from optimal buckets into the CBs. Diffusion is *controlled* by restricting the number of CBs into which elements from a given optimal bucket are diffused. Given a maximum performance degradation factor K, for an optimal bucket B_i of size $|B_i|$, its elements diffuse into no more than $\frac{K*|B_i|}{f_{CB}}$ composite buckets, where f_{CB} is the size of data set D over bucket size M, i.e., $\frac{|D|}{M}$. The resulting set of M composite buckets is stored in encrypted form on the server, with CB-specific bucket IDs.

Yao et al. [15] proposed a Deviation Bucketization (DB) scheme that extends QOB by further reducing false positives at the cost of at most M^2 buckets. First, DB generates a set of M QOB buckets. For each QOB bucket, DB computes an array comprising the deviations of each distinct data point from the bucket mean. QOB then bucketizes the deviation arrays, creating a set of second level buckets by subdividing each QOB bucket according to its deviation values. Higher frequency values are more likely to be queried, and DB buckets tighten the grouping of these values, greatly reducing false hits over QOB.

On average, lowering bucket width (i.e., increasing the number of buckets) reduces false positives by allowing queries more granular access to bucket domains, but is not without risk. Bucketization is susceptible to estimation and linking attacks [1, 3], as well as query access pattern attacks [5]. A tighter estimate of the underlying data distribution does not ensure that an adversary can determine precise plaintext values, but may be damaging to the extent an adversary does have particular knowledge of data values [1, 3].

In the following section, we introduce a new metric for estimating the risk of an adversary discovering information about the value distribution of a bucketized data set. Later, we present experiments demonstrating (1) the efficacy of our metric and (2) when controlled diffusion is applied to the set of DB buckets, the advantages afforded by DB's decreased bucket width are diminished. The authors in [4] acknowledge that QOB buckets normally lack sufficient entropy and variance, so it is reasonable to expect that subdividing them, as DB does, may cause additional and possibly substantial loss of privacy.

2 Metrics for Evaluation

We propose a new metric that considers the risk of exposure as a function both of the distinctness of the data distribution within a bucket, and the frequency with which queries access the bucket. For a set of M buckets over data set D with query distribution Q, we express the risk R of an adversary reliably estimating the data distribution as:

$$R(D,Q,M) = \sum_{i=1}^{M}\left(1 - \frac{d_i}{|D|}\right) * \frac{q_i}{|Q|} * \left(1 - \frac{1}{M}\right),\tag{2}$$

where $|D|$ is the size of D and d_i is the number of distinct values in the ith bucket, yielding the proportion of distinct bucket values $\frac{d_i}{|D|}$; $|Q|$ is the size of query distribution Q, and q_i is the number of query values accessing the ith bucket, yielding the proportion of query values $\frac{q_i}{|Q|}$. We add a normalization term, $1 - \frac{1}{M}$, which imposes a penalty for increasing the number of buckets over D. The frequency of bucket access is important as it reveals something about the preference for values a bucket contains. As the number of buckets increases, the distinct values per bucket must decrease on average, and some values are likely to be queried more often than others. Thus, our metric conveys that fewer distinct values per bucket, relative to a high rate of query access, may disclose not only *intra*-bucket distribution, but data distribution across buckets. If an adversary has some knowledge of one or more buckets, the access pattern across all buckets may help the adversary to extrapolate *inter*-bucket probabilities.

To measure how well a bucketization method minimizes false hits, we use a query precision metric [4]. Query precision is the number of values in the set of buckets satisfying a range query (i.e., positives), over the total number (superset) of values in those buckets. Returning to our previous video clip illustration, a query for clips shot before 2008 yields a query precision, positives/superset $= 800/900$, of 0.89. An increase in query precision is equivalent to a decrease in false positives.

We also consider a well-known measure of bucket privacy: entropy [9]. Entropy entails distributing a bucket's values as widely and uniformly over a large domain as possible, in order to decrease the probability (increase the uncertainty) of estimating the true value distribution. Bucket entropy is given by:

$$H(X) = -\sum_{i=1}^{n} p(x_i) * \log_2 p(x_i),\tag{3}$$

where $p(x_i)$ gives the probability mass function of outcome x_i. While entropy as a measure of bucket security is relatively static (i.e., varies only as buckets change), our metric, R, affords a dynamic query access component, which provides a mechanism to account for bucket privacy in terms of access frequency.

3 Experiments

We created a data set comprising 10^5 integer values generated randomly from a uniform distribution over the domain [1, 1000]. The query set comprised 10^4 range queries corresponding with the range of the data set, where values were also drawn from a uniform distribution.

We conducted three experiments. For each, the data set was bucketized using both QOB and DB algorithms at bucket sizes $M = 2,..., 30$ (M^2 for DB). In experiment 1, we applied our exposure metric to respective bucket sets. For experiments 2 and 3, both QOB and DB bucket sets were rebucketized using controlled diffusion with degradation factors of $K = 2, 4, 6$. Experiment 2 calculated the average query precision, for optimal (QOB), deviation (DB), and composite buckets (controlled diffusion). Experiment 3 measured entropy for optimal, deviation, and composite buckets.

All experiments were conducted in a simulated database environment, i.e., with data structures representing the various database tables necessary for each algorithm to operate. We ran our experiments on a 2 GHz i3-CPU PC, with 2 GB RAM.

3.1 Performance Evaluation

Figure 1a shows the risk of exposure R for optimal buckets (QOB) and second level deviation buckets (DB). Bucket exposure increases proportionally with bucket size M (i.e., with a decrease in bucket width). In terms of R, exposure risk is notably higher for DB than QOB. Put in perspective, for $M = 10$ optimal buckets, DB has generated at most 100 s level buckets. Recall that the data set contains a maximum 1000 distinct values, meaning the deviation buckets contain approximately $1000/100 \approx 10$ distinct values ($1000/10 \approx 100$, for optimal). This also means that the query access pattern gives a more precise reading of which data are requested. Put differently, R, shows that as buckets decrease in width, they have a proportionally higher risk of losing one of the primary advantages that bucketization affords, namely privacy. QOB also shows relatively high exposure; around 64 %, for example, at $M = 5$. This is consistent with [4] who pointed out that optimal buckets may warrant privacy concerns due to low variance and entropy.

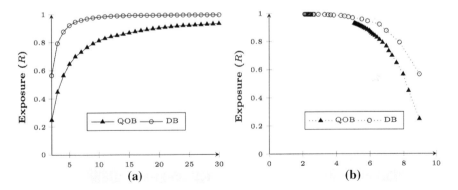

Fig. 1 **a** Estimated exposure risk for optimal (QOB) and deviation buckets (DB) and **b** comparison of R with entropy showing agreement of security assessment. **a** Size of M, **b** Average entropy

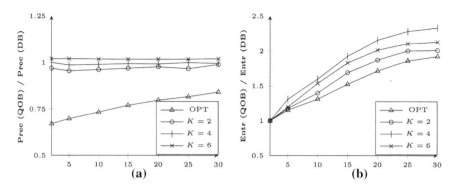

Fig. 2 Performance ratios for QOB, DB, and controlled diffusion bucket sets: **a** Average Query Precision. **b** Average Entropy; K = degradation factor for composite buckets. **a** Size of M, **b** Size of M

Figure 1b gives R as a function of average entropy for decreasing values of $M = 30,\ldots, 2$. Due to the nature of the comparison, data points for DB and QOB do not align one–to–one over the x-axis, with DB originating at entropy ≈ 2 ($M^2 = 900$), and QOB at entropy ≈ 4.5 ($M = 30$). The figure highlights a general agreement between R and entropy, i.e., exposure risk decreases as entropy (uncertainty) increases. Thus, our exposure metric, R, confirms that additional security mechanisms are necessary to increase bucket privacy.

In Fig. 2 we give results—query precision (a), entropy (b)—as ratios of QOB performance over DB performance for optimal and deviation buckets (OPT), and their corresponding composite buckets. QOB/DB composite bucket ratios are indicated by their respective degradation factors ($K = 2, 4, 6$). Thus, results with a value of 1.0 indicate no performance differences. We plot results for bucket sizes $M = 2, 5, 10, 15, 20, 25, 30$.

Figure 2a shows average query precision for optimal, deviation, and corresponding composite bucket sets. Consistent with previous findings [15], precision increases much faster for DB than QOB, as the number of DB buckets is increasing by a factor of M over QOB. This is indicated by ratio values less than 1.0. The increasing ratio (OPT) shows the advantages of DB decreasing as it approaches a ceiling on additional buckets around $M = 22$. That is, precision increases more slowly as bucket granularity rapidly approaches a maximum, while QOB precision is relatively constant increasing over sizes of M. When controlled diffusion is applied to each bucket set, however, advantages in query precision are minimized across all sizes of M. For a degradation factor of 2 ($K = 2$), DB composite buckets marginally outperform their QOB counterparts, while increasing factors of K reveal little to no difference (i.e., ratio is always close to 1.0).

Figure 2b shows average entropy for optimal, deviation, and corresponding composite bucket sets. Overall, QOB-based buckets have higher entropy than their DB counterparts, indicated by ratios greater than 1.0. The figure shows a relatively consistent relationship between QOB and DB. Even as entropy increases with controlled diffusion, QOB maintains its initial advantage in terms of bucket security due primarily to the greater width of its buckets.

Our results show that decreasing bucket width, even with controlled diffusion, may increase the risk of an adversary discovering information about the underlying data distribution. The diffusion of values from second level deviation buckets to composite buckets minimizes the advantages of DB; query precision substantially decreases, while minimal to no benefits over QOB are found in terms of entropy.

4 Concluding Remarks

In modern enterprise, it is critical to manage and protect the variety of sensitive multimedia data. In this paper, we introduced a new metric for estimating the risk of data exposure over a bucketized database. Our metric differs from established measures (e.g., entropy and variance) by accounting for the importance of bucket distinctness *relative to bucket access*. While entropy as a measure of bucket security is relatively static (i.e., varies only as buckets change), our metric, R, provides a mechanism to account for bucket privacy in terms of access frequency. Bucketized data that is never queried reveals only what can be discovered through standard inference attacks [3], while queries introduce information about the distribution of values across the data store. Our metric, then, may lend itself to modeling different scenarios of query access to determine whether the risk of exposure is acceptable for a given bucket set.

This paper also highlights the importance of bucket width in evaluating the security of bucketization methods. Our metric demonstrates that the advantage of decreasing bucket width, i.e., reducing false positives, can be offset due to a proportional loss of bucket privacy. In their presentation of Deviation Bucketization, Yao et al. [15] did not discuss the implications of bucket security. Thus, we used DB to both test its robustness with respect to data privacy concerns, as well as to highlight possible limitations of bucketization techniques that opt for a narrow bucket strategy, as this arguably undermines the purpose of bucketization.

As follow up to this work, we plan to investigate how the distribution of queries on a DAS server, relative to the bucketization method, impacts performance and security. QOB, for example, assumes that query distribution is uniform, which may be problematic. Our metric, R, shows that query access can enhance the likelihood of estimating the true data distribution. If queries are uniformly distributed, there is less risk of exposure. Optimal buckets may be less secure, however, if the query distribution is non-uniform, which is the most likely scenario for real-world databases. It may be that a more efficient bucketization strategy is one sensitive to the access pattern. Thus, we will explore bucketization based on the properties of query distributions more likely to reflect real-world access patterns, and their implications for existing bucketization methods. This may also provide insight as to the precise importance of the query distribution in modeling secure databases.

References

1. Agrawal R, Kiernan J, Srikant R, Xu Y (2004) Order preserving encryption for numeric data. In: 2004 ACM SIGMOD international conference on management of data, Paris, pp 563–574
2. Alwarsh M, Kresman R (2011) On querying encrypted databases. In: 2011 international conference on security and management, Las Vegas, pp 256–262
3. Damiani E, De Capitani di Vimercati S, Jajodia S, Paraboschi S, Samarati P (2003) Balancing confidentiality and efficiency in untrusted relational DBMSs. In: 10th ACM conference on computer and communication security, Washington, DC, pp 93–102
4. Hore B, Mehrotra S, Tsudik G (2004) A privacy-preserving index for range queries. In: 30th international conference on very large databases, Toronto, pp 720–731
5. Hore B, Mehrotra S, Canim M, Kantarcioglu M (2012) Secure multidimensional range queries over outsourced data. VLDB J 21(3):333–358
6. Huet B, Chua TS, Hauptmann A (2012) Large-scale multimedia data collections. IEEE MultiMedia 19(3):12–14 (IEEE Computer Society)
7. Li J, Omiecinski ER (2005) Efficiency and security trade-off in supporting range queries on encrypted databases. In: 19th annual IFIP WG 11.3 working conference on data and applications security, Storrs, CT, pp 69–83
8. Liu D, Wang S (2012) Programmable order-preserving secure index for encrypted database query. In: 2012 IEEE 5th international conference on cloud computing, Honolulu, pp 502–509
9. Shannon CE (1948) A mathematical theory of communication. Bell Syst Tech J 27:379–423
10. Smith JR, Döller M, Tous R, Gruhne M, Yoon K, Sano M, Burnett IS (2008) The MPEG query format: unifying access to multimedia retrieval systems. IEEE Multimedia 15(4):82–95
11. Sun W, Rane S (2012) A distance-sensitive attribute based cryptosystem for privacy-preserving querying. In: 2012 IEEE international conference on multimedia and expo, Melbourne, pp. 386–391
12. Wang J, Du X, Lu J, Lu W (2010) Bucket-based authentication for outsourced databases. Concurr Comput Pract Experience 22(9):1160–1180
13. Weis J, Alves-Foss J (2011) Securing database as a service: issues and compromises. IEEE Secur Privacy 9:49–55
14. Win LL, Thomas T, Emmanuel S (2011) A privacy preserving content distribution mechanism for DRM without trusted third parties. In: 2011 IEEE international conference on multimedia and expo, pp 1–6, Barcelona
15. Yao Y, Guo H, Sun C (2008) An improved indexing scheme for range queries. In: 2008 international conference on security and management, Las Vegas, pp 397–403

Mobile User Authentication Scheme Based on Minesweeper Game

Taejin Kim, Siwan Kim, Hyunyi Yi, Gunil Ma and Jeong Hyun Yi

Abstract The latest boom in the prevalence of smartphones has been encouraging various personal services to store and utilize important data such as photos and banking information. Thus, the importance of user authentication has also been growing rapidly. Nevertheless, many problems have arisen as a result of the common method of using a four-digit personal identification number (PIN) because of its potential for being breached by a brute force attack or shoulder-surfing attack. Various authentication schemes have been developed to overcome these problems. In this paper, we also propose a new password-based user authentication scheme that utilizes the well-known Minesweeper game, providing better usability as well as greater security. The proposed scheme provides its users a simple method for memorizing their passwords and usable security by allowing them to enter calculated values rather than the password itself.

Keywords Password · Usable security · Authentication · Shoulder-surfing attack

---

T. Kim · S. Kim · H. Yi · G. Ma · J. H. Yi (✉)
School of Computer Science and Engineering, Soongsil University, Seoul, Korea
e-mail: jhyi@ssu.ac.kr

T. Kim
e-mail: tjkim@ssu.ac.kr

S. Kim
e-mail: kimsiwan@ssu.ac.kr

H. Yi
e-mail: hyunyiyi@ssu.ac.kr

G. Ma
e-mail: gima@ssu.ac.kr

J. J. (Jong Hyuk) Park et al. (eds.), *Multimedia and Ubiquitous Engineering*,
Lecture Notes in Electrical Engineering 240, DOI: 10.1007/978-94-007-6738-6_29,
© Springer Science+Business Media Dordrecht(Outside the USA) 2013

1 Introduction

The recent increase in the use of smartphones has been replacing PCs in handling applications as their scope and area of application expands, providing their users with greater comfort. However, there are still many concerns about personal information leakage, viruses, and malware. Thus, there has been a gradual increase in the importance of secure user authentication methods to protect the personal data stored in smartphones. The current password-based authentication measures are user-friendly, but highly vulnerable to shoulder-surfing attacks, brute force attacks, and smudge attacks.

Much research has been conducted to resolve these concerns. However, the most of the methods developed [1–4] have been unsuitable for mobile devices, have compromised user-friendliness, or have continued to remain vulnerable to shoulder-surfing attacks. In this paper, we propose a user-friendly security method for mobile devices that provides defense against shoulder-surfing attacks. In addition, we conduct a security analysis through a comparison with previous techniques against brute force and shoulder-surfing attacks.

This paper is organized as follows. In Sect. 2, we discuss the suggested password authentication method, which is followed by a safety analysis in Sect. 3. Section 4 concludes the paper.

2 Proposed Scheme

Current password authentication schemes do not satisfy both security and usability requirements at the same time. For example, the PIN-Entry [1], DAS [2], and Passfaces [4] methods are secure against shoulder-surfing attacks but are vulnerable to recording attacks, while the Dementor-SGP scheme [3] has good security but lacks usability. Thus, in this paper, we propose a new scheme [5] that satisfies both the security and usability requirements by applying the idea of the Minesweeper game.

The password in the proposed scheme involves the locations and number of mines. First, the user will be prompted to set the mine locations at will. Then, the set-up is completed by pressing the OK button. After this, every time authentication is required, the user will be prompted to fill randomly selected cells in the entry interface with the number of mines in the 3×3 grid with the selected cell as the center. Authentication is carried out by entering the correct number of mines around each entry cell and then pressing the OK button. If each of the entries matches the correct number of mines, the user is successfully authenticated. The example shown in Fig. 1 depicts the mechanism with four entries in a 4×5 grid interface. Let us assume that we arrange the mines in an H shape, as shown in Fig. 1(left). Let us denote a coordinate in the grid as (x, y), where the grid consists of columns from 1 to y and rows from 1 to x. First, in the user authentication

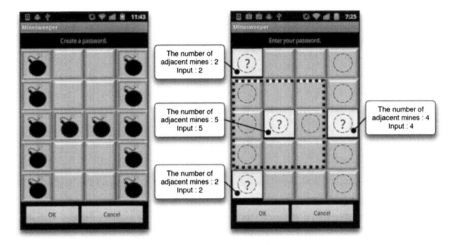

Fig. 1 The proposed scheme: password set-up (*left*) and user authentication (*right*)

process, the password is entered via randomly selected cells in the entry interface, as shown in Fig. 1(right).

The user fills in the entry interface cell located at (1,1) with the number '2' which is the total of the mines located at (1,1) and (1,2). For the entry at (2,3), the user fills in the number '5' because mines are planted at (1,2), (1,3), (1,4), (2,3), and (3,3). For the entry at (4,3), the number '4' is entered for the mines at (3,3), (4,2), (4,3), and (4,4). Finally, for the last entry at (1,4), the number '2' is entered because there are two mines around it. Once all the requested entries are filled, the user can confirm with the OK button for authentication. The authentication succeeds if all of the entered numbers are correct.

3 Security Analysis

This section describes a security analysis of the proposed password authentication scheme.

3.1 Password Space

The password space indicates the number of possible combinations for a given password length. Table 1 lists the password composition types and space values of the proposed and previous schemes.

The password space of the proposed scheme is equal to 2^{XY}, which is greater than those of the PIN-based password schemes. Similar to the Dementor-SGP and

Table 1 Password composition, space, and probability of a successful brute force attack

Scheme	Password composition	Password space	Probability of A successful brute force attack
PIN	N-digits	10^N	10^{-N}
DAS	N-digits	10^N	10^{-N}
Dementor-SGP	N user images (incl. hole image)	$_TP_N$	$1/(_TP_N)$
Passfaces	N facial images	$_TP_N$	9^{-N}
PIN-entry	N-digits	10^N	10^{-N}
Proposed sceme	N mine locations	2^{XY}	$1/(2^{XY})$

Note T number of elements in the password; N password length,
X X-axis units/length of grid; Y Y-axis units/length of grid,
M entry number; S number of entries exposed to attacker,
$_TP_N$ permutation to obtain an ordered subset of k elements from a set of n elements

Passfaces schemes that are allowed to expand the password space by increasing T, the new scheme can also expand the space by increasing X and Y to increase security against brute force attacks.

3.2 Brute Force Attack

A brute force attack attempts to hack a password by guessing every possible combination of the password. Table 1 lists the probabilities of successfully carrying out a brute force attack on the proposed and existing schemes.

The probability of a successful brute force attack depends on the password space. In a case where $X = 4$ and $Y = 5$, the probability of success is $1/(2^{20})$. This indicates that the proposed scheme is about 1.59 times more secure than Dementor-SGP with $T = 30$ and $N = 4$; 104 times more secure than PIN, DAS, or PIN-Entry with $N = 4$; and approximately 159 times more secure than Passfaces with $N = 4$.

3.3 Shoulder-Surfing Attack

The user authentication method proposed in this paper can conceal the locations of the mines because the user enters the number of adjacent mines instead of the actual password. Thus, even if an attacker obtains the correct entry values, the probability of success for an attacker is very low because the locations of the entry interface cells randomly change at each authentication attempt. In order to calculate the probability of success, the entry range of each grid coordinate must be known. Figure 2 depicts the division of a 3×3 grid into A, B, and C cells. In this

Fig. 2 Grid division

A	B	A
B	C	B
A	B	A

figure, the four corners are marked as A. The other outer cells are B, and C is used for the cells that are not marked as A or B. Counting itself and the adjacent cells, the total number of cells associated with an A cell is 4, with 6 for B and 9 for C, leading to digit ranges of 0–4, 0–6, and 0–9, respectively.

Thus, the probability of success of a brute force attack is 1/5 for A, 1/7 for B, and 1/10 for C. Equation (1) depicts the relationship between $X \times Y$ and the number of cells for each area:

$$\begin{aligned} A &= 4 \\ B &= 2(X + Y) - 8 \\ C &= XY - 2(X + Y) + 4 \end{aligned} \quad (1)$$

Consequently, the probability of entering the correct value in one random entry cell from an $X \times Y$ grid, based on Eq. (1), is as follows.

$$\frac{4}{5XY} + \frac{2(X + Y) - 8}{7XY} + \frac{XY - 2(X + Y) + 4}{10XY} \quad (2)$$

Let us assume that the number of cells seen by the attacker is S, and M is the number of random entries in the authentication when the attack occurs. In this case, only a portion of M can be identical to the cells seen by the attacker. Let us define the number of unidentified cells from M as n. The number of combinations for n can be calculated by $_{(XY-S)}C_n$, and for the identified cells within M, the number can be calculated by $_{S}C_{(M-n)}$. Equation (3) shows the proportion of selected combinations that include n among the possible combinations M:

$$\frac{_{S}C_{(M-n)}\,_{(XY-S)}C_n}{_{XY}C_M} \quad (3)$$

Only the correct combination for the randomly selected entry interface cells of n would allow the attacker to succeed in their attack. Finally, the overall probability of a successful brute force attack can be obtained by combining all cases of n. Equation (4) shows the attacker's probability of success in authentication.

$$\sum_{n=0}^{M} \left\{ \frac{_{S}C_{(M-n)}\,_{(XY-S)}C_n}{_{XY}C_M} \left(\frac{4}{5XY} + \frac{2(X + Y) - 8}{7XY} + \frac{XY - 2(X + Y) + 4}{10XY} \right)^n \right\} \quad (4)$$

As discussed earlier, if we assume that the parameters are $X = 4$, $Y = 5$, $M = 4$, and $S = 4$, then in a single attempt, the proposed scheme will allow a probability of success of as little as approximately 6.51×10^{-3}. Thus, the proposed scheme provides considerably more security against a shoulder-surfing attack than previous methods.

3.4 Recording Attack

A recording attack is a type of shoulder-surfing attack where the entire authentication process may be recorded using an electronic device such as a camera. In the case of the proposed scheme, the number of entries acquired by each recording attack is M. However, because the locations of the entry interface cells are randomly selected for each trial, the attacker may need the original mine locations or the correct digits for all of the entry cells. For an attacker, the latter option seems more feasible than the former. This is because the locations of the password mines can be relocated by only a change of one digit for every entry value, and the original mine locations can vary for the same digit entry combination [6].

In order to obtain the correct digits for all of the entries, the recording needs to be performed at least XY/M times. However, if a recording attack is performed twice or more, there will be considerable duplication in the obtained information. Thus, the chance of obtaining every entry from XY/M recording attacks is extremely low. Equation (5) shows the average of the numbers of newly obtained entries from recording attacks when S is the number of exposed entries.

$$\begin{cases} f_{(0)} = M \\ f_{(S)} = \sum_{n=0}^{M} n \dfrac{{_S}C_{(M-n)}\,{_{(XY-S)}}C_n}{{_{XY}}C_M} \end{cases} \tag{5}$$

Equation (6) shows the number of obtainable entries from multiple recording attacks based on Eq. (5).

$$\begin{cases} g_{(0)} = f_{(0)} \\ g_{(a)} = g_{(a-1)} + f_{(g_{(a-1)})} \end{cases} \tag{6}$$

Thus, $g(a) > XY$ has to be met for the number recording attacks to obtain the correct digits for all of the entries. For example, with the proposed scheme, an average of 17 recording attacks are required to obtain all of the entry values with parameters $X = 4$, $Y = 5$, and $M = 4$. Because, in reality, a recording attack can rarely be performed more than 10 times on the same user, the proposed scheme is secure against recording attacks.

4 Conclusion

This paper proposes a new password authentication method that adopts the Minesweeper game and can satisfy both security and usability requirements. Its suitability was proven using both a model test and user test with implementation in actual Android phones. The results proved that the suggested scheme can ensure user security against a shoulder-surfing attack, brute force attack, and especially, a recording attack. Although not very remarkable, the user test results showed improved usability over former schemes, which might be because it was inspired by the well-known Minesweeper game. We conclude that the presented authentication scheme ensures considerably better security than current mobile authentication schemes while providing proper usability for its users.

Acknowledgments This work was supported by a grant from the KEIT funded by the Ministry of Knowledge Economy (10039180).

References

1. Roth V, Richter K, Freidinger R (2004) A PIN-entry method resilient against shoulder surfing. In: Proceedings of the 11th ACM conference on computer and communications security, USA, pp 236–245
2. Park SB (2004) A method for preventing input information from exposing to observers. Patent application no.: 10-2004-0039209, Korea
3. MinInfo Co., http://www.mininfo.co.kr
4. Passfaces, http://www.passfaces.com
5. Yi JH, Kim T, Ma G, Yi H, Kim S (2012) Method and apparatus for authenticating password. Patent application no.: US 13/623,409
6. Kaye R (2000) Minesweeper is NP-complete. Math Intell 22:9–15
7. Olson JR, Olson GM (1990) The growth of cognitive modeling in human-computer interaction since GOMS. Hum Comput Interact 5:221–265
8. Lee S, Myung R (2009) Modified GOMS-model for mobile computing. J Soc Korea Ind Syst Eng 32:85–93

Design and Evaluation of a Diffusion Tracing Function for Classified Information Among Multiple Computers

Nobuto Otsubo, Shinichiro Uemura, Toshihiro Yamauchi and Hideo Taniguchi

Abstract In recent years, the opportunity to deal with classified information in a computer has increased, so the cases of classified information leakage have also increased. We have developed a function called "diffusion tracing function for classified information" (tracing function), which has the ability to trace the diffusion of classified information in a computer and to manage which resources might contain classified information. The classified information exchanged among the processes in multiple computers should be traced. This paper proposes a method which traces the diffusion for classified information among multiple computers. Evaluation results show the effectiveness of the proposed methods.

Keywords Prevention of information leaks · Network security · Log management

1 Introduction

The improvement in computer performance and propagation in various services has increased the opportunity to deal with classified information, such as customer information. According to the analysis [1] of personal information leakage incidents, it has been reported that leaks often happen by inadvertent handling and mismanagement, which account for approximately 57 % of all known cases of information leakage. In addition, several employees often share classified information. To trace the status of classified information in a computer and to manage the resources that contain classified information, we proposed a diffusion

---

N. Otsubo · S. Uemura · T. Yamauchi (✉) · H. Taniguchi
Graduate School of Natural Science and Technology, Okayama University, Okayama, Japan
e-mail: yamauchi@cs.okayama-u.ac.jp

H. Taniguchi
e-mail: tani@cs.okayama-u.ac.jp

J. J. (Jong Hyuk) Park et al. (eds.), *Multimedia and Ubiquitous Engineering*,
Lecture Notes in Electrical Engineering 240, DOI: 10.1007/978-94-007-6738-6_30,
© Springer Science+Business Media Dordrecht(Outside the USA) 2013

tracing function for classified information (tracing function), which manages any process that has the potential to diffuse classified information [2].

In this paper, we propose a method that uses the tracing function to trace the classified information being exchanged among multiple computers in internal network and to prevent information leakage outside internal network.

2 Requirements of Diffusion Tracing Function for Classified Information Among Multiple Computers

By tracing how classified information is diffusion, a computer can know which resources contain classified information. However, the tracing function [2] only traces the status of classified information in a computer. Thus, we propose a method that uses tracing function to trace classified information being exchanged among multiple computers in internal network and to prevent information leakage outside internal network.

In order to prevent information leakage outside the internal network, function needs to centrally manage the classified information that exists in the client computers in network. Moreover, it is necessary to determine whether the client computers are installed the tracing function or not, in order to prevent the diffusion of classified information to not installed computers.

We split the computers in network into a managed network and a non-managed network. Computers that need to handle classified information would be in the managed network, and the tracing function would be installed on them. A diffusion tracing function for classified information among multiple computers (tracing function for networks) must meet the following requirements:

(1) Computers that are installed the tracing function can be distinguished from computers that are not.
(2) The location and the flow of classified information in the managed network can be managed.
(3) The diffusion of classified information in the managed network can be instantly traced.
(4) Leakage of classified information out of the managed network can be detected.
(5) Leakage of classified information out of the managed network can be stopped in advance.

The tracing function for networks must also meet the following requests:

(1) Accurately trace the diffusion of classified information in a managed network.
(2) The processing overhead of tracing function for networks should be small.

3 Design

3.1 Overview of the Proposed Function

Figure 1 shows an overview of the function design to handle classified information in a managed network. A managed network consists of a single management server and multiple client computers installed the tracing function. The following describes what is achieved by using the tracing function for networks.

(1) Before sending data to another client computer, a process managed by tracing function (hereafter, managed process) in the client computer queries the management server to determine whether the tracing function is installed on the receiving computer. If the tracing function is installed, the transmission of classified information is allowed. If the function is not installed, the transmission is disallowed. This makes it possible to satisfy function requirements (1), (4), and (5).

(2) The tracing function writes a log in the client computer and transfers that log to the management server. The collected logs in the management server are used to manage the location and diffusion of classified information in the managed network. This satisfies function requirement (2).

(3) Before a managed process sends classified information, a managed process in the sending computer notifies sending classified information to the receiving computer. Then tracing function in the receiving computer mark a process receiving the classified information as managed process. This satisfies function requirement (3).

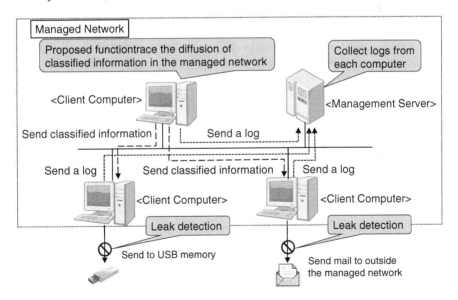

Fig. 1 Overview of diffusion tracing function for classified information among multiple computers

In order to satisfy requirements (1) and (3), communication functions must ensure that the receiving client computer is installed the tracing function before transmitting classified information. In the following sections, we propose a design of the function.

3.2 Communication Method to Pass the Information to be Managed

In order to control the communication of classified information, an application program, which monitors the communication of classified information (hereafter, communication monitor), is developed. Before computers exchange classified information, the communication monitor in the sending computer sends the destination port number and IP address of the receiver socket to the receiving computer. The receiving computer uses port number to mark a receiver socket as managed socket. Then the receiving computer marks a process receiving classified information from managed socket as managed process.

The management server maintains the list of IP address of client computers installed the tracing function. The communication monitor in the sending computer queries whether a receiving computer has been registered in the management server list, and, if so, the transmission process is initiated.

Figure 2 shows the flow of process using the communication monitor to send the classified information from the sending computer to the receiving computer as described below:

(1) Send system call of managed process is invoked.
(2) The processing of the send system call is suspended, and the IP address and port number of the receiver socket is sent to the communication monitor.
(3) The communication monitor queries the management server whether the receiving computer has been installed tracing function.

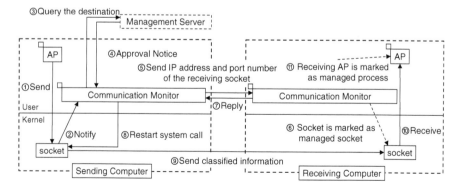

Fig. 2 Sending classified information to other computer AP

(4) If the receiving computer has been installed, the management server notifies the communication monitor that it has been allowed to send classified information. If it is not installed, the management server notifies the communication monitor that it is not allowed to send classified information.

(5) If transmission of classified information is allowed, the communication monitor in the sending computer sends the port number of the receiver socket to the receiving computer. If transmission of classified information is not allowed, the function terminates the processing of the send system call.

(6) The communication monitor in the receiving computer marks a socket using the receiving port number as managed socket.

(7) The communication monitor in the receiving computer replies to the communication monitor in the sending computer that it is ready to receive the classified information.

(8) After receiving the confirmation, the communication monitor in the sending computer restarts the send system call that was suspended in step (2).

(9) Classified information is sent.

(10) The application program in the receiving computer initiates a receiving system call and receives the classified information from the managed socket.

(11) The receiving AP is marked as a managed process.

4 Evaluation

4.1 Overhead

In order to evaluate overhead of the diffusion tracing function for classified information among multiple computers, we measured the time required to transfer managed files to other computers. We measured the time it takes to upload a managed file (1–10 MB) from an FTP client to an FTP server.

We used the following configuration as our measurement environment: the computer of FTP client has a Celeron D 2.8 GHz CPU and 768 MB memory. The computer of FTP server has a Pentium 4 3.0 GHz CPU and 512 MB memory. The two computers are connected using Ethernet 100Base-TX. OS used on both computers is Linux-2.6.0.

The measurement results are shown in Table 1. We found that the transfer time using the managed kernel function is a maximum of about 2.8 times of the transfer time using the original kernel function. This is because the communication monitor communicates with the management server when the send system call is called. Therefore, in order to reduce overhead, it is necessary to reduce the exchange of information between the communication monitor and the management server.

Table 1 File transfer time by FTP (ms)

	Data size of a transmitting file	
	1 MB	10 MB
Function before implementation	83	885
Function after implementation	233	2010
Overhead	150 (181 %)	1125 (127 %)

4.2 Evaluation of Diffusion Tracing of Classified Information from Logs

The diffusion of classified information among multiple computers can be analyzed by tracing the flow of classified information, using data in collected logs during FTP file transfer of the classified information. Logs from two the computer of FTP client and the computer of FTP server were evaluated after uploading files with classified information from the client to the server. ProFTP was used for the FTP server; LFTP was used for the FTP client.

Figure 3 shows the log output from the computer of FTP client. From the first line of the log, we find that the process PID 2907 of the FTP client "lftp" is a managed process, because it read the classified information file secret.txt. From lines 2 and 4–6 of the log, we see that the process PID 2907 has sent the data to the computer with the IP address 192.168.8.201.

Figure 4 shows the log output from the computer of FTP server. From the first line of the log, we find that a socket receiving classified information from the computer with the IP address 192.168.8.166 is marked as managed socket. From the second line of the log, the process PID 2796 of the FTP server "proftp" is marked to be managed. Lines 2–4 and 6 indicate that the managed process PID 2796 will exchange data with a computer with the IP address 192.168.8.166. Line 7 of the log shows that the process PID 2796 writes data to a managed file called secret.txt; this file is marked as a managed file.

```
1: Thu Feb  4 15:40:41 2010 PID: 2634 Socket Marked (by remote)  DOMAIN:INET
   SRC-IP:192.168.8.166 SRC-Port:32769, SRC-PID:2907
2: Thu Feb  4 15:40:41 2010 PID: 2796 Process Marked PID:2796
   PNM:/usr/local/sbin/proftpd (socket)
3: Thu Feb  4 15:40:41 2010 PID: 2796 Send to remote machine.  DOMAIN:INET
   PID:2796 PNM:/usr/local/sbin/proftpd DST-IP:192.168.8.166 DST-Port:32769
4: Thu Feb  4 15:40:41 2010 PID: 2796 Send to remote machine.  DOMAIN:INET
   PID:2796 PNM:/usr/local/sbin/proftpd DST-IP:192.168.8.166 DST-Port:32769
5: Thu Feb  4 15:40:41 2010 PID: 2634 Socket Marked (by remote)  DOMAIN:INET
   SRC-IP:192.168.8.166 SRC-Port:32771, SRC-PID:2907
6: Thu Feb  4 15:40:41 2010 PID: 2796 Send to remote machine.  DOMAIN:INET
   PID:2796 PNM:/usr/local/sbin/proftpd DST-IP:192.168.8.166 DST-Port:32769
7: Thu Feb  4 15:40:41 2010 PID: 2796 File Marked. FILE:/home/s-uemura/secret.txt
   INODE:426211 PID:2796 PNM:/usr/local/sbin/proftpd MODE:2 NO:3 DEV:300004
```

Fig. 3 Log output from the computer of FTP client

Design and Evaluation of a Diffusion Tracing Function

1: Thu Feb 4 15:40:41 2010 PID: 2907 <u>Process Marked PID:2907 PNM:/usr/bin/lftp</u>
<u>FILE:secret.txt</u>
2: Thu Feb 4 15:40:41 2010 PID: 2907 <u>Send to remote machine</u>. DOMAIN:INET
<u>PID:2907 PNM:/usr/bin/lftp DST-IP:192.168.8.201 DST-Port:21</u>
3: Thu Feb 4 15:40:41 2010 PID: 2871 Socket Marked (by remote) DOMAIN:INET
SRC-IP:192.168.8.201 SRC-Port:21, SRC-PID:2796
4: Thu Feb 4 15:40:41 2010 PID: 2907 <u>Send to remote machine</u>. DOMAIN:INET
<u>PID:2907 PNM:/usr/bin/lftp DST-IP:192.168.8.201 DST-Port:21</u>
5: Thu Feb 4 15:40:41 2010 PID: 2907 <u>Send to remote machine</u>. DOMAIN:INET
<u>PID:2907 PNM:/usr/bin/lftp DST-IP:192.168.8.201 DST-Port:21</u>
6: Thu Feb 4 15:40:41 2010 PID: 2907 <u>Send to remote machine</u>. DOMAIN:INET
<u>PID:2907 PNM:/usr/bin/lftp DST-IP:192.168.8.201 DST-Port:32777</u>

Fig. 4 Log output from the computer of FTP server

From these logs, the proposed methods can trace the flow of classified information from the computer with the IP address 192.168.8.166 to the computer with the IP address 192.168.8.201.

5 Related Work

By using another connection to send address range of the data when sending tainted data, received data is to be tainted, to trace the diffusion of classified information among multiple computers [3]. However, there is a problem that unable to trace the diffusion of classified information by UDP. On the other hand, by storing the address range of the data to improve the header of the packet, taint is imparted to the header, to trace the diffusion of classified information among multiple computers [4]. However, there is a problem that when send the packet to a computer that does not eliminate the header of the packet, improved header is treated as data. Multiple virtual machines can be processed on a trusted virtual machine monitor, providing isolated virtual environments on a per-machine basis to serve as mechanisms that prevent other users from accessing files [5]. To limit access to files by isolating environmental in the program unit, the mechanism used is to assign the domain name of the group to the files, and same domain user only access the files [6].

6 Conclusion

We proposed a diffusion tracing function of classified information among multiple computers to prevent information leakage outside internal networks. The classified information exchanged among the processes in multiple computers should be traced. Therefore, before computers exchange classified information, the proposed function sends IP address and the port number of the receiver socket. Then, the diffusion tracing function for classified information in the receiving computer traces the receiving classified information. An evaluation of logs shows that the proposed function traces a diffusion of classified information among multiple computers.

In future work, we will reduce the overhead of the proposed function.

References

1. Japan Network Security Association (2008) Information Security Incident Survey Report, http://www.jnsa.org/result/incident/data/2008incident_survey_e_v1.0.pdf
2. Tabata T, Hakomori S, Ohashi K, Uemura S, Yokoyama K, Taniguchi H (2009) Tracing classified information diffusion for protecting information leakage. IPSJ J 50(9):2088–2012 (in Japanese)
3. Kim CH, Keromytis DA, Covington M, Sahita R (2009) Capturing information flow with concatenated dynamic taint analysis. 2009 International conference on Availability, Reliability and Security (ARES 2009), pp 355–362
4. Zavou A, Portokalidis G, Keromytis DA (2011) Taint-Exchange: A generic system for cross-process and cross-host taint tracking. The 6th International Workshop on Security (IWSEC 2011), vol 7038. LNCS, pp 113–128
5. Garnkel T, Pfaff B, Chow J, Rosenblum M, Boneh D (2003) Terra: A virtual machine-based platform for trusted computing. In: Proceedings of 19th ACM SIGOPS Symposium on Operating System Principles (SOSP 2003), pp 193–206
6. Katsuno Y, Watanabe Y, Furuichi S, Kudo M (2007) Chinese-Wall process confinement for practical distributed coalitions. Proceedings of 12th ACM Symposium on Access Control Models and Technologies (SACMAT2007), pp 225–234

DroidTrack: Tracking Information Diffusion and Preventing Information Leakage on Android

Syunya Sakamoto, Kenji Okuda, Ryo Nakatsuka and Toshihiro Yamauchi

Abstract An app in Android can collaborate with other apps and control personal information by using the Intent or user's allowing of permission. However, users cannot detect when they communicate. Therefore, users might not be aware information leakage if app is malware. This paper proposes DroidTrack, a method for tracking the diffusion of personal information and preventing its leakage on an Android device. DroidTrack alerts the user of the possibility of information leakage when an app uses APIs to communicate with outside. These alerts are triggered only if the app has already called APIs to collect personal information. Users are given the option to refuse the execution of the API if it is not appropriate. Further, by illustrating how their personal data is diffused, users can have the necessary information to help them decide whether the API use is appropriate.

Keywords Android · Malware · Preventing information leakage · API control

1 Introduction

In recent years, adoption of the smartphone has been rapidly spreading, and Android [1] is one of the popular operating systems (OS) for smartphones. An app developer can make the app available through a Web site, such as Google Play Store [2]. However, an app [3] can hijack administrative privileges in order to exploit vulnerability in the Android OS and send out illegally collected personal information.

S. Sakamoto (✉) · K. Okuda · R. Nakatsuka · T. Yamauchi
Graduate School of Natural Science and Technology, Okayama University,
Okayama, Japan
e-mail: sakamoto@swlab.cs.okayama-u.ac.jp

T. Yamauchi
e-mail: yamauchi@cs.okayama-u.ac.jp

J. J. (Jong Hyuk) Park et al. (eds.), *Multimedia and Ubiquitous Engineering*,
Lecture Notes in Electrical Engineering 240, DOI: 10.1007/978-94-007-6738-6_31,
© Springer Science+Business Media Dordrecht(Outside the USA) 2013

Malware that target the Android OS are usually intended to illegally collect personal information. A mobile device contains a large amount of personal information, such as name, address, phone number, etc. and their information can be easily obtained by apps using the Android API. In addition, many users are unaware that mobile phones are not secure and usually do not come with any anti-malware software. For this reason, there is a possibility of information leakage while user did not notice the infection of malware.

An Android app is executed in sandbox, and communication with other apps is severely restricted, except using Intent [4]. Key features such as external communications and the acquisition of personal information require permissions from the user. However, the user cannot detect when the personal information is obtained by the app and whether that personal information was leaked.

In this paper, we propose DroidTrack: a method for tracking information leakage diffusion and preventing information leakage on Android, tracks information diffusion after the app has obtained personal information. DroidTrack alerts the user if there is a possibility of information leakage, and allows the user to limit the use of the API. DroidTrack monitors any app that uses the information-gathering API and displays a warning when the app also uses the API that sends information outside of the device. Personal information can also be leaked when one app obtains personal information and then sends it to another app, which sends the information out of the device. For this reason, DroidTrack manages both apps using the Intent. In addition, the user is allowed to decide whether to limit the use of the API, which sends information out of the device, thereby preventing information leakage.

2 Android Component and Security Issues

2.1 Android Component

In the Android OS, all apps operate on the Applications layer. If an app requires resources, it must use the API provided by the Application Framework. Android apps have individual user IDs (UID), and communication with an app with a different UID is highly restricted, except when using the Intent.

Apps cannot use the Android API to access a protected resource or to gather personal information without user's permission. It is necessary to obtain the user's permission [5] for app to use these APIs. An app can request specific permissions, e.g., permission to connect to the Internet (INTERNET) or permission to read the status of the unit (READ_PHONE_STATE). The safety of the resources is preserved by granting only minimum permissions required by the app.

Each Android app runs in sandbox. By default, apps cannot communicate with another app because they are strictly separated from each other. However, it is possible to enable communication between apps by using the Intent, which allows

DroidTrack: Tracking Information Diffusion   245

an app to communicate with another app and receive the results of process. In addition, an app can pass data both as a string or as an object.

2.2 Security Issues in Android

The following are the problems associated with malware and malware infections in the main security areas in the Android OS:

(1) Problems obtaining administrator authority.
(2) Problems with development tools (Android Debug Bridge (ADB) [6]).
(3) Permission abuse.
(4) Difficult detection of information leakage.
(5) WebKit abuse.

Problems (1), (2), and (3) are cause of the infection to malware. Problems (4) and (5) are related to malware behavior. Problem (4) makes it very difficult to inform the user when the app gathers information and what kind of personal information it gathers. Therefore, if a user installs malware by mistake, user cannot detect the leakage of personal information. In this work, we deal with problem (4) by preventing the transmission of information out of a smartphone.

3 Design Principles of the Proposed Method

3.1 Requirements and Challenges

In order to deal with the problem, we propose the following requirements:

(1) Detect all APIs with the possibility of information leakage.
(2) User can judge the risk of information leakage.
(3) Information leakage can be prevented by disallowing the execution of API.

In order to satisfy these three requirements, we propose the following challenges:

(A) Clarifying the condition of information leakage.
(B) Detecting all uses of APIs that have a possibility of information leakage.
(C) Controlling the operations of apps that have a possibility of information leakage.
(D) Allowing the user to decide whether app can receive sensitive information.

3.2 Solution

3.2.1 Solution for Challenge (A)

Information leakage can occur when an app uses the Android API to send information to out of the device (diffuse information) after obtaining personal information. The app can also obtain personal information without using the information-gathering API by using the Intent instead. In another scenario of information leakage, one app uses the information-gathering API and then communicates with another app that uses the information-diffusing API.

In this work, we address the following scenarios of information leakage:

(1) A single app uses the information-gathering API and the information-diffusing API or the Intent.
(2) One app uses the information-gathering API and then communicates with another app using either the Intent or the information-diffusing API.

3.2.2 Solution for Challenges (B) and (C)

As mentioned in Sect. 3.2.1, information leakage can occur when an app uses the API that obtains personal information, the API that diffuses the information, or the Intent. Therefore, to deal with challenges (B) and (C) (detecting all use of the API and controlling the operations of apps), we propose a method to control the behavior of the app as follows:

- by intercepting calls to the information-gathering API, information-diffusing API, or Intent,
- by determining the user's preference to control the use of API if either scenario described in Solution for challenge (A) is true, and
- by controlling the use of the APIs or the Intent based on the user's preference.

4 Method for Tracking Diffusion of Information and Preventing Information Leakage on Android

4.1 Design Principle

To address challenges (B) and (C), we propose the following requirements:

(1) To inform the user if there is potential for information leakage.
(2) To limit the use of APIs and the Intent with accuracy based on user's preference.

4.2 Basic Method

We change the framework of the Android as follows:

(1) "Hook" or intercept calls to the information-gathering API and the information-diffusing API and inform the user of both the name of the app using the API and the name of the API used.
(2) "Hook" or intercept calls to the Intent and inform the user of both the name of the app that uses the Intent and the name of the app called by the Intent.
(3) Execute a process based on the user's preference regarding the use of the APIs or the Intent.

The API is used differently depending on the type of personal information that is gathered by the app. Therefore, because of the change described in (2), the user can be informed when and what kind of personal information which the app obtains and attempts to transmit out of the device. Furthermore, the change described in (3) can be used to prevent information leakage according to the user's preferences. In the following section, we describe a method to track and prevent information diffusion by using the modified framework described above.

4.3 Control of API in the Framework

Figure 1 shows the flow of control of the API in the framework. DroidTrack consists of two "Control Aps" at the Applications layer and one "Calling Control AP Unit" at the Application Framework layer.

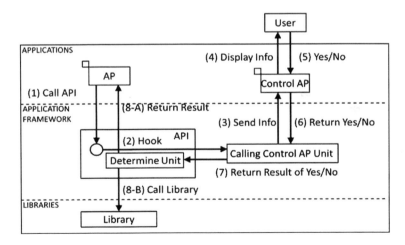

Fig. 1 Control of APIs

In Fig. 1, "Control AP" is an app that provides information about the API to the user and prompts the user to choose whether to limit the use of the API. The "Calling Control AP Unit" informs the "Control AP" whenever an app calls the information diffusing APIs and transmits back to the API engine the user's preferences regarding the use of the API.

The following describes details of the flow of the process in the framework:

(1) The app "AP" calls the information-diffusing API.
(2) The "Hook" intercepts the call to the information-diffusing API in the framework.
(3) The "Calling Control AP Unit" passes the information about the intercepted call to the "Control AP."
(4) The "Control AP" displays a warning dialog to the user if it suspects that an information leak is possible.
(5) The user replies to the dialog to indicate whether user allows the use of the API.
(6) The "Control AP" forwards the user's preference to the "Calling Control AP Unit."
(7) The "Calling Control AP Unit" returns the result to the "Determine Unit."
(8) The "Determine Unit" handles API based on the user's preference, as follows:

 (A) Error handling is carried out if the user disallows the API to be called by the app.
 (B) API process returns to normal mode if the user permits the API to be called by the app.

Like the above-mentioned procedure, we satisfy the requirement 4.1-(2) by requiring user's preference before use of APIs and processing according to the preference.

4.4 Control AP

4.4.1 Basic Mechanism

Figure 2 illustrates the mechanism of "Control AP." The Search-Leakage function triggers the Info Diffusion Manage Unit to check the possibility of information leakage. If there is a possibility, it passes the process to the Control-Write-Out function. If there is no possibility, it allows the API calls to be processed. The Info Diffusion Manage Unit updates the Information Diffusion data structure, examines the possibility of information leakage, and returns the result of the test to the Search-Leakage function. The Control-Write-Out function displays a warning dialog, and then accepts and returns the user's preference regarding the use of the APIs.

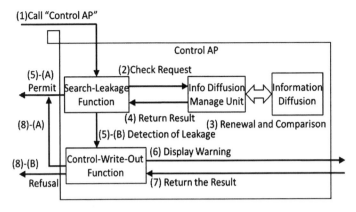

Fig. 2 Basic mechanism of control AP

4.4.2 Information Leakage Determination Method

Control AP monitors apps that use the information-gathering API by wrapping the app in a managed object. If the managed AP communicates with another app that uses the Intent, the second app is added to the managed object. Control AP also warns the user of the risk of information leakage if the managed app uses the information-diffusing API, thereby satisfying the requirement described in Sect. 4.1-(1).

5 Experiment of the Operation of DroidTrack

Prevention of information leakage by apps was tested using the following procedure:

(1) Obtain the phone number of the mobile device by using "getLine1Number," which serves as the information-gathering API.
(2) Transmit the personal information out of the device by using "sendText-Message," which serves as the information-diffusing API.

Figure 3 shows the dialog displayed by DroidTrack when the example app runs. The user can detect the use of the API by various information from the dialog. In this case, the user presses "Yes" to allow the use of the API and "No" to disallow the use of the API. DroidTrack could prevent information leakage by pressing "No".

Fig. 3 Warning dialog

6 Related Work

MockDroid [7] allows the user to provide fake or "mock" data to prevent a real personal information leakage. This method needs user's set up of permissions for each apps. Therefore, DroidTrack needs no user's set up, but needs user's input only when there is a possibility of information leakage. Furthermore, DroidTrack tracks all transmitted information, including transmissions without leakage, in order to check all API communications in and out of the device. AppFence [8] that using TaintDroid [9] provides a similar approach, but it modifies framework and Dalvik VM, and uses the policy which is made except on Android. On the other hand, DroidTrack modifies only the framework of the Android. Therefore, DroidTrack is easier to implement.

7 Conclusion

We have proposed DroidTrack, a method to warn of the risk of information leakage by monitoring apps that obtain personal information and by keeping track of information diffusion. DroidTrack prevents personal information leakage by controlling the behavior of the API, based on the user's preference, which they indicate when a warning is displayed. DroidTrack can also detect information leakage in scenarios where the Intent is used and where the information-diffusing API is used. In addition, DroidTrack can inform the user of the risk of information leakage and display a list of information-gathering APIs that could have diffused information to other apps.

References

1. Android. http://www.android.com/
2. Google play store. https://play.google.com/store
3. Droid dream. Google android market kills droid dream malware in Trojans. http://blogs.computerworld.com/17929
4. Intent. http://developer.android.com/reference/android/content/Intent.html
5. Access permissions. http://developer.android.com/intl/ja/reference/android/Manifest.permission.html
6. Android debug bridge. http://developer.android.com/guide/developing/tools/adb.html
7. Beresford AR, Rice A, Skehin N, Sohan R (2011) MockDroid: trading privacy for application functionality on smartphones. In: Proceedings of the 12th workshop on mobile computing systems and applications, pp 49–54
8. Hornyack P, Han S, Jung J, Schechter S, Wetherall D (2011) These aren't the droids you're looking for: retrofitting android to protect data from imperious applications. In: Proceedings of the 18th ACM conference on computer and communications security (CCS2011)
9. Enck W, Gilbert P, Chun B, Cox LP, Jung J, McDaniel P, Sheth AN (2010) TaintDroid: an information-flow tracking system for realtime privacy monitoring on smartphones. In: Proceedings of the 9th USENIX symposium on operating systems design and implementation (OSDI'10)

Three Factor Authentication Protocol Based on Bilinear Pairing

Thokozani Felix Vallent and Hyunsung Kim

Abstract Secure authentication mechanism is a pre-requisite to remote access of server's resources particularly when done over the Internet. This paper presents a three factor authentication protocol which is based on verification of user's: biometrics, knowledge proof of a password and possession of token to pass authentication. The proposed protocol utilizes bilinear mapping for session key establishment and elliptic curve discrete logarithm problem for security.

Keywords Authentication · Bilinear pairing · Three factor authentication

1 Introduction

Information security is concerned with the assurance of confidentiality, integrity and availability of information in all forms. There are many tools and techniques that can support the management of information security one of which is the use of tokens that store client identifying information like smart card [1–5]. Smart card authentication falls short of password sharing among colleagues, password guessing and smart card breaching [3]. In applications with strict user identification smart card flaws can be dealt with by employing the biometric authentication besides the password. Biometrics is hard to forge hence outworks impersonation attack resilience thus provides a reliable means of authentication [3–5].

T. F. Vallent
Department of IT Convergence, Kyungil University, Kyungpook,
Kyungsansi 712-701, Korea
e-mail: tfvallent@gmail.com

H. Kim (✉)
Department of Cyber Security, Kyungil University, Kyungpook,
Kyungsansi 712-701, Korea
e-mail: kim@kiu.ac.kr

J. J. (Jong Hyuk) Park et al. (eds.), *Multimedia and Ubiquitous Engineering*,
Lecture Notes in Electrical Engineering 240, DOI: 10.1007/978-94-007-6738-6_32,
© Springer Science+Business Media Dordrecht(Outside the USA) 2013

A biometric system is a pattern recognition system that extracts an individual's unique features set for authentication by comparing these features' template pre-stored in the database [2, 4, 6]. Three factor authentication involves knowledge proof by checking user's knowledge of correct password, token possession and biometrics matching before authentication. Biometric authentication can be applied for identification and non-repudiation and for preserving the integrity like in passport, medical records access control among others [2, 3].

In 1981, Lamport first proposed a remote password authentication scheme for insecure communication. The protocol uses verification table hence it's at the edge of a huge security risks once the system is compromised [7]. Hwang and Li (2000) proposed a remote user authentication scheme using smart cards based on ElGamal's public key cryptosystem the protocol suffers from man-in-the middle attack [2]. In 2010 Li and Hwang proposed another remote user authentication based on biometrics verification, smart card, one-way hash function but still bears a problem of man-in-the-middle attack [2, 4].

2 Preliminaries

This section introduces mathematical background necessary for the proposed protocol's description.

2.1 Bilinear Pairing

Consider two groups G_1 and G_2 both of order q, a cyclic additive group and cyclic multiplicative group respectively. A map $\hat{e}: G_1 \times G_1 \rightarrow G_2$ has following properties [6].

- Bilinearity; $\forall\ P,\ Q \in G_1$ and $\forall a,\ b \in Z^*_q$
$$\hat{e}(aP,\ bQ) = \hat{e}(P,\ Q)^{ab}.$$
- Non-Degeneracy; $\hat{e}(P,\ P) \neq 1$, where $P \neq 0$ to avoid everything totally mapped to the identity element.
- Computability: $\hat{e}: G_1 \times G_1 \rightarrow G_2$ is efficiently computable.

Notice that $\hat{e}(aP,\ bQ) = \hat{e}(bP,\ aQ)$ by the bilinear property.

2.2 Computational Discrete Logarithm Problem

Given $Q = kP$, where $P,\ Q \in G_1$, it is relatively easy to compute Q given k and P. However, it is relatively hard to determine k given Q and P [8–10].

3 Three Factor Authentication Protocol

The major contributions of the proposed protocol are: (1) achieving mutual authentication and session key agreement in an efficient way; (2) no need of directory or password verification table (3) offline password guessing attack resilience (4) insider attack resilience and replay attack resilience. The network environment involves one registration server, many users subscribing to different service servers under the same registration server. The users access services from the service servers over an insecure channel by means of handy devices. Entities mutual authentication is a pre-requisite before granting access to service server's resources. The protocol takes four phases: *set up phase, registration phase, login phase and password change phase* as discussed below.

3.1 Set Up Phase

The registration server, RS selects two groups G_1 and G_2, a bilinear mapping $\hat{e}$: $G_1 \times G_1 \rightarrow G_2$ and a cryptographic function $H():\{0, 1\}^* \rightarrow G_1$. RS selects a master secret key s and computes a corresponding public key as $P_s = sP$, where P is the generator of the group G_1. Then RS publishes $\{G_1, G_2, q, P, P_s, \hat{e}, H()\}$ while keeping the master key s secret.

3.2 Registration Phase

Any user registers with a registration server (RS) and the procedures is as follows.

R1. A user, U_i, submits $H(PW_i\|b)$, $H(f_i\|ID_i)$ and ID_i to RS where PW_i is a chosen password, b is a random number, f_i is hashed biometrics (B_i) and ID_i is identity.

R2. RS computes U_i's public key $Q_u = H(ID_i)$ and U_i's private key $P_u = sQ_u$.

R3. Further RS computes $V_i = P_u\oplus y_i$ and $t_i = H(P_u)$, where $y_i = H(PW_i\|b)\oplus H(f_i\|ID_i)$.

R4. RS sends the smart card to U_i securely stored with $\{ID_i, H(), t_i, V_i\}$.

R5. Upon receipt of the smart card, U_i inserts the random number b on the smart card so the information stored in the card is $\{ID_i, H(),Q_u, t_i, V_i, b\}$.

In a similar manner, a service server, SS_j registers with RS and is issued a pair of a public key, $Q_{ssj} = H(ID_{ssj})$ and a private key, $P_{ssj} = sH(ID_{ssj})$.

3.3 Login and Verification Phase

To login to the service server SS_j user inserts the smart card into a terminal and inputs identity ID_i, password PW_i and his/her biometrics B_i into a device and then:

A1. The smart card computes $H(PW_i\|b)$, $f_i = H(B_i)$ and $H(f_i\|ID_i)$ then cross-checks the correctness of ID_i, f_i and $H(PW_i\|b)$ and if found valid the processes continues as below otherwise terminates.
A2. The smart card computes $P_u^* = V_i \oplus y_i$.
A3. Then the smart card checks if $t_i = H(P_u^*)$. If the result holds, it means U_i inputted correct password PW_i, identity ID_i and the biometrics B_i hence is authenticated.
A4. Now the smart card selects a random number r_i and computes r_iP.
A5. Smart card computes $K_a = H(r_iP\|P_s\|Q_u\|Q_{ssj}\|\hat{e}(sQ_uP, r_iQ_{ssj}))$, $SK = H(K_a\| ID_i\|ID_{ssj})$ and $auth = H(r_i\|y_i\|ID_i\|ID_{ssj}\|SK)$ and sends $\{E_{Ka}(r_i\|auth\|ID_i), r_iP, y_i\}$ to SS_j.
A6. In turn SS_j computes $K_a = H(r_iP\|P_s\|Q_u\|Q_{ssj}\|\hat{e}(Q_ur_iP, sQ_{ssj}))$ and uses it to decrypt $E_{Ka}(r_i\|auth\|ID_i)$, the message from smart card $\{E_{Ka}(r_i\|auth\|ID_i), r_iP, y_i\}$.
A7. Then SS_j cross-checks if ID_i and $auth$ are valid then continues to compute the session key as $SK = H(K_a\|ID_i\|ID_{ssj})$ and verifies $auth = H(r_i\|y_i\|ID_i \|ID_{ssj}\| SK)$. When it computations hold SS_j's accepts U_i's request otherwise rejects.

3.4 Password Change Phase

At will U_i has the right to change his/her password from PW_i to PW_i^{new} for security reasons without involving RS. Below is the procedure for password change phase.

C1. U_i submits identity, ID_i, current password, PW_i and his/her biometrics, B_i into a device before submitting the new password PW_i^{new}.
C2. Smart card computes $H(PW_i\|b)$, $f_i = H(B_i)$ and $H(f_i\|ID_i)$ and checks the validity of the results. The process continues if the calculations holds otherwise terminates.
C3. Then smart card computes $y_i = H(PW_i\|b) \oplus H(PW_i^{new}\|b)$.
C4. Smart card replaces y_i with $y_i^{new} = H(PW_i^{new}\|b) \oplus H(f_i\|ID_i)$.

4 Security and Performance Analysis

This section points out the security properties and computational performance.

4.1 Security Analysis

This section shows how the proposed protocol achieves notable security requirements.

Proposition 1 *The proposed protocol supports mutual Authentication.*

Proof Mutual authentication is satisfied in the sense that each party verifies the counterpart's legitimacy. This property is implicitly embedded in the ability to establish a session key SK between the two. When U_i sends the message $\{E_{Ka}(r_i\|auth\|ID_i), r_iP, y_i\}$ to SS_j, h/she is assured that only the intended service server will be able to compute the session key $SK = H(r_iP\|P_s\|Q_u\|Q_{ssj}\|\hat{e}(sQ_uP, r_iQ_{ssj}))$, because no one else could compute $\hat{e}(Q_ur_iP, sQ_{ssj}) = \hat{e}(Q_u\ r_iP, sQ_{ssj})$ but SS_j only by bilinear pairing property. While on the hand SS_j knows only U_i could form a valid pair $\hat{e}(sQ_u\ r_iP, Q_{ssj})$.

Proposition 2 *The proposed protocol supports offline password guessing attack resilience.*

Proof Offline password guessing attack resilience is met by the fortification of the smart card stored information, $\{ID_i, H(), Q_u, t_i, V_i, b\}$. Even if an adversary skims the information the tough task of guessing PW_i awaits. The attempt would mean computing $H(PW_i\|b)\oplus H(f_i\|ID_i) = P_u\oplus V_i$, which would require knowledge of the private key P_u, the random umber b and the hashed biometrics f_i. So guessing the four parameters, PW_i, P_u, b and f_i is extremely hard besides b is a high entropy number.

Proposition 3 *The proposed protocol supports replay attack resilience.*

Proof Replay attack is resisted by usage of fresh random number r_i in the message $\{E_{Ka}(r_i\|auth\|ID_i), r_iP, y_i\}$. Even if an adversary would try to replay an old authenticated message $\{E_{Ka}(r_i\|auth\|ID_i), r_iP, y_i\}^*$, h/she will fail to get authenticated obviously because $auth \neq auth^*$ as a result of the randomness of r_i in each session.

Proposition 4 *The proposed protocol supports insider attack resilience.*

Proof Insider attack is overcome by securing the message $\{H(PW_i\|b), H(f_i\|ID_i)$ and $ID_i\}$ during registration phase. In order to fabricate a correct login message, $P_u^{**} = V_i^* \oplus y_i^*$ an adversary has to guess PW_i, b, and f_i simultaneously, which is impossible within polynomial time. Thus the protocol copes up with insider attack.

4.2 Performance Analysis

This section gives computational cost of the protocol in comparison with related protocols depicted in Table 1.

Table 1 Comparison with related protocols on computational cost

Protocols	Computational operations				
	No. of pairing	No. of hash function	No. of point addition in EC	No. of scalar multiplication in EC	No. of symmetric encryption or decryption
Ours	2	9	1	2	1
Juang et al's	2	10	0	3	2
Das et al's	2	2	1	2	0

Table 1 is gives communication cost of the shown protocols with an exception of symmetric encryption/decryption, the operations' computational load in ascending order are: bilinear pairing operation is heaviest followed by hash function then EC point addition operation and finally EC scalar multiplication with reference to [6]

Compared with communication load of other smart card based authentication protocols shown in Table 1 and presented in [6, 11] our protocol is moderately heavier but achieves high security level. Therefore the proposed protocol is relatively efficient and strong against smart card well known attacks.

5 Conclusion

A secure three factor smartcard biometric authentication protocol has been proposed that is facilitated by the properties of bilinear pairing and computational discrete logarithm problem. The proposed protocol does not require a directory or password verification table and supports offline password change without involving the server. Further the protocol resists lost or stolen password guessing attack. Though has comparatively higher computation load the proposed protocol achieves high security.

Acknowledgments This work was supported by the National Research Foundation of Korea Grant funded by the Korean Government (MEST) (NRF-2010-0021575).

References

1. Sanchez-Reillo R (2001) Smart card informations and operations using biometrics. IEEE AESS systems magazine
2. Jeon IS, Kim HS, Kim MS (2011) Enhanced biometrics-based remote user authentication using smart cards. J Secur Eng 16(1):9–19
3. Arkantonakis KM, Tunstall M, Hancke G, Askoxylakis I, Mayes K (2009) Attacking smart cards systems: theory and practice. Inf Secur Tech Rep 14:46–56
4. Li X, Niu JW, Ma J, Wang WD (2011) Cryptanalysis and improvement of a biometric-based remote user authentication scheme. J Comput Appl 34:73–79

5. Chen TH, Hsiang HC, Shih WK (2011) Security enhancement on an improvement on two remote user authentication schemes using smart cards. Furth Gener Comput Syst 27:377–380
6. Juang WS, Nien WK (2008) Efficient password authenticated key agreement using bilinear pairing. Math Comput Model 47:1238–1245
7. Lao YP, Hsiao CM (2012) A novel multi-server remote user authentication scheme using self-certified public keys for mobile clients. Furth Gener Comput Syst 29(3):886–900
8. Giri D, Srivastava PD (2006) An improved remote user authentication scheme with smart cards using bilinear pairings. In: IACR Cryptology ePrint Archive, 2006, p 274
9. Lao YP, Hsiao CM (2011) The improvement of ID-based remote user authentication scheme using bilinear pairings. In: Consumer electronics, communications and networks international conference (CECNet), pp 865–869
10. Sarier ND (2010) Improving the accuracy and storage cost in biometric remote scheme. J Netw Comput Appl 33:268–274
11. Das ML, Saxena A, Gulati VP, Phatak DB (2006) A novel remote user authentication scheme using bilinear pairings. J Comput Secur 25:184–189
12. Wang D, Ma CG, Wu P (2012) Secure password-based remote user authentication scheme with non-tamper resistant smartcards. In: Proceedings of the 26th annual IFIP WG 11.3 conference on data and application security and privacy, pp 112–121
13. Pippal RS, Jaidhar CD, Tapaswi S (2012) Security issue in smart card authentication scheme. Int J Comput Theory Eng 4(2):206–211
14. Chang CC, Lee CY, Chiu YC (2009) Enhanced authentication scheme with anonymity for roaming service in global mobility networks. Comput Commun 32:611–618
15. Li X, Ma J, Wang W, Xiong Y, Zhang J (2012) A novel smart card and dynamic ID based remote user authentication scheme for multi-server environment. Math Comput Model. doi: 10.1016/j.mcm.2012.06.033

A LBP-Based Method for Detecting Copy-Move Forgery with Rotation

Ning Zheng, Yixing Wang and Ming Xu

Abstract Copy-move is the most common tampering manipulations, which copies one part of the image and pastes into another part in the same image. Most existing techniques for detecting tampering are sensitive to rotation and reflection. This paper proposed an approach to detect Copy-Move forgery with rotation. Firstly the suspicious image is divided into overlapping blocks, and then LBP operator are used to produce a descriptor invariant to the rotation for similar blocks matching. It is effective to solve the mismatch problem caused by the geometric changes in duplicated regions. In order to make the algorithm more effective, some parameters are proposed to remove the wrong matching blocks. Experiment results show that the proposed method is not only robust to rotation, but also to blurring or noise adding.

Keywords Copy-move · Image forgery · LBP · Rotation invariant

1 Introduction

With the wide application of powerful digital image processing software, such as Photoshop, digital image tampering becomes increasingly easy. At the same time, digital image forensic cause more and more attention. Copy-move is the most common tampering manipulations, which copies one part of the image and pastes into another part in the same image. The first method for detecting copy-move

---

N. Zheng · Y. Wang (✉) · M. Xu
College of Computer, Hangzhou Dianzi University, Hangzhou 310018, China
e-mail: star_rats@163.com

N. Zheng
e-mail: nzheng@hdu.edu.cn

M. Xu
e-mail: mxu@hdu.edu.cn

J. J. (Jong Hyuk) Park et al. (eds.), *Multimedia and Ubiquitous Engineering*,
Lecture Notes in Electrical Engineering 240, DOI: 10.1007/978-94-007-6738-6_33,
© Springer Science+Business Media Dordrecht(Outside the USA) 2013

forgery was proposed by Fridrich [1]. They lexicographically sorted quantized discrete cosine transforms (DCT) coefficients of small blocks and then checked whether the adjacent blocks are similar or not. In [2] Farid proposed a new method by adopting the PCA-based feature, which can endure additive noise, but the detection accuracy is low. Nevertheless, these methods are sensitive to the geometric changes in the copied part.

To solve the above problem, log-polar transform (LPT) may be performed on image blocks followed by wavelet decomposition [3]. There are some approaches [4] that extracted interest points on the whole image by scale-invariant feature transform (SIFT). However, these schemes still have a limitation on detection performance since it is only possible to extract the key points from particular points of the image. This paper presents a comprehensive novel method to detect duplicated regions that have undergone geometric changes, particularly reflection and rotation.

2 Rotation Invariant Feature

2.1 LBP Operator

LBP operator is an effective texture description operator. It has been successfully applied in image processing areas these years. Next, introduce how to calculate the LBP value. In 3×3 window, the gray value of the center point of the window as a threshold value, other pixels in the window do binarized processing, generates an 8-bit binary string. Then, according to the different positions of the pixels, get the LBP value of the window by weighted summing. It can be computed by

$$LBP = \sum_{i=0}^{7} s(g_i - g_c)2^i, where\ s(x) = \begin{cases} 1 & x \geq 0 \\ 0 & x < 0 \end{cases} \tag{1}$$

Here g_c is the center pixel of the window, g_i represents surrounding pixels. Generally the order of the neighboring pixels is started by the pixel to the right of the center pixel, counterclockwise marked. The LBP value can reflect the texture information for the region. LBP can be expanded to a circular neighborhood. Using (P, R) to describe the neighborhood, where P represents the number of sampling points, R is the radius of the neighborhood. The gray values of neighbors which do not fall exactly in the center of pixels are estimated by interpolation.

2.2 Rotation Invariance

The $LBP_{P,R}$ operator produces 2^P different output values, corresponding to 2^P different binary patterns that can be formed by the P pixels in the neighbor set. When the image is rotated, the gray values g_i will move along the perimeter of the circle. After rotation, a particular binary pattern results in a different $LBP_{P,R}$ value. This does not apply to patterns comprising of only 0 (or 1) which remain constant at all rotation angles. To remove the effect of rotation, assign a unique identifier to each rotation invariant local binary pattern, we define:

$$LBP_{P,R}^{ri} = \min\{ROR(LBP_{P,R}, i), \ i = 0, 1, \ldots P - 1\} \tag{2}$$

where $ROR(x,i)$ performs a circular bit-wise right shift on the P-bit number x i times, superscript ri means rotation invariant. $LBP_{P,R}^{ri}$ quantifies the occurrence statistics of individual rotation invariant patterns corresponding to certain features in the image, hence, the patterns can be considered as feature detectors. In the case of $P = 8$, $LBP_{P,R}^{ri}$ will generate 36 different values or 36 patterns. Let vector V represents the occurrence number of individual patterns. When block is rotated, V' is extracted. It is expected that V and V' are similar, the correlation coefficients between them is close to 1. Compare the similarity between V and V', it is easy to identify the duplicated blocks.

$$corr2(V, V') = \left(\sum_m \sum_n (V_{mn} - \overline{V})(V'_{mn} - \overline{V'})\right) \Big/ \sqrt{\left(\sum_m \sum_n (V_{mn} - \overline{V})^2\right)\left(\sum_m \sum_n (V'_{mn} - \overline{V'})^2\right)} \tag{3}$$

3 The Proposed Method

The framework of the proposed method is given as follow: (1) Dividing the suspicious image into blocks; (2) Extracting appropriate features from each block; (3) Searching similar block pairs; (4) Finding correct blocks and output them. The following is the implement details.

Step1: An $M \times N$ color image is first split into overlapping blocks of $B \times B$ pixels. Adjacent blocks have one different row or column. Thus $(M - B + 1)(N - B + 1)$ blocks would be getting. Let A_i denote the i-th block of pixels, $i = 1,2\ldots,N_{blocks}$, where $N_{blocks} = (M - B + 1)(N - B + 1)$.

Step2: Then feature vectors are extracted from each block A_i. The mean of each of three channels has been proved effective against JPEG compression and blurring. The fourth feature is entropy. It can be calculated as, $v = -\sum_{i=0}^{255} p_i \log_2 p_i$ where p_i is the proportion of the number of pixels which gray value is i to total pixels numbers. In previous detection methods, uniform areas in the image may lead to false matches. The entropy can be used to identify blocks with insufficient textural information. Thus, blocks whose entropy is lower than a defined threshold e_{min} could be discarded. However, this could prevent the system from detecting duplicates in areas with scarce textural information. Finally, a 1×4 eigenvectors v can be gotten from each block.

Step3: The extracted eigenvectors are arranged a matrix L with the size of $(M - B + 1)(N - B + 1) \times 4$, then L is lexicographically sorted. To search for the similar block pairs, the $corr2(V_i, V_j)$ is computed using (3), for every $|j - i| \leq \varphi_a$, when the following conditions are satisfied: $\left|v_i^k - v_j^k\right| \leq \varphi_b, k = 1, 2, 3, \left|v_i^4 - v_j^4\right| \leq \varphi_c$ and $d_{ij} \leq \varphi_d$. Where $\varphi_a, \varphi_b, \varphi_c, \varphi_d$ are threshold discussed in next section and d_{ij} is the Euclidean distance between two blocks. It can be calculated as $d_{ij} = \sqrt{\left(x_i - x_j\right)^2 + \left(y_i - y_j\right)^2}$, (x_i, y_i) is the top-left corner's coordinate of the i-th block.

Let c_{ip} be the higher correlation coefficient computed for V_i. If c_{ip} is greater than a user-defined similarity threshold φ_s, records the offsets between i-th block and p-th block as well as their locations. Whenever we find a pair of matching blocks, create a record. Finally, form a list Q with all the created records.

Step4: Initialize a black map image P with the size $M \times N$. According to the Q, mark the suspicious blocks in P. Morphologic operations are applied to P to fill the holes in the marked regions and remove the isolated points, then output the final result.

4 Experimental Results and Analysis

The experiments were carried out on the Matlab R2009a. All tampered images are generated from three datasets. The first dataset are the Uncompressed Color Image Database (UCID) [5]. The second is several uncompressed color PNG images of size 768×512 pixels released by the Kodak Corporation [6]. The last is a tampered image sets including original color images and their copy-move forged versions [7]. Two standard metrics detection accuracy rate (DAR) and false positive rate (FPR) will be adopted to quantify the accuracy and robustness. These two parameters indicated how precisely our method could locate the duplicated regions. The more *DAR* is close to 1 and *FPR* is close to 0, the more precise the method would be. Unless otherwise noted, all the thresholds in the experiment are set as: $B = 8, \varphi_a = 35, \varphi_b = 2, \varphi_c = 0.3, \varphi_d = 40, \varphi_s = 0.95, e_{min} < 4$. The specific thresholds would be given in case of the default parameters are not well working.

4.1 Effectiveness Testing

In order to test the effectiveness of the proposed method, for the first experiment, we choose some images with the size of 200×200 pixels from the third dataset. The detection results can be seen from Fig. 1. The first line is forgery images, the second line is detection results. All the duplication regions are non-regular and the detection algorithm can find the tampered regions precisely, though each image has complex texture background.

Fig. 1 The detecting results for non-regular copy-move forgery

The basic motivation of our scheme is to detect regions that have undergone geometric changes, particularly reflection and rotation. Following experiments are designed to detect the duplicative region when it is rotated with different degrees. This kind of forgery can not be detected by the other existing methods. In Fig. 2 the duplicative regions have been horizontal reflection, vertical reflection and rotated. Detecting result shows the method can effectively solve this situation.

Fig. 2 The detecting results when the regions are reflected and rotated

4.2 Robustness and Accuracy Test

In real life, some evil people often handle the tampered images with post-processing operation, such as noise adding, blurring or mixture operations. Experiments on images that distorted by blur and additive white Gaussian noise were also performed to test how precise our method was in these two cases. 100 images were chosen from the datasets, for each image, copying a square region at a random location and pasting onto a non-overlapping region. The square region's size was fixed as 48×48 in this part and the parameters were set as: $B = 16, \varphi_a = 30, \varphi_b = 2, \varphi_c = 0.3, \varphi_d = 40, \varphi_s = 0.98, e_{min} < 3$ for AWGN distortion, while default parameters were used for blurred images. Results of tampered images distorted by AWGN with different power and blur were shown in Table 1. The values indicated that the algorithm had the ability to locate tampering regions in the case of processing distorted image. All tampering images had been detected and the detecting precision was good.

4.3 Computational Complexity

The number of blocks and the dimension of the vector are two important aspects of reducing computational complexity. Assuming a 256×256 image and the size of block is 8×8. Table 2 displays the comparison results between the different methods. Table 2 shows the time complexity of algorithm [1, 2, 8] has higher than our method.

Table 1 The result of additive white Gaussian noise and Gaussian blurring

SNR	20 dB	30 dB	40 dB
DAR	0.924	0.961	0.982
FPR	0.085	0.092	0.087
w, σ	5, 0.5	5, 1	5, 1.5
DAR	0.988	0.928	0.895
FPR	0.041	0.106	0.213

Table 2 Computation complexity comparisons

Methods	Numbers of blocks	Feature dimension	Detection time (s)
Fridrich's	58081	64	55.6
Farid's	58081	32	50.8
Wang's	58081	4	20.2
Our proposed	58081	4	4.3

5 Conclusions

This paper present an automatic duplication image region detection algorithm based on LBP. It works in the absence of digital watermarking and does not need any prior information about the tested image. Compared with previous works, our algorithm used less features to represent each blocks, and was more effective. The experiment results prove that the proposed method have nice robustness to post-processing and rotation. Thus, our method could be useful in some areas of forensic science.

Acknowledgments This paper is supported by NSFC (No. 61070212, No.61003195), NSF of Zhejiang Province, China (No. Y1090114), the State Key Program of Major Science and Technology (Priority Topics) and the science and technology search planned projects of Zhejiang Province, China (No 2010C11050, No. 2012C21040).

References

1. Fridrich J, Soukal D, Lukas J (2003) Detection of copy–move forgery in digital images. In: Proceedings of digital forensic research workshop, IEEE Computer Society, Cleveland, OH, USA, August, pp 55–61
2. Popescu A, Farid H (2004) Exposing digital forgeries by detecting duplicated image regions, Technical Report TR2004-515, Department of Computer Science, Dartmouth College
3. Myna AN, Venkateshmurthy MG, Patil CG (2007) Detection of region duplication forgery in digital images using wavelets and log-polar mapping. In: Proceedings of the international conference on computational intelligence and multimedia applications, Washington, DC, USA, pp 371–377
4. Huang H, Guo W, Zhang Y (2008) Detection of copy–move forgery in digital images using sift algorithm. In: Proceedings of the 2008 IEEE Pacific-Asia workshop on computational intelligence and industrial application, IEEE Computer Society, Washington, DC, USA, pp 272–276
5. Schaefer G, Stich M (2004) UCID-An uncompressed colour image database. In: Proceedings of the SPIE, storage and retrieval methods and applications for multimedia, pp 472–480
6. http://r0k.us/graphics/kodak/
7. http://faculty.ksu.edu.sa/ghulam/Pages/ImageForensics.aspx
8. Wang JW, Liu GJ (2009) Detection of image region duplication forgery using model with circle block. In: International conference on multimedia information networking and security

Attack on Recent Homomorphic Encryption Scheme over Integers

Haomiao Yang, Hyunsung Kim and Dianhua Tang

Abstract At CDCIEM 2012, Yang et al. proposed a new construction of somewhat homomorphic encryption scheme over integers, which is quite efficient in the perspective of the key size. In this paper, we present an effective lattice reduction attack on Yang et al.'s scheme, where it is easy to recover the plaintext by applying LLL algorithm.

Keywords Homomorphic encryption · LLL algorithm · Lattice · Cloud computing

1 Introduction

Fully homomorphic encryption (FHE) can operate the arbitrary plaintext information homomorphically, just by operating ciphertexts, without decryption. However, how to construct an efficient FHE scheme has been still an open problem for over 30 years. In 2009, the old open problem was solved by the breakthrough work of Gentry [1]. At the same time, Gentry still gave a construction framework that a fully homomorphic scheme could be transformed from a "somewhat" homomorphic scheme.

H. Yang
College of Computer Science and Engineering, UEST of China, Chengdu 610054, China

H. Kim (✉)
Department of Cyber Security, Kyungil University, Kyungpook, Kyungsansi 712-701, Korea
e-mail: kim@kiu.ac.kr

D. Tang
Science and Technology on Communication Security Laboratory, Chengdu 610041, China

J. J. (Jong Hyuk) Park et al. (eds.), *Multimedia and Ubiquitous Engineering*, Lecture Notes in Electrical Engineering 240, DOI: 10.1007/978-94-007-6738-6_34, © Springer Science+Business Media Dordrecht(Outside the USA) 2013

Gentry's somewhat scheme originally worked with ideal lattices. At 2010, Dijk et al. proposed a very simple somewhat homomorphic scheme only over the integers, which had owned merit of conceptual simplicity [2]. However, this simplicity came at the cost of public key size in $O(\lambda^{10})$. Although at 2011, Coron et al. reduced the public key size to $O(\lambda^7)$, it was still too large for practical applications [3]. At 2012, Yang et al. further reduce the public key size to $O(\lambda^3)$ by encrypting with a new form [4].

In this paper, based on LLL algorithm, we present an effective attack on Yang et al.'s scheme. Our attack shows that it is easy to recover the plaintext by using lattice reduction: it is a matter of applying LLL in a lattice of dimension 3.

2 Recent Somewhat Homomorphic Encryption Scheme

For convenience, the same notations are used as in [4]. The construction of Yang et al.'s somewhat homomorphic encryption scheme is as follows.

- **KG**(λ): Choose randomly an odd η-bit integer $p \in [2^{\eta-1}, 2^{\eta})$. Choose randomly four integers $l_0, l_1 \in \mathbb{Z} \cap (0, 2^{\gamma}/p), h_0, h_1 \in (-2^{\rho}, 2^{\rho})$. Compute $x_i = pl_i + 2h_i$, $i = 0, 1$. Assume that $|x_0| > |x_1|$. Set public key $pk = \langle x_0, x_1 \rangle$, and secret key $sk = p$.
- **Enc**(pk, m): To encrypt a bit $m \in \{0, 1\}$, choose randomly two integers $r \in (-2^{\rho'}, 2^{\rho'}), r_1 \in (-2^{\rho}, 2^{\rho})$ and compute the ciphertext $c = m + 2r + r_1 x_1 \bmod x_0$.
- **Eval**$(pk, C, \vec{c})$: The function is the same as in [2].
- **Dec**(sk, c): Output $m = (c \bmod p) \bmod 2$, where $(c \bmod p)$ is the integer in $(-p/2, p/2)$.

To foil various attacks, a convenient parameter set is $\rho = \lambda, \rho' = 2\lambda$, $\eta = O(\lambda^2), \gamma = O(\lambda^3)$, where λ is a security parameter.

Remark 1 According to the parameter set in Yang et al.'s scheme, we have $r = O(2^{2\lambda}), r_1 = O(2^{\lambda})$ and $c = O(2^{\lambda^3})$.

Remark 2 In Yang et al.'s scheme, the message is encrypted with a new form. However, the new form of encryption results in the lattice reduction attack by LLL algorithm. Before describing the attack, we first give a brief introduction of LLL algorithm.

3 Lattice and LLL Algorithm

Definition 1 (*Lattice*) Let $B = (b_1, b_2, \ldots, b_n)^{\mathrm{T}}$, where $b_1, b_2, \ldots, b_n \in \mathbb{R}^m$ are n linearly independent row vectors, the lattice generated by B is

$$\mathcal{L}(B) = \{x_1b_1 + x_2b_2 + \ldots x_nb_n | x_i \in \mathbb{Z}\},$$

where, we refer to $b_1, b_2, \ldots, b_n \in \mathbb{R}^m$ as a *basis* of the lattice and m as its *dimension* [5].

One basic parameter of a lattice is the length of the shortest nonzero vector in the lattice. This parameter is denoted by λ_1. By *length* of a vector v, the Euclidean norm of v, or the $\|v\|$ norm, defined as following

$$\|v\| = \sqrt{\sum_{i=1}^{n} |v_i|^2}.$$

Definition 2 (*LLL Algorithm of Lattice*) LLL algorithm can transform a basis to a reduced basis [6].

INPUT: $b_1, \ldots, b_n \in \mathbb{Z}^n$
OUTPUT: δ-LLL reduced basis for $\mathcal{L}(B)$
Start: Compute Gram-Schmidt orthonormal basis $\tilde{b}_1, \tilde{b}_2, \ldots, \tilde{b}_n$

Reduction Step:

for $i = 2$ to n do
for $j = i - 1$ to 1 do
$b_i \leftarrow b_i - c_{i,j}b_j$ where $c_{i,j} = \lceil \langle b_i, \tilde{b}_j \rangle / \langle \tilde{b}_j, \tilde{b}_j \rangle \rfloor$

Swap Step:

if $\exists i$ s.t. $\delta \|\tilde{b}_i\|^2 > \|\mu_{i+1,i}\tilde{b}_i + \tilde{b}_{i+1}\|$ then

$b_i \leftrightarrow b_{i+1}$

goto **start**

Output: $b_1, \ldots, b_n$

One important property of LLL-reduced basis is that its first vector is relatively short, as shown in the next claim. Our attack is based on the following claim.

Claim 1. Let $b_1, b_2, \ldots, b_n \in \mathbb{R}^n$ be a δ-LLL reduced basis. Then

$$\|b_1\| \leq \left(\frac{2}{\sqrt{4\delta - 1}}\right)^{n-1} \lambda_1(\mathcal{L}).$$

4 Attack to Yang et al.'s Somewhat Scheme

This section provides an effective lattice reduction attack on Yang et al.'s scheme, where it is easy to recover the plaintext by applying LLL algorithm.

4.1 Basic Idea

A ciphertext in Yang et al.'s scheme is of the form: $c = m + 2r + r_1 x_1 \mod x_0$, where m is the message (0 or 1), x_0 and x_1 are known large integers, and r and r_1 are random small integers. But it is very easy to recover the unknown integers and hence m using lattice reduction: it is a matter of applying LLL algorithm in a lattice of dimension 3.

4.2 Construction of Lattice

By using **Enc** algorithm, we have $c = m + 2r + r_1 x_1 \mod x_0$. Since $|x_0| > |x_1|$, we have that

$$c = m + 2r + r_1 x_1 + a x_0, \quad |a| < r_1.$$

Let $k_1 = m + 2r$, $k_2 = r_1$, $k_3 = a$ and $k = (k_1, k_2, k_3)$, we have that

$$(1, k_2, k_3) \begin{pmatrix} c & 0 & 0 \\ -x_1 & 1 & 0 \\ -x_0 & 0 & 1 \end{pmatrix} = (k_1, k_2, k_3),$$

where $k_1 = O(2^{2\lambda+1})$, $k_2 = O(2^{\lambda})$ and $k_3 = O(2^{\lambda})$. Let

$$B = \begin{pmatrix} c & 0 & 0 \\ -x_1 & 1 & 0 \\ -x_0 & 0 & 1 \end{pmatrix} = \begin{pmatrix} b_1 \\ b_2 \\ b_3 \end{pmatrix}.$$

Therefore, we have a lattice of dimension 3: $\mathcal{L}(B)$, and the lattice basis is $b_1 = (c, 0, 0), b_2 = (-x_1, 1, 0)$ and $b_3 = (-x_0, 0, 1)$.

4.3 Attack by Using LLL Algorithm

On one hand, we would like to show how can find a short vector v in this lattice. First of all, we compute

$$\det(\mathcal{L}(B)) = c.$$

Then by using the LLL algorithm, we find a short lattice vector v whose length satisfies

$$\|v\| \leq O(\lambda_1(\mathcal{L}(B))) \leq O\left((\det(\mathcal{L}(B)))^{1/3}\right) = O(c^{1/3}) = O\left(2^{i^3/3}\right),$$

Attack on Recent Homomorphic Encryption Scheme 273

where the second inequality follows from Minkowski's theorem [6] and the second equality follows from Yang et al.'s parameter set.

On the other hand, we have $\|k\| = O(2^{2\lambda+1})$, which is much smaller than $O\left(2^{\lambda^3/3}\right)$. So the vector k is just the shortest vector. We know that for a lattice of dimension 3, there exists a probabilistic polynomial time algorithm [7] to find the shortest vector k. Therefore, from $k_1 = m + 2r$, we have $m = k_1 \bmod 2$, and the message m is recovered.

5 Experiment of Attack

We run it on a Thinkpad Notebook, featuring an Intel CPU P8400 (2.26 GHz), with 3 GB of RAM. Our implementation uses Shoup's NTL library [8] version 5.5.2 for high-level numeric algorithms.

The parameters $\rho = \lambda, \rho' = 2\lambda, \eta = O(\lambda^2), \gamma = O(\lambda^3)$ is set, where λ is security parameter. For convenience, first let $\lambda = 10$. The running result is as follows.

[The secret key]
$p = 8141053646305563512 40736280183$

[The public key]
$x_0 = 783469587406863538604408155646589147564269731 01293533788204$
2138651189443564537158008237864259649318100268560099902558 69548506
0297920323382382578970214587143935820829059846396671970726 88343270
3172873371164071755804678677123387852389765041755479346673 09204186
472998295482528397099779210082237396490 15191
$x_1 = 4990470515888642578274456262245152160068752355116743 0388079$
0956209112354280471455436125919891346709467357161260487473 86820180
7484979143164058979908650925071252417291186628400604894569 88543515
0077652233270569526538944715177949342689224931128439962275 17667057
683506047282708117582681286435026226 18273548

[The random numbers for encrypting]
$r = 101223, r_1 = 17$

[The ciphertext of 0 bit]
$c = 134365584464806541581914066295721951090088037443825550 729017$
2696081738564418939956764758687032484381578824196706410898 90904936
0324781234172057102252947323720029351694856275531084703076 65362183
5815191172051073626894052903321275505705916301267934547260 09060705
83378446501774369191989441509165516283 14339

Table 1 Attack Yang' et al. scheme with LLL algorithm (Intel CPU P8400 (2.26 GHz) with 3 GB of RAM)

Security parameter λ	Number of run	Number of success	Ratio of success (%)	Average time per run (s)
6	100	96	96	0.00078
10	100	95	95	0.02481
20	100	98	98	6.7111
30	100	92	92	210.962

As a result, the lattice basis: $b_1 = (13436558446480654158191406629572$ 195109008803744382555072901726960817385644189399567647586870324843 8157882419670641089890904936032478123417205710225294732372002935 16 948562755310847030766536218358151911720510736268940529033212755057 05916301267934547260090607058337844650177436919198944150916551 6283 14339, 0, 0), $b_2 = (-4990470515888642578274456262245 15216006875235511$ 67430388079095620911235428047145543612591989134670946735716126048 7 473868201807484979143164058979908650925071252417291186628400604894 569885435150077652233270569526538944715177949342689224931128439962 27517667057683506047282708117582681286435026226182735 48, 1, 0), $b_3 =$ $(-78346958740686353860440815564658914756426973101293533788204 2138$ 6511894435645371580082378642596493181002685600999025586954850 60297 9203233823825789702145871439358208290598463966719707268834327031 72 8733711640717558046786771233878523897650417554793466730920418647 29 982954825283970997792100822373964901519 1, 0, 1).

The first vector of LLL-reduced basis: $\tilde{b}_1 = (202446, -17, 11)$. It is easy to verify $k = (k_1, k_2, k_3) = (202446, -17, 11) = (2r + 0, k_2, k_3)$. So, the message 0 is recovered. Then we run experiments with different security parameters. The result is shown in Table 1.

6 Conclusion

In this paper, we pointed out that the somewhat homomorphic encryption scheme over integers presented by Yang et al. is vulnerable to lattice reduction attack. By using LLL algorithm, we could recover the plaintext easily. As shown in experiments, there was more than 90 % of recovering plain texts for different security parameters.

Acknowledgments This work was supported by National Research Foundation of Korea (NRF) under "2011 Korea-China Young Scientist Exchange Program" and National Science Fund of China under Grant No. 61103207 and also was partially supported by the National Research Foundation of Korea Grant funded by the Korean Government (MEST) (NRF-2010-0021575).

References

1. Gentry C (2009) Fully homomorphic encryption using ideal lattices. In: STOC 2009, pp 169–178
2. Dijk M, Gentry C, Halevi S, Vaikuntanathan V (2010) Fully homomorphic encryption over the integers. In: Advances in cryptology—EUROCRYPT 2010. LNCS, vol 6110, pp 24–43
3. Coron JS, Mandal A, Naccache D, Tibouchi M (2011) Fully homomorphic encryption over the integers with shorter public keys. In: Advances in cryptology—CRYPTO 2011. LNCS, vol 6841, pp 487–504
4. Yang H, Tang D, Xia Q, Wang X (2012) A new somewhat homomorphic encryption scheme over integers. In: Proceedings of CDCIEM 2012, pp 61–64
5. Regev O (2005) On lattices, learning with errors, random linear codes, and cryptography. In: Proceedings of STOC 2005, pp 84–93
6. Nguyen PQ, Vallée B (2009) The LLL algorithm: survey and applications., Information security and cryptographySpringer, Heidelberg
7. Lenstra HW, Lenstra AK, Lovasz L (1982) Factoring polynomials with rational coefficients. Math Ann 261:515–534
8. Shoup V. NTL: A library for doing number theory. http://shoup.net/ntl/, Version 5.5.2

A New Sensitive Data Aggregation Scheme for Protecting Data Integrity in Wireless Sensor Network

Min Yoon, Miyoung Jang, Hyoung-il Kim and Jae-woo Chang

Abstract Since wireless sensor networks (WSNs) are resources-constrained, it is very essential to gather data efficiently from the WSNs so that their life can be prolonged. Data aggregation can conserve a significant amount of energy by minimizing transmission cost in terms of the number of data packets. Many applications require privacy and integrity protection of the sampled data while they travel from the source sensor nodes to a data collecting device, say a query server. However, the existing schemes suffer from high communication cost, high computation cost and data propagation delay. To resolve the problems, in this paper, we propose a new and efficient integrity protecting sensitive data aggregation scheme for WSNs. Our scheme makes use of the additive property of complex numbers to achieve sensitive data aggregation with protecting data integrity. With simulation results, we show that our scheme is much more efficient in terms of both communication and computation overheads, integrity checking and data propagation delay than the existing schemes for protecting integrity and privacy preserving data aggregation in WSNs.

Keywords Sensor network · Data aggregation · Integrity · Data privacy · Signature

---

M. Yoon · M. Jang · H. Kim · J. Chang (✉)
Deptartment of Computer Engineering, Chonbuk National University, Jeonju,
Republic of Korea
e-mail: jwchang@jbnu.ac.kr

M. Yoon
e-mail: myoon@jbnu.ac.kr

M. Jang
e-mail: brilliant@jbnu.ac.kr

H. Kim
e-mail: melipion@jbnu.ac.kr

J. J. (Jong Hyuk) Park et al. (eds.), *Multimedia and Ubiquitous Engineering*,
Lecture Notes in Electrical Engineering 240, DOI: 10.1007/978-94-007-6738-6_35,
© Springer Science+Business Media Dordrecht(Outside the USA) 2013

1 Introduction

Recently, due to the advanced technologies of mobile devices and wireless communication, wireless sensor networks (WSNs) have increasingly attracted much interest from both industry and research. Since a sensor node has limited resources (i.e., battery and memory capacity), data aggregation techniques have been proposed for WSNs [1]. Another issue of WSNs is how to preserve sensitive measurements of everyday life where data privacy becomes an important aspect. In many scenarios, confidentiality of transported data can be considered critical, for instance, data from sensors might measure patients' health information such as heartbeat and blood pressure details. So, maintaining data privacy of a sensor node even from other trusted participating sensor nodes of the WSN is critical issue [2]. Although the existing data aggregation schemes have been proposed to preserve data privacy, they have the following limitations. First, the communication cost for network construction and data aggregation and data integrity is considerably expensive. Secondly, the existing schemes do not support data integrity due to communication loss. However, since the existing privacy-preserving schemes do not support privacy preservation and integrity protection simultaneously, it is required to carefully design a good data aggregation scheme for recent applications of WSNs, where both the privacy of sensed data and the integrity of the data should be provided [3].

To reserve these problems, in this paper, we propose a new and resource efficient scheme that can aggregate sensitive data protecting data integrity in WSNs. Our scheme utilizes complex numbers, which is an algebraic expression and can use arithmetic operations, such as addition (+), to aggregate and hide data (for data privacy) from other sensor nodes and adversaries during transmissions to the data sink. In our scheme, the real unit of a complex number is used for concealing sampled data whereas the imaginary unit is used for providing data integrity checking. Thus, our scheme not only prevents recovering sensitive information even though private data are overheard and decrypted by adversaries or other trusted participating sensor nodes but also provides data integrity checking. For data security, our scheme can be built on the top of the existing secure communication protocols like [4].

The rest of the paper is organized as follows. In Sect. 2, we present some related work. Section 3 describes our integrity protecting sensitive (private) data aggregation scheme in detail. Simulation results are shown in Sect. 4. Along with some future research directions, we finally conclude our work in Sect. 5.

2 Related Work

In this section, we present the existing data aggregation schemes for supporting data privacy and data integrity in WSNs. There are privacy preserving data aggregation schemes, such as *i*PDA and *i*CPDA. He et al. proposed *i*PDA [3] and *i*CPDA [5] schemes for WSNs to support integrity. In the iPDA scheme, they protect data integrity by designing node disjoint two aggregation trees rooted at the query server where each node belongs to a single aggregation tree. In this technique, first, every sensor node slices its private data randomly into L pieces and $L-1$ pieces are encrypted and sent to the randomly selected sensor nodes of the aggregation tree keeping one piece at the same sensor node. The same process is independently done for each sensor node using another aggregation tree. Then, all the sensor nodes which received data slices from multiple sensor nodes decrypt the slices using their shared keys and sum the received data slices including its own. After that, each sensor node sends the sum value to its parent from the respective aggregation tree. In the same way, the sum data from another set of sensor nodes are transmitted to the query server through another aggregation tree. In the end, the aggregated data from two node-disjoint aggregation trees reach to the base station where the aggregated data from both aggregation trees are compared. If the difference of the aggregated data from the two aggregation trees doesn't deviate from the predefined threshold value the query server accepts the aggregation result, otherwise, it rejects the aggregated result by considering them as polluted data. However, there are some shortcomings in the *i*PDA. First of all, during protecting data privacy it generates high traffics in the WSN. As a result, communication cost is significantly increased in the *i*PDA. Secondly, all sensor nodes use secret keys to encrypt their all data slices before sending to their respective $2(L-1)$ number of sensor nodes. So, every sensor node has computation overhead of decrypting all the slices they received before aggregating them. In the *i*CPDA, three rounds of interactions are required: Firstly, each node sends a seed to other cluster members. Next, each node hides its sensory data via the received seeds and sends the hidden sensory data to each cluster member. Then, each node adds its own hidden data to the received hidden data, and sends the calculated results to its cluster head which calculates the aggregation results via inverse and multiplication of matrix. To enforce data integrity, cluster members check the transmitted aggregated data of the cluster head. There are some disadvantages of *i*CPDA. Firstly, the communication overhead of *i*CPDA increases quadratically with the cluster size. Secondly, the computational overhead of CPDA increases quickly with the increase of the cluster size which introduces large matrix, whereas lower cluster size introduces lower privacy-preserving efficacy.

3 Integrity-Protecting Sensitive Data Aggregation Protocol

To overcome previously mentioned shortcomings of the *i*PDA and *i*CPDA, in this section, we propose a new, efficient scheme in order to support data privacy and integrity in data aggregation for WSNs. Our scheme is based on the algebraic properties of the complex numbers and it not only ensures that no trend about sensitive data of a sensor node is released to any other nodes and adversaries but also provides data integrity of the aggregated value of sensor data. In particular, we apply the additive property of complex number for data aggregation. We know that other aggregation functions, such as Average, Count, Variance, Standard Deviation and any other Moment of the measured data, can be reduced to the additive aggregation function Sum [4].

Our privacy and integrity preserving scheme is performed through five step. In the first step, we assign a special type of positive integer 2^n (where, $n = 0$ to $Bn \times 8-1$, such that Bn is the number of free bytes available in the payload) to every sensor node as node ID. This is because the binary value of every integer of 2^n type has only one high bit (1). In addition, the position of the high bit for all integers of this type is unique. The sink node knows a data contributing sensor node through the signature of Node-ID. The Node-ID of a sensor node is used to generate a signature of a fixed length. A signature is a fixed size bit stream of binary numbers for a given integer. Signature of a senor node ID can be generated by using the technique presented in the work [6]. We can determine the length of the signature based on the size of a given WSN. When the size of the WSN increases we can increase the length of the signature up to the Bn bytes. In other words, different size WSNs can have signatures of different lengths. The detail of using signatures has been presented in our previous work [7].

When the network receives a SQL like query for SUM aggregation function, in the second step, the sampled sensitive data ds of each sensor node is, first, concealed in a by combining with a unique seed (sr) which is a private real number. The seeds can be selected from an integer range (i.e., space between lower bound– upper bound). By increasing the size of the range, we can further increase the level of the data privacy. Hence, our approach can support data privacy feature strongly. To support data integrity, an integer value b—*the difference of the previous sensed value* and the current sensed value of the sensor node—with i is appended to the a by using *genCpxNum()* function to form a complex number $C = a + bi$. We assumed that any sensor node cannot be compromised before sending first round data to the sink node. Every source sensor node keeps the original sensed value d of the current round to deduce b in the next round which is updated in each round of data transmission. Next, the source node encrypts the customized data R'_1, i.e., $R_1 = a + bi$, and the signature of the node by using a secret key $K_{x,y}$ [8] and transmits the cipher text C_j to its parent. The term $K_{x,y}$ denotes a pairwise symmetric key shared by nodes x and y where the node x encrypts data by using a key $K_{x,y}$ and the node y decrypts the data by using the key $K_{x,y}$. In this way, our algorithm converts the sampled data into an encrypted complex number form. Hence, it not only protects the transmitting trend of private data but also doesn't let neighboring sensor nodes and adversaries to recover sensitive data even though they overheard and decrypted the sensitive data.

In the third step, the parent sensor node (i.e., data aggregator) decrypts the received data by using respective pairwise symmetric keys of its child sensor nodes. For each child node, the parent node computes the difference value (b') of the two real units by using the stored previous data and received current data of the child node. For the first round, the value of b' is also zero. For this, the parent node always keeps the record of the previously received data from each of the child nodes and it updates the previous data by current one in every round. To support local integrity checking, the parent node first compares just computed difference value with the currently received difference value (imaginary unit) from the child node and then compares the difference value with local threshold δ. If the imaginary unit of the child's current data is equal to the computed difference value and the imaginary unit is not greater than δ then the parent node accepts the data of the child node. Otherwise, the parent node rejects the data of the child sensor node considering as polluted data. After that the parent node adds the data of child nodes including its own by using additive property of complex number to produce an intermediate result R'. At the same time, it superimposes signatures (*SSig*) of the contributed nodes by performing bitwise *OR* operation on the bit-streams of the node IDs and forwards the encrypted intermediate result '*Cr*' towards the sink node. Since this approach needs just one bit to carry an ID of a sensor node it is 16 times scalable than the existing work CMT [4] where plaintexts (2-byte each) are used for carrying IDs of sensor nodes by simply concatenating them. Note: Different types of application can have different value for the threshold δ. Thus, our algorithm supports local integrity checking which enforces to provide consistent data from child nodes. Above process continues at all nodes of the upper levels of the network until the whole partially aggregated data of the network reach to the sink node.

In the fourth step, when the sink node receives all intermediate result sets C_{rs} (partially aggregated encrypted customized data with superimposed signature) from the 1-hop child nodes, it decrypts them by using respective pairwise symmetric keys and computes the final aggregation SUM_2 from C_{rs}. Since SUM_2 is of complex number form and the sensed data has been concealed in the real unit by using private seeds identifying the information of the contributed sensor nodes is necessary to deduce actual SUM value.

In the last step, the sink node first knows data contributing nodes by checking the high bits (1 s) of the received superimposed signature by performing bitwise AND operation with the pre-stored signature files or superimposed signature of the Node-IDs of the all nodes of the network. For this, it separates SUM_2 into real unit SUM_{2R} and imaginary unit SUM_{2IM}. Because the sampled data of sensor nodes has been concealed within the real unit, the sink node computes the actual aggregated result SUM by subtracting (an inverse operation of masking, step 2) SUM_{1R} (a freshly computed sum value of the private seeds of the contributed source nodes) from SUM_{2R}. The final result SUM is always accurate and reliable because of the following two reasons. First, a complex number is an algebraic expression and hence the underlying algebra gives the accurate result of the aggregated sensor data. Second, since the private seeds are fixed integer values (i.e., seeds are not random numbers) after collecting data by the sink node it subtracts exactly the same values that have been added to the sensor data during data hiding process by every source node. At the same time, before accepting the SUM, the sink node performs global integrity checking of SUM to assure whether the SUM_2 has been polluted by an adversary in transit or not. For this, like parent nodes, the sink node also computes the difference value (B') of the two real units by using the stored previous data and received current data from the network. The sink node first compares just computed difference value $B'i$ with the currently received difference value i.e., SUM_{2IM}, from the network and then compares the difference value (SUM_{2IM}) with global threshold Δ (for every application, the maximum value for $\Delta = \delta \times N$, where N is the total number of nodes in a network). If the imaginary unit SUM_{2IM} of the current data from the network is equal to the just computed difference value B'_i and the SUM_{2IM} is not larger than Δ then the sink node accepts the data of the network and returned the actual SUM to the query issuer. Otherwise, the sink node rejects the SUM considering it as forged/polluted data by adversary or other nodes.

4 Performance Evaluation

In this section, we present simulation results of our scheme by comparing it with iPDA and iCPDA schemes in terms of communication overhead and integrity checking. For this, we use TOSSIM simulator running over TinyOS operating system and GCC compiler. We consider 100 sensor nodes distributed randomly in 100 × 100 m area.

Figure 1 shows communication overhead in terms of energy dissipation by the iPDA, iCPDA and our schemes with respect to varying number of sensor nodes in the WSN. The power consumption by our scheme is always lower than that of iPDA and iCPDA schemes. The reason is that the iPDA and iCPDA schemes generate too many unnecessary messages in the WSN while achieving integrity protecting and privacy preservation in data aggregation. And Fig. 2 compares integrity checking feature of all the three schemes. It is shown that our scheme can detect every polluted message but the iPDA and iCPDA has very low rate of polluted message detection. The reason is that every node in our scheme performs local integrity checking of the coming data from the lower level nodes. But, only sink node checks the integrity in iPDA and so does the cluster heads in iCPDA.

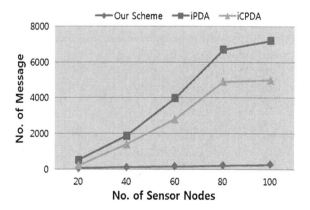

Fig. 1 Energy consumption

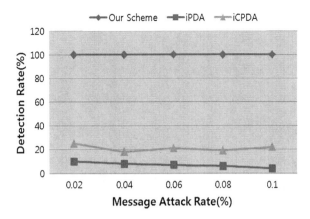

Fig. 2 Integrity checking

5 Conclusion

In this paper, we proposed an efficient and general scheme in order to aggregate sensitive data protecting data integrity for private data generating environments such as patients' health monitoring application. For maintaining data privacy, our scheme applies the additive property of complex numbers where sampled data are customized and given the form of complex number before transmitting towards the sink node. As a result, it protects the trend of private data of a sensor node from being known by its neighboring nodes including data aggregators in WSNs. Moreover, it is still difficult for an adversary to recover sensitive information even though data are overheard and decrypted. Meanwhile, data integrity is protected by using the imaginary unit of complex-number-form customized data at the cost of just two extra bytes. Through simulation results, we have shown that our scheme is much more efficient in terms of communication and computation overheads, data propagation delay and integrity checking than the iPDA and iCPDA schemes.

As future work, we will provide more simulation results by designing data integrity and sensitive data preserving scheme under collusive attacks.

Acknowledgments This research was supported by Basic Science Research program through the National Research Foundation of Korea (NRF) funded by the Ministry of Education, Science and Technology (grant number 2010-0023800).

References

1. Considine J, Li F, Kollios G, Byers J (2004) Approximate aggregation techniques for sensor databases. In: Proceedings of ICDE, April 2004
2. Conti M, Zhang L, Roy S, Di Pietro R, Jajodia S, Mancini L-V (2009) Privacy-preserving robust data aggregation in wireless sensor networks. Secur Commun Netw 2:195–213

3. He W, Liu X, Nguyen H, Nahrstedt K, Abdelzaher T (2008) iPDA: an integrity-protecting private data aggregation scheme for wireless sensor networks. In: Proceedings of the IEEE MILCOM
4. Castelluccia C, Mykletun E, Tsudik G (2005) Efficient aggregation of encrypted data in wireless sensor networks. In: The second annual international conference on mobile and ubiquitous systems: networking and services, pp 109–117
5. He W, Liu X, Nguyen H, Nahrstedt K (2009) A Cluster-based protocol to enforce integrity and preserve privacy in data aggregation. In: Proceedings of the 29th IEEE international conference on distributed computing systems workshops, pp 14–19
6. Zobel J, Moffat A, Ramamohanarao K (1998) Inverted files versus signature file for text indexing. ACM TDS 23(4):453–490
7. Bista R, Chang JW (2010) Energy efficient data aggregation for wireless sensor networks. Sustainable wireless sensor networks, ISBN, pp 978–995
8. Blaß E-O, Zitterbart M (2006) An efficient key establishment scheme for secure aggregating sensor networks. In: Proceedings of the 2006 ACM symposium on information, computer and communications security, March 2006, pp 303–310

Reversible Image Watermarking Based on Neural Network and Parity Property

Rongrong Ni, H. D. Cheng, Yao Zhao, Zhitong Zhang and Rui Liu

Abstract Reversible watermarking can recover the original cover after watermark extraction, which is an important technique in the applications requiring high image quality. In this paper, a novel image reversible watermarking is proposed based on neural network and parity property. The retesting strategy utilizing the parity detection increases the capacity of the algorithm. Furthermore, the neural network is considered to calculate the prediction errors. Experimental results show that this algorithm can obtain higher capacity and preserve good visual quality.

Keywords Reversible watermarking · Neural network · Parity property · Retesting strategy

1 Introduction

Reversible watermarking can completely restore the original digital contents after data extraction. For this characteristic, reversible watermarking is very useful for some applications where the availability of the original data is essential, such as military image processing and medical image sharing.

Early reversible watermarking algorithms mainly focus on lossless compression until the difference expansion algorithm is proposed by Tian [1]. The method divides the image into pairs of pixels and uses each legitimate pair for hiding one bit of information. It has high embedding capacity and high quality, and becomes the basic idea of some reversible watermarking methods. Later, prediction error

R. Ni (✉) · Y. Zhao · Z. Zhang · R. Liu
Institute of Information Science, Beijing Jiaotong University, Beijing 100044, China
e-mail: rrni@bjtu.edu.cn

H. D. Cheng
Department of Computer Science, Utah State University, Logan, UT, USA

J. J. (Jong Hyuk) Park et al. (eds.), *Multimedia and Ubiquitous Engineering*,
Lecture Notes in Electrical Engineering 240, DOI: 10.1007/978-94-007-6738-6_36,
© Springer Science+Business Media Dordrecht(Outside the USA) 2013

expansion (PEE) method is proposed by Thodi and Rodriguez [2]. Their method uses PEE to embed data, and suggests incorporating expansion embedding with histogram shifting to reduce the location map. Then, several PEE-based methods have been proposed [3–6]. In [6], Sachnev et al. propose a method which combines sorting and two-pass-testing with prediction error expansion method. The algorithm has higher capacity and lower distortions than most of other existing reversible watermarking methods.

In this paper, a novel image reversible watermarking is proposed based on neural network and parity property. Because the real embedded data is not always identical with the testing bit, some ambiguous pixel cells are generated. A retesting strategy utilizing the parity detection activates the capacity of the ambiguous pixel cells. As a result, the capacity is increased. Furthermore, considering the global feature, the neural network is used to predict the prediction errors. The experimental results show that the proposed algorithm can obtain higher capacity and preserve good visual quality.

2 Proposed Algorithm Based on Neural Network and Parity Detection

In the proposed algorithm, all pixels of the image are divided into two sets: the "Cross" set and the "Dot" set (Fig. 1) as suggested in [6]. The watermark bits are embedded in the "Cross" set first, and then embedded in the "Dot" set.

2.1 Prediction Based on Neural Network

During "Cross" embedding, the "Cross" set is used for embedding data while the "Dot" set works as the reference signals. And vice versa.

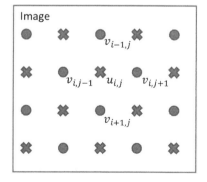

Fig. 1 "*Cross*" set and "*Dot*" set

Fig. 2 Structure of neural network

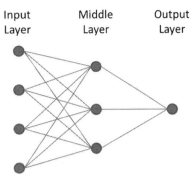

A center pixel of a cell is predicted by the four neighboring pixels. In this paper, neural network is used to predict the pixel values considering the global feature. Since four pixels in the neighboring region are utilized to calculate the prediction values, a neural network with four inputs is designed here. As shown in Fig. 2, the input layer has four neurons and the middle layer is created with three neurons. The output layer has one neuron which refers to the central pixel value.

After the construction of the neural network, the corresponding weights and parameters can be determined by training. Considering the global influence and generalization, a great number of pixel cells from many natural images are input to the neural network for obtaining a common model of prediction. This model can be shared by encoder and decoder in advance. Thus, the prediction value $u'_{i,j} = \lfloor nnpredict(v_{i,j-1}, v_{i-1,j}, v_{i,j+1}, v_{i+1,j}) \rfloor$. Where, $nnpredict(.)$ is the prediction model based on neural network.

2.2 Data Embedding and Extraction

The combination of difference expansion and histogram shifting method [2] is also utilized in this paper.

If the prediction error $e_{i,j} = u_{i,j} - u'_{i,j}$ inside the region $[T_n, T_p]$, $e_{i,j}$ is expanded to $E_{i,j} = 2 \times e_{i,j} + b$. T_n is the negative threshold and T_p is the positive threshold. Otherwise, the pixel does not carry any data and the prediction error is simply shifted. That is,

$$E_{i,j} = \begin{cases} 2 \times e_{i,j} + b & \text{if } e_{i,j} \in [T_n, T_p] \\ e_{i,j} + T_p + 1 & \text{if } e_{i,j} > T_p \text{ and } T_p \geq 0 \\ e_{i,j} + T_n & \text{if } e_{i,j} < T_n \text{ and } T_n < 0 \end{cases} \quad (1)$$

The watermarked value is computed by $U_{i,j} = u'_{i,j} + E_{i,j}$.

During extraction, if $E_{i,j} \in [2T_n, 2T_p + 1]$, the watermark can be extracted. Otherwise, shifting is used to recover the image. That is,

$$
e_{i,j} = \begin{cases} [E_{i,j}/2] & \text{if } E_{i,j} \in [2T_n, 2T_p + 1] \\ E_{i,j} - T_p - 1 & \text{if } E_{i,j} > 2T_p + 1 \\ E_{i,j} - T_n & \text{if } E_{i,j} < 2T_n \end{cases} \tag{2}
$$

Then, $u_{i,j} = u'_{i,j} + e_{i,j}$.

2.3 Improved Classification Using Parity Property

To ensure $U_{i,j}$ without overflow or underflow problems, two-pass-testing [6] is used here. If a pixel can be modified twice based on Eq. (1), it belongs to Class A; if the pixel is modifiable once owing to overflow or underflow errors during the second embedding test, it belongs to Class B; and if the pixel cannot be modified even once, the pixel belongs to Class C. During the testing process, bit "1" is used as an embedding bit for positive prediction errors, and bit "0" is for negative prediction errors. The locations of Class B and Class C are marked in a location map, which is also embedded with the payload.

In the decode phase, use once-embedding-test to distinguish Class A, and Class B (or Class C). And further discriminate Class B and Class C using the location map. However, some pixel cells belonging to Class B will be misclassified to Class A if the actually embedded bit does not coincide with the testing bit.

We utilize the parity characteristic and retesting strategy to activate the capacity of Class B. After once-embedding-test during the extraction phase, the pixel cells are assigned into two parts: Part one contains the cells without overflow or underflow, and Part two contains the overflow or underflow cells. As a result, Part one is the set consisting of Class A and partial Class B, while Part two is the set containing Class C and part of Class B. It is obvious that the elements of Class B which are attributed in Part one are problem pixel cells. Since they will cause the wrong localization in the location map, these problem pixel cells should be identified further.

For the cells in Part one, a retesting detection is designed to distinguish the ambiguous cells belonging to Class B. As for the expandable pixel cells, $U_{i,j} = u'_{i,j} + e_{i,j} = u'_{i,j} + 2e_{i,j} + b$. Thus, $U_{i,j} - u'_{i,j} = 2e_{i,j} + b$. Due to $2e_{i,j}$ is an even number, $b = \text{LSB}(U_{i,j} - u'_{i,j})$, here $\text{LSB}(x)$ means the LSB of x. For the positive prediction errors, if $U_{i,j} - u'_{i,j}$ is an even number, the embedded bit is not consistent with the testing bit. Thus, add one to the pixel value $U_{i,j}$ and retest the corresponding prediction error using the testing bit "1". For the negative prediction errors, if $U_{i,j} - u'_{i,j}$ is odd, subtract one from the pixel value $U_{i,j}$ and retest the corresponding prediction error using the testing bit "0". If the retesting result shows the pixel value is overflow, it belongs to Part two. Otherwise, it still belongs to Part one. After the retesting, Part one only contains Class A, and Part two contains Class B and Class C. Further classification is conducted to distinguish Class B and Class C with the help of the location map.

3 Encoder and Decoder

3.1 Data Embedding

We first embed data in "Cross" set, then embed in "Dot" set. For recovering data, threshold values T_n(7 bits), T_p(7 bits), payload size $|P_{cross}|$(17 bits) or payload size $|P_{dot}|$(17 bits), and the length of location map (7 bits) should be known first. We will embed these 38 bits into the first 38 pixels' LSB. The original 38 LSB should be recorded with the payload. The "Cross" embedding method is designed as follows:

Step 1: Calculate the prediction errors. For each pixel $u_{i,j}$, compute the prediction value and the corresponding prediction error $e_{i,j}$ based on the common neural network.

Step 2: Sort the prediction errors. For each pixel $u_{i,j}$, compute the variance $Var_{i,j}$ of the four neighbor pixels which is used as the sorting parameter. Skip the first 38 pixels. Sort the pixel cells according to the ascending order $\{Var_{i,j}\}$ to produce a sorted row of prediction errors e_{sort}.

Step 3: Determine the threshold. According to the two-pass-testing, all pixels are classified in one of classes A, B and C. Although the shiftable pixels can be modified, they cannot carry watermark bits. Only the expandable pixels in Class A and Class B are capable of carrying data. Let set of expandable pixels in class A be EA. Let set of expandable pixels in class B be EB.

In the sorted vector e_{sort}, create the location map L. If a pixel belongs to Class B, the corresponding element in the location map is marked as "0"; while if the pixel belongs to Class C, it is marked as "1". If $|P_{cross}| \leq |EA| + |EB| - |L| - 38$ and $|EA| \geq |L|$ are satisfied, the to-be-embedded bits can be successfully embedded. Otherwise, increase the threshold T_p or decrease T_n, and repeat Step 3.

Step 4: Embed data. The location map L, the true payload P_{cross}, and the first 38 LSBs will be embedded in the image by using the embedding method described in Sect. 2.2. The location map L is first embedded in Class A. The elements belonging to Class A and Class B are all used to improve the capacity. Use the auxiliary data to modify the first 38 LSB values of the pixels by simple binary replacement. If the last to-be-embedded bit is processed, the "Cross" embedding phase is completed.

After 4 steps, the "Cross" embedding process is finished. The "Dot" embedding scheme uses the modified pixels from the "Cross" set for computing the predicted values. The original pixels from the "Dot" set are used for embedding data, and the embedding procedure is similar to the "Cross" embedding. After the "Dot" embedding, the watermarked image is obtained.

3.2 Data Extraction

Double decoding scheme is the inverse of the double encoding scheme. We only describe the "Cross" decoding method.

Step 1: Calculate the prediction values. For each pixel $U_{i,j}$, compute the prediction value based on the common neural network. Then, the prediction errors $E_{i,j}$ are obtained afterwards.

Step 2: Sort the prediction errors. Skip the first 38 pixels. Sort the pixels according to $Var_{i,j}$ to get the set of sorted prediction errors E_{sort}. Read the first 38 LSB values to recover the values of T_n, T_p, payload size P_{cross}, and the length of location map.

Step 3: Extract the watermark. Skip the first 38 sorted cells. Test every pixel cell to classify it into Class A, Class B and Class C according to Sect. 2.3. Extract location map from Class A firstly. Further classification is conducted to distinguish Class B and Class C based on the location map. Then, extract data from Class A and Class B, meanwhile recover the original prediction errors using the method in Sect. 2.2. The extracted data is the cascading of the true payload, and the 38 LSBs.

Step 4: Restore the original image. Computer the original pixel values based on $u_{i,j} = u'_{i,j} + e_{i,j}$.

Step 5: Recover the rest pixels. Replace the first 38 LSB values of the pixels with the extracted 38 LSBs.

When the "Dot" and "Cross" decoding are both finished, the entire watermark is obtained and the original image is restored.

4 Experimental Results

Several 8-bit gray images "Lena", "Baboon" and "Plane" with size 512×512 are used in the experiments. Figure 3 shows the watermarked images with payload 50000 bits:

Figure 4 shows the performances of Capacity versus Visual quality in terms of payload and Peak Signal-to-Noise Ratio (PSNR). The horizontal axis represents the capacity in terms of bpp (bits per pixel). The vertical axis represents the corresponding PSNR. The results show that our method has both high visual

 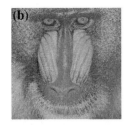 

Fig. 3 The watermarked *gray* images. **a** Lena (PSNR = 51.4 dB). **b** Baboon (PSNR = 43.0 dB). **c** Plane (PSNR = 53.5 dB)

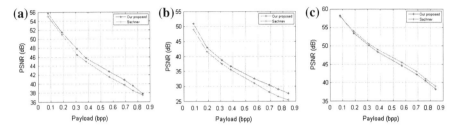

Fig. 4 Capacity versus PSNR for testing images. **a** Results for Lena. **b** Results for Baboon. **c** Results for Plane

quality and high capacity. Compared with [6], our method can achieve better results for "Lena" and "Baboon", and get comparable result for "Plane". The reason is that the neural network is not trained enough.

5 Conclusions

A high capacity image reversible watermarking based on neural network and parity property is proposed. A retesting strategy utilizing the parity detection activates the capacity of the ambiguous pixel cells. In addition, the prediction errors are obtained by using the neural network to consider the global feature. The experimental results show that this algorithm can obtain higher capacity and preserve good visual quality. In the future, we will further research and discuss the effectiveness of the neural network.

Acknowledgment This work was supported in part by 973 Program (2011CB302204), National Natural Science Funds for Distinguished Young Scholar (61025013), National NSF of China (61073159, 61272355), Sino-Singapore JRP (2010DFA11010), Fundamental Research Funds for the Central Universities (2012JBM042).

References

1. Tian J (2003) Reversible data embedding using a difference expansion. IEEE Trans Circuits Syst Video Technol 8:890–896
2. Thodi DM, Rodriguez JJ (2007) Expansion embedding techniques for reversible watermarking. IEEE Trans Image Process 3:721–730
3. Tsai WL, Yeh CM, Chang CC (2009) Reversible data hiding based on histogram modification of pixel differences. IEEE Trans Circuits Syst Video Technol 6:906–910
4. Tsai PY, Hu C, Yeh HL (2009) Reversible image hiding scheme using predictive coding and histogram shifting. IEEE Signal Process Mag 6:1129–1143
5. Luo LZ, Chen N, Zeng X, Xiong Z (2010) Reversible image watermarking using interpolation technique. IEEE Trans Inf Forensics Secur 1:187–193
6. Sachnev V, Kim HJ, Nam J, Shi YQ, Suresh S (2009) Reversible watermarking algorithm using sorting and prediction. IEEE Trans Circuits Syst Video Technol 7:989–999

A Based on Single Image Authentication System in Aviation Security

Deok Gyu Lee and Jong Wook Han

Abstract An image protection apparatus includes an information collecting unit for collecting personally identifiable information to be embedded in images captured by an image capturing instrument; and an information processing unit for extracting personal information from the ·collected personally identifiable information. Further, the image protection apparatus includes an information embedding unit for embedding the extracted personal information into a captured image; and an image signature unit for writing a signature on the captured image by using the extracted personal information.

Keywords Aviation Security · Surveillance System · Authentication · Authorization

1 Introduction

As the national airspace system grows increasingly interconnected to partners and customers both within and outside the Rep. of Korea government, the danger of cyber-attacks on the system is increasing. Because of low-cost computer technology and easier access to malware, or malicious software code, it is conceivable

---

This research was supported by a grant (code# 07aviation-navigation-03) from Aviation Improvement Program funded by Ministry of Construction & Transportation of Korean government.

D. G. Lee (✉) · J. W. Han
Electronic and Telecommunications Research Institute, 161 Gajeong-dong, Yuseong-gu, Daejeon, Republic. of Korea
e-mail: deokgyulee@etri.re.kr

J. W. Han
e-mail: hanjw@etri.re.kr

J. J. (Jong Hyuk) Park et al. (eds.), *Multimedia and Ubiquitous Engineering*, Lecture Notes in Electrical Engineering 240, DOI: 10.1007/978-94-007-6738-6_37, © Springer Science+Business Media Dordrecht(Outside the USA) 2013

for individuals, organized crime groups, terrorists, and nation-states to attack the Rep. of Korea air transportation system infrastructure.

Consider an airport in which passengers and employees can enter common areas, like transportation facilities, and waiting areas. However, secured areas, like luggage transport and service stations, are available for authorized employees only. The highest security areas, such as the air traffic control room, are accessible to specialized personnel who are appropriately authorized. The keyword here is "uthorization", meaning that people who are not authorized to access a physical location should not be allowed physical or electronic access to that location. In the surveillance world, the exact same rules apply and the potential recipient of the surveillance data must have the same authorization that an ordinary person of any trade would have to be physically or electronically present at that location. However, during emergency operations, controlled dissemination of sensitive data may become necessary in order to obtain support services or to prevent panic. It has been shown that during crisis people require clear instructions so that their maximum cooperation is obtained. However, these instructions should not release unauthorized information or reveal the existence of such information.

This paper relates to an apparatus and a method for processing image information, and more particularly, to an image information processing apparatus and method capable of adding information on an image capturing device and signature information to image data and storing the image data to maintain security of the image data and use the image data as digital proof.

2 Related Work

With the development of image photographing technology, techniques for maintaining security of image data captured by an image capturing device and protecting copyright are proposed. For example, captured images are transmitted to a limited image information output device and reproduced or identification information such as watermarking is embedded in image data to protect copyright of image information [1].

In the case of embedding watermarking in image information, it is possible to confirm the copyright holder of the image information even though the image information is displayed at or transmitted to an undesired place and prevent the image information from being illegally copied. Furthermore, users can watch the image information without having any difficulty and track the source of the image information and image information copy routes when watermarking is embedded in the image information.

However, watermarking does not have legal force capable of preventing the image information from being illegally copied or transmitted although it can show the copyright holder or the source of the image information and allow users to confirm image information copy routes and the source of the image information. Accordingly, security of image information cannot be efficiently maintained only with watermarking when the image information includes personal information related to privacy or data requiring the maintenance of security [2. 3].

A distributed architecture for multi-participant and interactive multimedia that enables multiple users to share media streams within a networked environment is presented in "An architecture for distributed, interactive, multi-stream, multi-participant audio and video". In this architecture, multimedia streams originating from multiple sources can be combined to provide media clips that accommodate look-around capabilities. SMIL has been the focus of active research "The use of smil: Multimedia research currently applied on a global scale" and "About the semantic verification of SMIL documents", and many models for adaption to real world scenarios have been provided. A release control for SMIL formatted multimedia objects for pay-per-view movies on the Internet that enforces DAC is described in "Regulating access to smil formatted pay-per-view movies". The cinematic structure consisting of acts, scenes, frames of an actual movies are written as a SMIL document without losing the sense of a story. Here access is restricted to the granularity of an act in a movie. A secure and progressively updatable SMIL document "Sputers: A secure traffic surveillance and emergency response architecture" is used to enforce RBAC and respond to traffic emergencies. In an emergency response situation, different recipients of the live feeds have to be discriminated to people playing different roles [1–7].

While most models addresses the need of multimedia, their approach does not incorporate semantics of multimedia. None of the approaches are completely satisfactory for surveillance multimedia. They primarily address textual documents and exploit the granular structure of XML documents. Multimedia for various reasons as discussed above has to be treated differently because there is a sense of temporal synchrony and continuity involved. Synchronization and integration of different and diverse events to produce sensible information is nontrivial when compared to textual data. The process of retrieval without losing the sense of continuity and synchronization needs sophisticated techniques and algorithms which all of the above models do not completely address. Although our approach to provide controlled information flow in real-time multimedia systems is based in concepts similar to MLS, the developed methods and techniques are also applicable in other security models, like Role-Based or Discretionary Access Control models.

3 ACRS (Aviation Control Room Surveillance)

It is an object of the paper to provide an image information processing apparatus and method for adding information on an image capturing device and predetermined signature information to image data obtained using the image capturing device to protect the image data from infringement of security such as illegal copy and transmission and adding information on the place and time at which the image data is obtained to the image data to use the image data as digital proof.

An apparatus for processing image information according to the paper comprises: an image capturing unit for generating image data and collecting information on the image capturing unit; an image processing unit for adding at least one of the information on the image capturing unit and signature information to the image data using the image data and the information on the image capturing unit transmitted from the image capturing unit; and an image storage unit for storing the image data output from the image processing unit.

A method for processing image information according to the paper comprises: an image capturing step of generating image data and collecting information on the image capturing step; an image processing step of adding at least one of the information on the image capturing step and signature information to the image data; and an image storing step of storing the image data.

3.1 Security Framework for Physical Environment

The Security framework for physical environment contains a few essential components, such as an authentication, an authorization, and a security policy. They work at each smart door and often cooperate with a smart surveillance established by a smart image-unit in the physical environment. Since the smart door is installed at the border of each physical domain and every physical environment must pass through it, it is supposed to be a core component and suitable in providing security functions described in security framework for physical environment. Whenever a new access to physical environment is found, it should be able to authenticate and authorize it and enforce security policy based on security rules set by the corresponding smart security administrator [8].

Figure 1 depicts the overall architecture of secure physical environment.

Figure 1 is a view illustrating a configuration of an image photographing system to which an image information processing apparatus according to the paper is applied.

In view of the above, the present invention provides an image protection apparatus that embeds personal information and signature information in an image to thereby protect the image from others. In accordance with an embodiment of the present invention, there is provided an image protection apparatus including: an information collecting unit for collecting personally identifiable information to be embedded in images captured by an image capturing instrument; an information processing unit for extracting personal information from the collected personally identifiable information; an information embedding unit for embedding the extracted personal information into a captured image; and an image signature unit for writing a signature on the captured image by using the extracted personal information. It is preferable that the information processing unit extracts device information of the image capturing instrument, and the information embedding unit embeds the device information into the captured image. Further, it is preferable that the information embedding unit verifies validity of the extracted

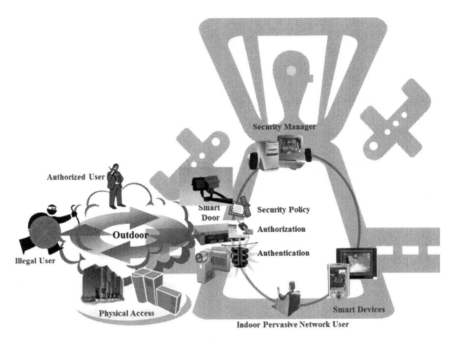

Fig. 1 Architecture of secure physical environment

personal information in cooperation with an external personal information server through a network, and embeds the verified personal information in the captured image. It is preferable that the information collecting unit collects personally identifiable information by using one or more method of image recognition barcode, personal identification tag, radio frequency identification (RFID), sensor, face recognition, and iris recognition. It is preferable that the image signature unit writes signatures on frames in the captured image by using the personal information. Further, it is preferable that the image signature unit writes signatures on each or parts of the units of preset number of frames in the captured images. Furthermore, it is preferable that the information collecting unit and information processing unit are installed in the image capturing instrument, and the information embedding unit and image signature unit are installed in a server connected through a network to the image capturing instrument. In a feature of the present invention, because both personal information and signature information are embedded in images, a user cannot identify an image of another user without validating relevant personal information and additionally access control is enforced on the administrator administrating the images. When a particular case occurs, personal information embedded in an associated image can be used as digital evidence.

Further, because signature information are embedded into images in the form of various cryptographic signatures including public key certificates, particular types of facts such as place and time can be verified using the signature information.

Therefore, illegal acts related to the images can be prevented before the fact, and legal measures can be taken after the fact as the image source and illegal act associated with the images can be identified. Hereinafter, an embodiment of the present invention will be described below with reference to the accompanying drawings. The image capturing device and the image information server illustrated in Fig. 1 according to an embodiment of the present invention, which shows a case in which the image processor adds the information on the image capturing unit or the signature information to the image data. The image photographing device includes the image capturing unit and the image processor. The image processor includes an information receiver, a device information processor, a signature information processor, and an image transmitter. The image information server includes an image receiver and a storage unit. The image capturing unit is a device capable of capturing an image, obtaining image data from the captured image and collecting information on the image capturing unit. For example, a CCTV, a digital camera, a video camera or a communication terminal including a camera module can be used as the image capturing unit. The image data obtained by the image capturing unit is transmitted to the image processor and undergoes a data processing operation of adding signature information or the information on the image capturing unit thereto. The image processor adds the information on the image capturing unit or the signature information to the image data generated by the image capturing unit. The information receiver included in the image processor receives the image data generated by the image capturing unit and the information on the image capturing unit and the device information processor adds the information on the image capturing unit to the image data transmitted from the information receiver. The signature information processor adds the signature information to the image data transmitted from the device information processor and the image transmitter transmits the image data having the information on the image capturing unit or the signature information added thereto to the image information server. Here, positions of the device information processor and the signature information processor can be changed each other. That is, the signature information can be added first, and then the information on the image capturing unit can be added.

The information receiver receives the image data generated by the image capturing unit and transmits the information to the device information processor. The information on the image capturing unit can include an identifier given to the image capturing unit or information on the place and time at which the image capturing unit obtains the image data. When the image capturing unit is a CCTV, for example, the information on the image capturing unit can include the place where the CCTV is installed and the time when the CCTV records the image. When the image capturing unit is a communication terminal, the information on the image capturing unit can include an identification number given to the communication terminal. The device information processor adds the information on the image capturing unit to the image data transmitted from the information receiver. That is, the place and time at which the image data is obtained or the identification number of the image capturing unit that captures the image data can be added to the image data as the information on the image capturing unit. The image data can be used as digital proof of a specific event when the place and time at which the image data is captured is added thereto and the source of the image data can be easily detected when the identification number of the image capturing unit is added thereto. The device information processor transmits the image data having the information on the image capturing unit added thereto to the signature information processor. The signature information processor can embed signature information including a predetermined encryption key in the image data transmitted from the device information processor. According to an embodiment, the signature information processor can add public key based signature information, symmetric key based signature information or public key and symmetric key based signature information to the image data.

The public key based signature information can be generated according to Rivest Shamir Adleman (RSA) algorithm and the symmetric key based signature information can be generated according to Vernam or data encryption standard (DES) algorithm. The symmetric key based signature information requires transmission of an additional secret key and is difficult to authenticate with safety although it is encrypted at a high speed. On the other hand, the public key based signature information does not require transmission of the additional secrete key and is easily authenticated with safety while it is encrypted at a low speed. Accordingly, an algorithm of generating the signature information can be selected according to a degree to which maintenance of security of the image data is required. When the signature information is added to the image data, the image data can be accessed only using a predetermined decryption key. Accordingly, the possibility that the image data is exposed to hacking or illegal copy according to arbitrary access can be reduced when the signature information is added to the image data. The application server that provides the image data to communication subscribers can provide the decryption key to only an authenticated communication subscriber through a text message to maintain security of the image data. The signature information can be added to the image data at regular intervals.

The image data having the information on the image capturing unit and the signature information added thereto is transmitted to the image transmitter. That is, at least one of the information on the image capturing unit and the signature information is added to the image data transmitted from the image capturing unit and transmitted to the image transmitter. The image transmitter transmits the image data received from the signature information processor to the image information server.

The image receiver receives the image data from the image transmitter. The storage unit stores the image data transmitted from the image receiver. The image data transmitted from the image receiver has at least one of the information on the image capturing unit and the signature information added thereto. The image information server can determine whether the decryption key transmitted from the application server corresponds to the encryption key embedded in the image data, extract the image data stored in the storage unit and transmit the image data to the application server when the application server requests the image information server to transmit the image data through the communication network.

The image capturing device and the image information server illustrated in Figure 1 according to another embodiment of the present invention, which shows a case in which the image information server adds the information on the image capturing unit or the signature information to the image data. The image capturing device includes the image capturing unit and the image processor and the image information server includes a receiver, a device information processor, a signal information processor and a storage unit. The image capturing unit is a device capable of recognizing an object through a lens and a sensor, obtaining image data from the recognized object and collecting information on the image capturing unit. The image processor transmits the image data and the information on the image capturing unit received from the image capturing unit to the image information server. The receiver included in the image information server receives the image data and the information on the image capturing unit transmitted from the image processor and the device information processor receives the image data and the information on the image capturing unit from the receiver and embeds the information on the image capturing unit in the image data. The signature information processor adds predetermined signature information to the image data having the information on the image capturing unit added thereto and the storage unit stores the image data including the signature information. The receiver receives the image data and the information on the image capturing unit from the image processor.

The device information processor embeds the information on the image capturing unit in the image data and transmits the image data to the signature information processor. The information on the image capturing unit depends on the type of the image capturing unit. For example, when the image capturing unit is a CCTV, the information on the image capturing unit can include the place where the CCTV is installed and the time when the CCTV obtains the image data. When the image capturing unit is a communication terminal including a camera module, the information on the image capturing unit can include the identification number of the communication terminal.

The source of the image data can be easily searched when the identification number of the image capturing unit is embedded in the image data and the image data can be used as digital proof when the place and time at which the image capturing unit captures the image data is added thereto. The information on the image capturing unit can be recorded in a meta-data region of the image data. The signature information processor can embed signature information including a predetermined encryption key in the image data transmitted from the device information processor. The signature information processor can generate the signature information according to a predetermined algorithm and embed the signature information in the image data. The signature information can be generated according to a public key based algorithm or a symmetric key based algorithm. In general, the case that the image information server embeds the information on the image capturing unit and the signature information in the image data requires a data processing speed and available capacity greater than the data processing speed and available capacity required for the case that the image photographing device embeds the information on the image capturing unit and the signature information in the image data. Accordingly, it is desirable to generate the signature information using the public key based algorithm that easily performs safe authentication and does not require an addition secret key to be transmitted while having a low encryption speed. The image data to which the signature information has been added is stored in the storage unit of the image information server.

The signature information can be added to the image data at regular intervals. The storage unit stores the image data transmitted from the signature information processor. The image data transmitted from the signature information processor has at least one of the information on the image capturing unit and the signature information added thereto. The image information server can receive a predetermined decryption key from the application server and compare the decryption key with the encryption key included in the image data to determine whether the image data is transmitted when the application server requests the image information server to transmit the image data through the communication network.

4 Conclusion

We have presented a surveillance framework for audio–video surveillance of multi-level secured facilities during normal and pre-envisioned emergencies. This paper relates to an apparatus and a method for processing image information, and more particularly, to an image information processing apparatus and method capable of adding information on an image capturing device and signature information to image data and storing the image data to maintain security of the image data and use the image data as digital proof. However, it is also important to address data integrity and source authentication issues. These issues, along with the development of a complete and comprehensive prototype system are part of our future work.

References

1. Kodali N, Farkas C, Wijesekera D (2003) Multimedia access control using rdf metadata. In: workshop on metadata for security, WMS 03
2. Kodali N, Wijesekera D (2002) Regulating access to smil formatted pay-per-view movies. In: 2002 ACM workshop on XML security
3. Rutledge L, Hardman L, Ossenbruggen J (1999) The use of smil: multimedia research currently applied on a global scale
4. Kodali N, Wijesekera D, Michael. Sputers: A secure traffic surveillance and emergency response architecture. J Intell Transp Syst

5. Pihkala K, Cesar P, Vuorimaa P (2002) Cross platform smil player. In: International conference on communications, internet and information technology
6. Rutledge L, Ossenbruggen J, Hardman L, Dick CA (1999) Bulterman. Anticipating SMIL 2.0: the developing cooperative infrastructure for multimedia on the Web. Computer Networks, Amsterdam, Netherlands, 31(11–16):1421–1430
7. Schmidt BK (1999) An architecture for distributed, interactive, multi-stream, multi-participant audio and video. In: Technical report no CSL-TR-99-781, stanford computer science department
8. Kodali N, Wijesekera D, Farkas C (2004) SECRETS: a secure real-time multimedia surveillance system. In: Proceedings of the 2nd Symposium on intelligence and security informatics
9. Damiani E, di Vimercati SDC (2003) Securing xml based multimedia content. In: 18th IFIP international information security conference
10. Damiani E, di Vimercati SDC, Paraboschi S, Samarati P (2000) Securing XML documents. Lect Notes Compt Sci 1777:121–122
11. FAA'S NEXTGEN AIR TRAFFIC CONTROL SYSTEM A CIO's Perspective on Technology and Security Georgetown University Institute for Law, Science, and Global Security & Billington CyberSecurity, 28 Feb 2011
12. Damiani E, di Vimercati SDC, Paraboschi S, Samarati P (2002) A fine grained access control system for xml documents. ACM Trans Info Syst Security 5:121–135
13. Gu X, Nahrstedt K, Yuan W, Wichadakul D, Xu D (2001) An xml-based quality of service enabling language for the web. Kluwer Academic Publishers, Norwell
14. Kodali N, Farkas C, Wijesekera D (2003) Enforcing integrity in multimedia surveillance. In: IFIP 11.5 working conference on integrity and internal control in information systems

Part VI
Multimedia and Ubiquitous Services

A Development of Android Based Debate-Learning System for Cultivating Divergent Thinking

SungWan Kim, EunGil Kim and JongHoon Kim

Abstract Six Thinking Hats which is designed by Edward de Bono enhances the excellence of the thinking and has an effect on cultivating divergent thinking. In particular, it is effective in seeking a reasonable solution by analyzing some problems from a variety of views in debate-learning. In this paper, we developed a system sharing voice and images based on Six Thinking Hats, using sensors of android device. We analyzed tools and guidelines by making design structural model for designing the system. We developed debate-learning system, verified its utility and analyzed improvements through a demonstration and a practice for educational experts.

Keywords Android · Divergent thinking · Six thinking hats · Debate-learning system · Mobile learning

1 Introduction

Thinking also can be improved through practicing and many methods were studied as a learning method. There are brainstorming, Six Thinking Hats, Attribute Listing, Morphological Synectics, Forced relation, Synectics in typical creative methods [1]. It has been studied that Six Thinking Hats makes people practice each field of thinking and a change in attitude by looking problems in a different ways and improve different abilities of thinking.

However, it does not guarantee time for activity of thinking which is a prerequisite of creative thinking method due to lots of contents of curriculum compared to time

---

S. Kim · E. Kim · J. Kim (✉)
Department of Computer Education, Teachers College, Jeju National University,
Jeju-si, Korea
e-mail: jkim0858@jejunu.ac.kr

S. Kim
e-mail: kswandrea@naver.com

E. Kim
e-mail: computing@korea.kr

J. J. (Jong Hyuk) Park et al. (eds.), *Multimedia and Ubiquitous Engineering*,
Lecture Notes in Electrical Engineering 240, DOI: 10.1007/978-94-007-6738-6_38,
© Springer Science+Business Media Dordrecht(Outside the USA) 2013

for each studying. In addition, it is difficult to study because there is a shortage of some examples of thinking methods and programs for students studying the method.

Thus, in this paper, we developed a debate-learning system based on android among smart devices which are embedded portable and different sensors. Using our developed system, students' thinking progress with sync function of phonic and image data and provide them other learners through the sharing server so that it is possible to share and evaluate the opinion. We analyzed the result by conducting expert evaluation to verify its benefits and find some future improving way.

2 Six Thinking Hats

Edward de Bono argued that thinking is a function to manipulate the intelligence and it can be improved through the process of practicing [2]. Six Thinking Hats, designed by him, is a method to intend people to think one thing at a time. If you decouple emotion and logic, and information and creativity, you can come up with more ideas. You can use some, not six at once according to the six thinking colors. You need to be aware of the rules that if you wear a certain color's hat, you have to think in a way of applying to the color. Mental activities by each hat's color are same as Table 1 as follows.

The purpose of this method is to simplify the thinking by dealing with each field at a time, and guide a change in attitudes by changing the color of hat. Simplification of thinking lessens difficulties of thinking from a variety of fields at the same time and improves the ability of having different perspectives and positions. A change in attitude also requires people to think in a broad ways including a negative way and a creative way.

People should take a neutral attitude using objective information to illuminate the problem and its background which is white hat supposed to discuss. Red hat provides an opportunity to express people's emotions and feelings. Emotions are supposed to be excluded in a process of solving the logic problems generally; however, it is very hard to obstruct the involvement of emotion. Therefore, we can expect some creative discoveries such as insight by suggesting emotions and feelings as the whole truth without the logical reasons. Black hat is based on critical thinking and it needs legitimate grounds. Critical thinking is one of important thinking abilities to prevent illegal or expected damages in advance and understand weakness of some countermeasures. Yellow hat serves as a role to find

Table 1 Mental activities by hat's colors

Types of hat	Mental activities
White hat	Facts and objective information
Red hat	Emotions and feelings, intuition and sixth sense
Black hat	Negative judgement, impossible reason
Yellow hat	Positive judgement, constructive decision and opinion
Green hat	New and creative ideas
Blue hat	Organization of all ideas

positive value and it requires people to be sensitive to an object's value all the time. Optimistic value can be distinguished from mental activity of red hat which presents simple emotions and feelings just founded on facts and truth. Green hat suggests new ideas. It makes an entirely new up-to-date idea or modifies or improves established ideas. Blue hat is a manager to comment other people's thinking and synthesize a final conclusion.

3 Design of Debate-Learning System

We present the design structural model of our debate-learning system using Six Thinking Hats with help of procedures of web debate-learning from Korea Education & Research Information Service [3] as follows as Table 2.

In the stage of pre-learning, students study characteristic and roles of Six Thinking Hats and look into concrete examples from examples. In the stage of discussion, it is necessary to choose the topic of debate and students collect opinion and grounded data according to the color of hat from online and offline. They think steadily presenting their data with spoken languages. They also have peer review about their opinion and this would make the atmosphere of discussion more active through choosing a great participant and rewarding in the stage of after-learning.

4 A Development of Debate-Learning System

The map of overall debate-learning system developed by suggested design structural model is as in the following Fig. 1.

It makes log-in function with procedures of user confirmation to distinguish one's opinion from others in the discussion. People would become a member of teacher and learner largely and teacher obtains certification to acquire the authority to register the topic of discussion. Learner can join without difficulty and use the system.

Table 2 Design structural model of debate-learning system

Stage	Details	Tool and guide line
Pre-learning	Six Thinking Hats	Guiding characteristics and roles of Six Thinking Hats looking into examples for each theme
Discussion activities	Choose the topic of discussion	Guiding themes in diverse criteria choosing a topic
Discussion activities	Wear the hat and collect the grounded data	Choosing one type of thinking hats individually (log-in function) collecting data for an argument (a camera, WebView)
Discussion activities	Discussion activities and peer review	Presenting opinion and its data with voice and images (voice and images sync) listening to other people's opinion and evaluation (a developed player and a separate point)
After-learning	Offer feedback	Selecting a good participant with peer review vitalizing atmosphere of a debate

Before discussion, people have to learn about the Six Thinking Hats. They examine the multi-media data about thinking method and study the opinion according to the hat's color from examples using the player developed from this system.

When you finish learning of the method, you participate in the discussion with the procedures of choosing the topic and the hat. You collect grounded data for your opinion from Internet, a embedded camera on smart device and so on. Learner also evaluates other learner's opinion besides debating.

4.1 Membership and User Authentication

Our developed application starts like Fig. 2.

When you start the android application, you can check the condition of communication through Thread in Main Activity and process the update of topics for discussion. Topics for discussion are defined as a language of XML and it updates when teacher register a topic or modify the topic. If there are communication problems, the application is closed after noticing the user.

Information of log-in and membership is written as a form of a request message in the mode of POST and transmitted to the server. User authentication in the process of log-in maintains and Session starts when it corresponds with the information from DB information.

4.2 A Guide to Six Thinking Hats

It is so essential to understand about Six Thinking Hats before a debate that we organize specific characteristics and roles of Thinking Hats to make people learn them from image data and examples. We present guided-materials using ScrollView to overcome the small screen of smart devices and examples with images and voice using our own player.

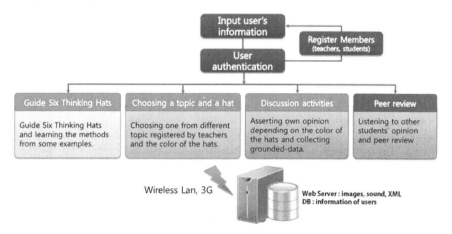

Fig. 1 Overall debate-learning system

Fig. 2 Topic update and user authentication

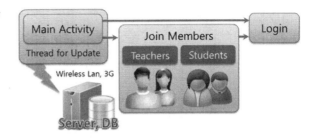

4.3 Choice of a Topic for a Debate and One of Thinking Hats

One of topic for a debate and classified opinion are defined as XML language and then they communicate with server. XML files are supposed to feed the relevant information to DB and renew themselves when teacher register a topic or learner record own opinion. A XML file which has been already downloaded need not to be reused because it has been expressed on ListView the first one time parsing. Therefore, we have intended that parsing would be fulfilled by SAX Parser which rarely use its memory, considering features of mobile device [4]. Information from parsing has re-defined to express the attributes of the topic and the opinion more clearly.

After parsing a XML file which all the contents about the topic for a debate is defined, it would be expressed on re-defined ListView. A user choose the topic for a debate on the ListView and download a XML file through Thread, which defines the registered contents of opinion according to the selected topic. Once you download it, Handler sends the message, summon serve activity from Intent of android and print it on the ListView by parsing the downloaded XML file. If the user click the Thinking Hats button arranged at the bottom, registered opinion would be renewed, and you can put your opinion using the registration of opinion button.

4.4 Discussion Activities

We record learner's argument in a voice using the microphone sensor of the smart device depending on Six Thinking Hats. The voice would be compressed in a AMR-NB way which has been developed from 3GPP (3rd Generation Partnership Project), a joint research project [5]. AMR-NB, designed for voice recorder and communication in mobile device, has a compressive force of 4.75–12.2 kbps capacity [6].

We also collect well-grounded data in a image form using WebView and the camera sensor. This improvement item aims for students who do not have good power of word-painting to make their thinking more easily. Collected image data would be printed on GalleryView at the bottom and we can check them moving from side to side with a user's touch input. Besides, if you choose something in the middle of recording an argument, it would be inserted into the opinion. This point of time that something inserted for Sync modulation between voice and images would be remembered by millisecond and we write it with XML elements.

4.5 Peer Review

Android provides verbal or video-typed playback player, however, we have developed our own player which has sync function of voice and images because a debate in our system can be progressed with them. Learner can give five grades after listening to other learner's opinion, judging validity and suitability of Six Thinking Hats. We provide interface which learner can give separate grades with touch input to overcome the limit of input in smart device.

5 Results

We have conducted expert evaluation to examine utility of our developed Six Thinking Hats-based debate-learning system and improvements. Evaluators are consisting of twenty teachers who have level-one qualification for a licensed teacher and career for ten years. Expert evaluation has been progressed by responding some questions through a demonstration and practice experience about the application and by checking level-five Likert criterion at intervals of 2.5 grades, choosing or describing opinion. Contents of the survey have been made in four fields like Table 3.

Ninety percentages of teachers have agreed that Six Thinking Hats is effective on debate-learning from the following assessment's results. However, few of them have used this method in the educational practice. After analyzing the reasons why the Six Thinking Hats rarely has been used, we can find out lack of guideline and activity programs for Six Thinking Hats consists fifty percentages and the lack of school hours in school forms thirty-five percentage, like Fig. 3. For such these reasons, we expect our debate-learning system would have high effectiveness because it does not have restrictions from time and space.

The result from expert evaluation about contents and construction of the debate-learning system is as follows, Fig. 4. We can see constructive results generally. In particular, there is a high percentage on the response that expressing learner's thinking into voice and images would be effective.

Viewed from the functional side, sync accuracy of voice and images of our developed player has acquired 8.5 points and the response that peer review in a touch input way would be positive in attracting people to learn also has formed same points (Fig. 5).

We have looked into benefits and improvements of our developed debate-learning system, comparing with existing web-based system and other system in offline. Above all, the response that expressing learner's thinking into voice and images is more effective than its of established text-centered, web-based debate-learning system acquires positive 7.375 points.

We can obtain some improvements in the narrative that it would be better to provide more materials with responded opinion and to improve a speed-control function of spoken languages. We consider these comments are due to the difference between a small screen and listening ability of speech.

A Development of Android Based Debate-Learning

Table 3 Expert evaluation contents for verification of the system

Field	Contents details
Values and utility	Is Six Thinking Hats effective on improving debate skills?
	Do you use the thinking method in the school?
	If you do not, why is the reason?
Contents and construction	Is a guide for Six Thinking Hats suitable for the level of learners?
	Is the method of participating in a debate appropriate to the level of learners?
	Is using sync function of voice and images effective for learners to express their opinion?
	Is overall user's interface construction convenient?
Functions	Is the sync function of the player accurate?
	Is peer review effective to attract learners to participate in a debate?
	Are you satisfied with operating time of the system?
Benefits and improvements	Is the new system more effective than text-centered and web-based debate-learning system?
	What is the merit of this system compared to the web-based debate-learning system and mental activities offline?
	What is the improvements of developed debate-learning system?

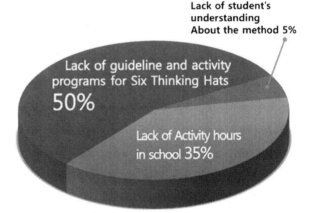

Fig. 3 Reasons of unused the method in the educational practice

6 Conclusion

This research is based on Edward de Bono's Six Thinking Hats and we developed android-based debate-learning system. Six Thinking Hats lessens difficulty of mental activities thinking one thinking field at a time and makes it possible to contemplate problems from different perspectives by experiencing various thinking fields. Therefore, thinking depending on Six Thinking Hats is one of effective learning methods for learners to solve diverse problems differently and reasonably.

We sometimes face into the limit that it is difficult to show thinking in case of learner whose language expression is not good. To solve this problem, it would be better to use multi-media as data to support learner's thinking. It is very convenient to make analog data to the digital and to collect enormous data from Internet.

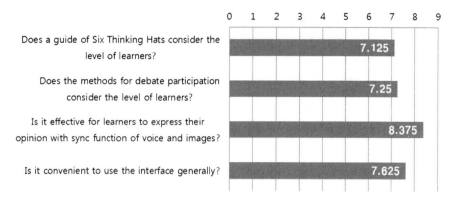

Fig. 4 Results of responding about contents and construction

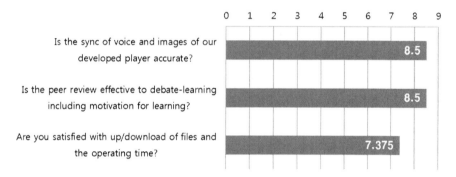

Fig. 5 Results in a functional part

Using smart device with high portability, we solve problems offline because we would gather data without restrictions from time and space and guarantee the time for learners to organize their thinking. As you see in the results of expert evaluation, we could find out that it is very positive to express leaner's thinking into voice and images with sync function in an educational way and, if we consult some future improvements, our system would serve as a debate-learning system with high effectiveness and it would be helpful to improve learner's mental activities.

References

1. Pike RW (1994) Creative training techniques handbook: tips, tactics, and how-to's for delivering effective training. Lakewoods Publications, Minneapolis
2. de Bono E (1985) Six thinking hats. Penguin Book, London
3. Korea Education & Research Information Service (2001) Guide for the use of information & communication technology education. Korea Education & Research Information Service, Seoul
4. SAX. http://www.saxproject.org
5. Google. http://developer.android.com/guide/basics/what-is-android.html
6. Wikipedia. http://en.wikipedia.org/wiki/Adaptive_Multi-Rate_audio_codec

Development of a Lever Learning Webapp for an HTML5-Based Cross-Platform

TaeHun Kim, ByeongSu Kim and JongHoon Kim

Abstract With the advent of smart devices, educational apps for smart learning are actively being developed, but the existing native apps run only on specific devices and are not compatible with other devices. Webapp, a new app development method, which is written using HTML5, supports a cross-platform. In this study, a Webapp for learning about levers as mentioned in elementary school textbooks was developed using HTML5. The proposed Webapp was tested by a group of incumbent expert elementary school teachers, revealing that the proposed contents and Webapp offer high educational value.

Keywords HTML5 · Webapp · Cross-platform · Smart learning

1 Introduction

Amid the new information technologies that are being developed at alarming speeds, diverse inventions that encompass new information technologies are being developed. Of them, representative inventions are smart devices such as smartphones and smart pads, the advent and spread of which have enormously changed modern-day people's lifestyles. This is true for education, for which smart devices have provided new paradigms in terms of methods. Smart-learning has the same root as e-learning, m-learning, and u-learning, but uses smart devices, which

---

T. Kim · B. Kim · J. Kim (✉)
Department of Computer Education, Teachers College, Jeju National University, Jeju, Korea
e-mail: jkim0858@jejunu.ac.kr

T. Kim
e-mail: gtranu@naver.com

B. Kim
e-mail: pigpotato79@naver.com

J. J. (Jong Hyuk) Park et al. (eds.), *Multimedia and Ubiquitous Engineering*,
Lecture Notes in Electrical Engineering 240, DOI: 10.1007/978-94-007-6738-6_39,
© Springer Science+Business Media Dordrecht(Outside the USA) 2013

differentiates it from other learning methods and is drawing attention to it as a new education method.

Current operating systems for smart devices are dominated by Apple's IOS and Google's Android. Large numbers of smart-device-driven apps are now hitting the market. IOS and Android are different operating systems, however, which require developers to make apps using these two platforms or give up on one OS-based app for the other. This inter-device incompatibility problem, if not resolved, will adversely affect digital textbook projects, app development, and other smart learning initiatives.

Recently, the web standardization organization W3C proposed the next-generation web-standard advanced HTML5 based on the web standard and Web 2.0. HTML5 pursues web standards and can support cross-platforms. A Cross-platform refers to the capability of a computer program, operating system, computer language, programming language, computer software, etc. to operate on various kinds of computer platforms. HTML5, which is enabling the implementation of the cross-platform feature, is drawing attention as a new method of developing smart device apps.

This study developed the lever learning Webapp that can support cross-platforms to eliminate inter-smart device incompatibility problems. It also investigated methods of implementing smart learning using HTML5.

2 Related Researches

2.1 Smart Learning

Smart learning is a method of electronic learning that the learner can access the learning content easily using smart device and its related technologies. It is self-initiated learning method that enables user-tailored learning and self-directed learning and the interaction among the learners and between the learners and the teachers [1, 2].

Smart learning is similar to electronic learning that encompasses e-learning (electronic learning), m-learning (mobile learning), and u-learning (ubiquitous learning), and is no different from these methods. Each learning method is able to classify by their characteristic.

E-learning is used to access the learning content using desktop computers and it is distributed in many different forms of educational applications including online courses and web-based learning. It can be usually accessed at fixed locations with internet connections such as computer labs or from homes [3].

M-learning is an advanced stage of e-learning where in the learner is equipped with handheld mobile device to access the learning content using various wireless technologies. M-learning has the benefits of mobility and its supporting platform, which can be summarized as being ubiquity, convenience, localization and personalization. The major advantage of m-learning is learning can happen anytime and anywhere, so it can support continuous learning [3–5].

U-learning is equivalent to some form of simple m-learning. But it is context aware and also provides anywhere, anytime learning using various mobile and sensor technologies. Besides the domains of e-learning or m-learning, u-learning may use more context awareness to provide most adaptive content for learners. The main characteristics of u-learning are permanency, accessibility, immediacy, interactivity, situating of instructional activities [3, 6].

2.2 HTML5

HTML5 changed the concept of Web from documents to a platform for Webapps.

The HTML5 specifications encompass the HTML5 grammar, available elements, attributes, and relevant APIs, and the various surrounding APIs are related to HTML5 but are basically individual, independent specifications. Although these APIs do not actually belong to the HTML5 specifications, all of them are generally referred to as "HTML5" in a broad sense, and HTML5, CSS, and Javascript are collectively called "Open Web Platform" [7, 8].

3 System Design and Content Selection for the Implementation of Webapp

3.1 System Design

The proposed Webapp system is illustrated in Fig. 1. Because it is intended to be compatible with all platforms, it was implemented using HTML5, CSS, and Javascript so that it could be used on diverse devices (PCs, smartphones, and smart pads). PCs or smartphones enable the use of the Internet, unlike smart pads, so Webapp was designed to store and load data using DB servers. Specifically, in a non-networked situation, data are internally stored; and in a networked situation, data are uploaded onto a DB server and downloaded from it to enable learning with different devices.

The proposed Webapp content comes from the Elementary School Science Subject Sixth-year Second Semester Unit 3 lesson "What are the benefits of using levers?" This unit helps students understand the concept, forms, and transformation of energy and the benefits of using levers and pulleys [9].

We proposed the integrated subject content shown in Table 1 that is suitable to theme-oriented project learning tailored to elementary students' cognitive levels and for performing experiments and tasks using STEAM-based apps, and that enables convergence of textbooks. Levers were selected as the learning content to

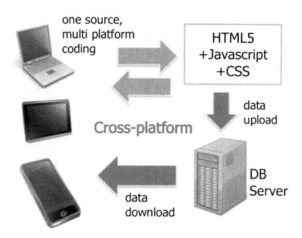

Fig. 1 Overall of the proposed Webapp system

Table 1 Analysis of textbooks encompassing the subject of steam learning

Subject (grade)	Unit	Description
Science (4)	Weighing	Weighing by balancing
		Weighing using the scales that I made
Science (6)	Energy and tools	Knowing about the benefits of levers
		Knowing about the benefits of pulleys
Mathematics (6)	Proportional expression	Knowing about and using the features of proportional expressions
Fine Arts (5 and 6)	Designs that deliver laughter	Making everyday products based on interesting ideas

enable STEAM-based learning using Webapp, in line with the design that encompasses weighing in the science subject, proportional formulae in mathematics, and laughter in fine arts.

4 Webapp Development

4.1 Development Environment

To implement native apps, the integrated development environment should be used to enable utilization of programming languages and SDK; but for Webapp, HTML5, CSS, and Javascript can all be written in general text editors and executed and confirmed in browsers. The web programming environment for storing and sharing learning data was crafted in line with Apache, PHP, and MySQL. To use the Webapp offline by installing smart devices such as native apps, Webapp was ported in the form of a hybrid app using Appspresso 1.0, and Android-SDK and JDK are needed to port it to hybrid apps for Android-based devices.

4.2 Implementation

To enable the lever learning Webapp to support diverse devices, a large view of letters on smartphones was enabled using the viewport function, in consideration of the small display of smartphones. The viewport-declared metatag does not affect desktop PCs and thus, can adjust the display according to the Webapp developer's intention.

The learning content offered units, problems for study, and activities to help students understand the learning activities. HTML5 has adopted new tags designed to divide the semantic markup into logical structures, and has eliminated mere modification tags, thereby expressing document structures more distinctively and data embedded in documents.

To help students understand the lever's point of force, supporting point, and point of action, a related game was developed using everyday items. The game, using HTML5 Canvas, enables the implementation not only of diagrams but also of animations without plug-ins.

Clicking the Start button will make three gray points appear on everyday items. The learner should guess what point each location refers to. When s/he answers correctly, s/he confirms the explanation of the Nos. 1, 2, and 3 levers, and can go to the next problem.

To receive the user inputs, event listeners were used. Desktops and smart devices have different input methods, namely, a mouse or a touchpad, so event listeners should be separately registered. The mouse-input-based desktop PCs used mousedown, mousemove, and mouseup events, and the touchpad-based input smart devices used touchstart, touchmove, and touchend events. The event listeners should be selected depending on the device used, so the method of receiving the userAgent character string that identifies the relevant browser that sent that string was used to determine the content to be sent to the system from the server.

To store learning content and share it with other learners, text-based note pads and image-based picture boards were used. The text-based note pads were designed to be stored in the local storage, among the Web storage types supported by HTML5. Data, once stored in the local storage, remain even when the window is closed, which allows the user to later access the page and continue to use the data. Data can also be stored in the local storage during offline learning and can be uploaded in an online environment to be effectively used for learning.

The image-based picture board was implemented using Canvas. If the canvas.toDataURL() method, which functions with the conversion of the canvas into png or jpg files, is used, the relevant image can be stored as a text file in the DB server to be shared and used as backup data and as learning data. All the note pads and picture boards were produced using the web program that ran on PHP and MySQL, in line with the databases.

5 Experts' Evaluation and Analysis of the HTML5 Webapp

5.1 Experts' Evaluation and Method

The education potential of the proposed Webapp was evaluated by eight experts who specialized in elementary school education and computer education. They had taught for more than 5 years and had experience of app development.

The evaluation items were developed and used to accurately assess the educational value and potential for use of the proposed Webapp, and to define what must be improved. The use of the Webapp was demonstrated to the experts, and the experts were allowed to use it on smartphones and desktop browsers. The evaluation items were rated on a five-step Likert interval scale with 2.5-point intervals, or else the experts were allowed to select or state their opinions.

5.2 Results of the Experts' Evaluation and Analysis

First, to determine the experts' knowledge of smart learning, their interest and development experience in smartlearning were analyzed. All of them showed much interest in smartlearning, and seven had experience in using smartphones and developing native apps.

On the question that pertained to the difficulty of developing educational apps using Android or IOS, 37.5 % of the experts highly rated the high-level maintenance difficulty and the inter-OS incompatibility (Fig. 2).

Regarding the educational value of Webapp and its possible use in the education field, 87.5 % of the experts gave it a high education value, and all of them stated that it is worth using in the education field (Fig. 3).

In addition, the experts indicated that diverse educational contents should be secured and that the browser compatibility should be improved (Fig. 4).

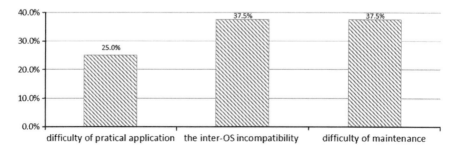

Fig. 2 Difficulty of developing educational native app

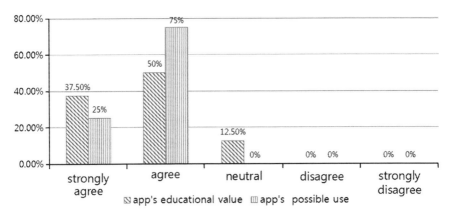

Fig. 3 Educational value and possible use in the education field of the lever learning Webapp

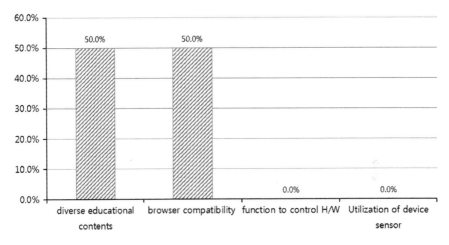

Fig. 4 Expert's advice for improvements

Thus, Webapp for lever learning is worth using in the actual educational field, and has excellent strengths compared with native apps, making it worthwhile to use for developing Webapps.

6 Conclusion

To develop e-learning as a new educational methodological approach, this study used the latest HTML5, developed the lever learning Webapp, and examined the possibility of its educational application.

The HTML5-based Webapp, similar to existing apps, have different strengths that enable the pursuit of diversified learning and promote motivation for learning.

Furthermore, it can support multiple platforms regardless of the device, which allows learners in diverse learning situations to easily access learning contents. Notably, it allows learners even without smartphones to use the same content on desktop PCs, thus enabling easy access to the latest education contents. Also, for teachers, Webapp can be developed in a short period with an easy method, and can be easily maintained in a short period, thereby boosting its educational value.

The HTML5 standard is still popular and is expected to be further improved, and yet it has to be perfectly supported by all browsers. Native apps have the advantage of controlling the hardware levels, which Webapps cannot do. Nonetheless, Webapps, if produced in the form of hybrid apps, enable access to hardware. If the requirements are accurately analyzed and if diverse HTML5 functions are designed and developed, Webapps can have tremendous education value.

References

1. Kim S, Yon YI (2011) A model of smart learning system based on elastic computing. In: Proceedings of 9th international conference on software engineering research, management and applications, Baltimore, USA, pp 184–185
2. Shin DH, Shin YJ, Choo H, Beom K (2011) Smartphones as smart pedagogical tools: implications for smartphones as u-learning devices. Comput Hum Behav 27(6):2207–2214
3. Mandula K, Meda SR, Jain DK, Kambham R (2011) Implementation of ubiquitous learning system using sensor technologies. In: Proceedings of IEEE international conference on technology for education, pp 142–148
4. Parsons D, Ryu H (2006) A framework for assessing the quality of mobile learning. In: Proceedings of international conference for process improvement, research and education, Kerkrade, The Netherlands
5. Seppälä P, Alamäki H (2003) Mobile learning in teacher training. J Comput Assist Lear 19(3):330–335
6. Ogata H, Yano Y (2004) Context-aware support for computer-supported ubiquitous learning. In: Proceedings of 2nd IEEE international workshop on wireless and mobile technologies in education, Taiwan, pp 27–34
7. Lubbers P, Albers B, Salim F (2010) Pro HTML5 programming. Appress, New York
8. Hogan BP (2010) HTML5 and CSS3: develop with tomorrow's standards today. Pragmatic Bookshelf, Dallas
9. Ministry of education science and technology of South Korea (2011) Elementary school science 6th 2nd semester teacher's guide. Kumsung, Seoul

Looking for Better Combination of Biomarker Selection and Classification Algorithm for Early Screening of Ovarian Cancer

Yu-Seop Kim, Jong-Dae Kim, Min-Ki Jang, Chan-Young Park and Hye-Jeong Song

Abstract This paper demonstrates and evaluates the classification performance of the optimal biomarker combinations that can diagnose ovarian cancer under Luminex exposed environment. The optimal combinations were determined by T Test, Genetic Algorithm, and Random Forest. Each selected combinations' sensitivity, specificity, and accuracy were compared by Linear Discriminant Analysis (LDA) and k-Nearest Neighbor (k-NN). The 8 biomarker data used in this experiment was obtained through Luminex-PRA from the serum of 297 patients (cancer 81, benign 216) of two hospitals. In this study, the results showed that selecting 2–3 markers with Genetic Algorithm and categorizing them with LDA shows the closest sensitivity, specificity, and accuracy to those of the results obtained through complete enumerations of the combination of 2–4 markers.

Keywords Biomarker · Ovarian cancer · Marker · T Test · Genetic algorithm · Random forest · LDA · Logistic regression

---

Y.-S. Kim · J.-D. Kim · C.-Y. Park · H.-J. Song (✉)
Department of Ubiquitous Computing, Hallym University, 1 Hallymdaehak-gil, Chuncheon, Gangwon-do 200-702, Korea
e-mail: hjsong@hallym.ac.kr

Y.-S. Kim
e-mail: yskim01@hallym.ac.kr

J.-D. Kim
e-mail: kimjd@hallym.ac.kr

C.-Y. Park
e-mail: cypark@hallym.ac.kr

M.-K. Jang
Department of Computer Engineering, Hallym University, 1 Hallymdaehak-gil, Chuncheon, Gangwon-do 200-702, Korea
e-mail: wscang@gmail.com

Y.-S. Kim · J.-D. Kim · M.-K. Jang · C.-Y. Park · H.-J. Song
Bio-IT Research Center, Hallym University, 1 Hallymdaehak-gil, Chuncheon, Gangwon-do 200-702, Korea

---

J. J. (Jong Hyuk) Park et al. (eds.), *Multimedia and Ubiquitous Engineering*,
Lecture Notes in Electrical Engineering 240, DOI: 10.1007/978-94-007-6738-6_40,
© Springer Science+Business Media Dordrecht(Outside the USA) 2013

1 Introduction

Ovarian cancer is a malignant tumor frequently arising in the age between 50 and 70. Early diagnosis is very closely associated with a 92 % 5-year survival rate, yet only 19 % of ovarian cancers are detected early [1]. Therefore, early detection of ovarian cancer has great promise to improve clinical outcome. It is evident that the development of a biomarker for early detection of the ovarian cancer has become paramount [2].

Biomarker consists of molecular information based on the pattern of a single or multiple molecules originating from DNA, metabolite, or protein. Biomarkers are indicators that can detect the physical change of an organism due to the genetic changes.

Along with the completion of the genome project, various biomarkers are being developed, providing critical clues for cancers and senile disorders (Fig. 1).

The early stages of research focused on a single biomarker for cancer diagnosis. Recent researches focus on combining multiple biomarkers to diagnose cancer more efficiently. Researches tend to focus especially on improving the sensitivity and specificity in order to increase the accuracy of the diagnosis, and the commercialization of multi-biomarkers seems to be close at hand. However, a new technology to find the right biomarker combinations is required, since the sensitivity and quantity has not yet reached a satisfactory level [3].

In this research, the mass value of the biomarkers was obtained using Luminex [4]. Luminex follows the panel reactive antibody (PRA), a solid phase-based method of Luminex corp. This paper determines the optimal marker combinations for ovarian cancer diagnosis from the combinations selected with T Test [5], Genetic Algorithm [6], and Random Forest [7] from all the possible combinations

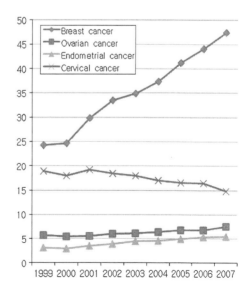

Fig. 1 Changes in gynecologic cancer causes in Korea (1999–2007)

of the 8 markers, based on their the florescence data measured. Linear Discriminant Analysis [5] and k-Nearest Neighbor [6] were used to evaluate the sensitivity, specificity, and classification accuracy of the optimal combinations. The research aims to determine the optimal marker combination and categorization algorithm by comparing the experimental results with all the possible combinations.

Methods for the collection of data are illustrated in Sect. 2, and the experimental details are demonstrated in Sect. 3. The results of the marker combinations and its classification performance are discussed in Sect. 4, and Sect. 5 presents the conclusion.

2 Data Collection

The serum samples from 81 patients with ovarian cancer, 216 patients with benign pelvic masses were used. Sera were provided by Hallym University Medical Center (HUMC) and ASAN Medical Center. These samples were reacted with Luminex-beads attached with 8 biomarkers, and the florescence from the antibodies on the beads was measured. In order to equalize the range of the biomarker florescence, the florescence values of each biomarker were normalized to 0–1 based on their maximum and minimum values.

3 Methods

This paper conducts two experiments: (1) determination of biomarkers with T Test, Genetic Algorithm (GA), and Random Forest (RF), and (2) performance comparison of the selected markers using Linear Discriminant Analysis (LDA) and k-Nearest Neighbor (k-NN).

The number of randomly created tree for the RF was 50, and the k for k-NN was 3. The combination of the biomarkers consisted of 2–4 markers, and leave-one-out cross validation was conducted for the evaluation.

The algorithms for the marker combination and combinations of evaluation algorithms used in the experiment are shown in Fig. 2. As a control, the optimal marker combination gained through a complete enumeration survey of 2–4 markers was used.

4 Results

The experiment compares the difference in performance of the selected 2–4 multi-biomarkers by T Test, GA, and RF to that of the optimal combination amongst the total possible combinations of the markers. The sensitivity, specificity, and

Fig. 2 The algorithms for the marker combination and combinations of evaluation algorithms

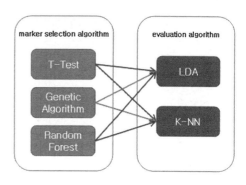

accuracy of each optimal combination for both cancer and benign group was measured and compared with LDA and k-NN.

The markers that ought to be combined was limited to four, because the high cost to combine more than 4 markers will make it difficult to realize and commercialize the use of multi-biomarkers. Also to avoid the infringement of patent, the names of the markers are concealed.

4.1 Optimum Combination Results According to Classification Algorithm

Table 1 shows the optimal marker combination and their performance when applying LDA and k-NN to the combine the 2–4 markers. M*number* means an individual bio marker. As seen in the Table 1, M1 and M7, M6 are most frequent, having the appearance of 6, 4, and 3 times respectively. The best accuracy of 80.5 % was seen in the 4-marker combinations.

Table 2 compares the accuracy of the selected markers through T Test with accuracy of optimum marker combinations. From the selected four markers through T test, M6 is the most probable marker as the high frequency marker. This might have been the cause of the large performance difference of about 5 %.

Table 1 Performance test of the classification algorithm through all the possible marker combinations of 2–4 markers and the formation of the optimal combination

Classifier	Marker 1	Marker 2	Marker 3	Marker 4	Sensitivity	Specificity	Accuracy
LDA	M1	M7			0.531	0.894	0.795
	M1	M4	M7		0.556	0.894	0.801
	M1	M2	M5	M7	0.543	0.903	0.805
k-NN	M1	M6			0.519	0.830	0.785
	M1	M6	M7		0.506	0.912	0.801
	M1	M2	M6	M8	0.494	0.921	0.805

Looking for Better Combination of Biomarker Selection

Table 2 Classification performance comparison of the marker combinations obtained through T test

Classifier	Marker 1	Marker 2	Marker 3	Marker 4	Sensitivity	Specificity	Accuracy
LDA	M8	M3			0.543	0.796	0.727
	M8	M3	M6		0.568	0.815	0.748
	M8	M3	M6	M5	0.531	0.815	0.737
k-NN	M8	M3			0.333	0.847	0.707
	M8	M3	M6		0.457	0.847	0.741
	M8	M3	M6	M5	0.469	0.889	0.774

Table 3 Classification performance comparison of the marker combinations obtained through Genetic Algorithm

Classifier	Marker 1	Marker 2	Marker 3	Marker 4	Sensitivity	Specificity	Accuracy
LDA	M1	M8			0.531	0.880	0.785
	M1	M8	M2		0.543	0.884	0.791
	M1	M8	M2	M7	0.556	0.866	0.781
k-NN	M1	M8			0.432	0.903	0.774
	M1	M8	M2		0.395	0.917	0.774
	M1	M8	M2	M7	0.407	0.889	0.758

The four markers selected with Genetic Algorithm (Table 3) and Random Forest (Table 4) includes the two most frequent markers (M1, M7). It shows 3 and 2 % difference for the accuracy.

4.2 Dot Plot and ROC Curve for Optimum Marker Combination

As illustrated in Fig. 3, the cancer and benign are not distinguishable in one dimension. The results were easier to analyze when it was projected in two dimensions using a marker combination. The ROC curve of Fig. 4 demonstrates the aforementioned statement.

Table 4 Classification performance comparison of the marker combinations obtained through Random Forest

Classifier	Marker 1	Marker 2	Marker 3	Marker 4	Sensitivity	Specificity	Accuracy
LDA	M1	M6			0.543	0.852	0.768
	M1	M6	M5		0.494	0.866	0.764
	M1	M6	M8	M5	0.543	0.866	0.778
k-NN	M1	M6			0.494	0.875	0.771
	M1	M6	M5		0.432	0.861	0.744
	M1	M6	M8	M5	0.407	0.857	0.734

Fig. 3 Dot plot of individual markers in optimum marker combination

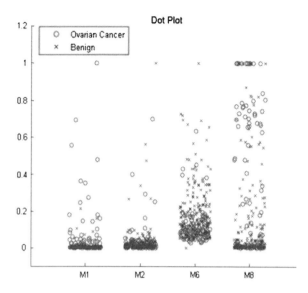

Fig. 4 ROC curve of individual markers and 4-markers combination

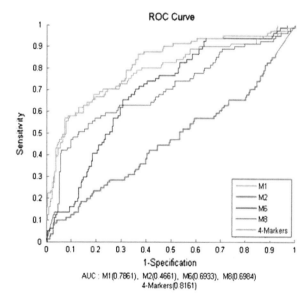

5 Conclusion

This paper searches for the biomarker combination that can easily distinguish between malignant and benign tumors. Comparing the classification performance of ovarian cancer, selecting 2 or 3 markers through GA and classifying with LDA shows the most similar sensitivity, specificity, and accuracy to that of the marker combinations of 2–4 markers derived from the complete enumeration survey.

However, except for the LDA 3-marker combination of GA, the marker combination of 2–4 markers obtained with the marker selection algorithm showed significantly low accuracy compared to the marker combinations obtained from the complete enumeration survey.

Acknowledgments The research was supported by the Research & Business Development Program through the Ministry of Knowledge Economy, Science and Technology (N0000425) and the Ministry of Knowledge Economy (MKE), Korea Institute for Advancement of Technology (KIAT) and Gangwon Leading Industry Office through the Leading Industry Development for Economic Region.

References

1. American Cancer Society (2012). http://www.cancer.org/Cancer/OvarianCancer
2. Brian N, Adele M, Liudmila V, Denise P, Matthew W, Elesier G, Anna L (2009) A serum based analysis of ovarian epithelial tumorigenesis. Gynecol Oncol 112:47–54
3. ChiHeum C (2008) Biomarkers related to diagnosis and prognosis of ovarian cancer. Korean J Obstet Gynecol 39:90–95
4. SunKyung J, EunJi O, ChulWoo Y, WoongSik A, YongGu K, YeonJun P, KyungJa H (2009) ELISA for the selection of HLA isoantibody and comparison evaluation of Luminex panel reactive antibody test. J Korean Soc Lab Med 29:473–480
5. David F, Roger P, Robert P (1998) Statistics, 3rd edn. W. W. Norton & Company, New York
6. Tom MM (1997) Machine learning. The McGraw-Hill, New York
7. Leo B (2001) Random forest. Mach Learn 45:5–32
8. Suraj DA, Greg PB, Tzong-Hao C, Katharine JB, Jinghua Z, Partha S, Ping Y, Brian CM (2009) Development and preliminary evaluation of a multivariate index assay for ovarian cancer. PLoS ONE 4:e4599

A Remote Control and Media Sharing System Based on DLNA/UPnP Technology for Smart Home

Ti-Hsin Yu and Shou-Chih Lo

Abstract The remote control and media sharing of consumer devices are key services for smart living. The involving of mobile devices into these services has become a technology trend. Existing solutions to these services restrict these devices to be located in the same local network. In this paper, we design and implement an integrated architecture that supports the outdoor remote control to home devices and the sharing of digital media among indoor and outdoor devices. By following the digital home related standards, we show our system design with the details of hardware and software components.

Keywords DLNA · UPnP · Smart home · Media sharing · Remote control

1 Introduction

With the popularity of digital consumer products, digital content can be seen everywhere. For the easy sharing of digital media such as videos, photos, and music between these consumer products, the Digital Living Network Alliance (DLNA) was initiated in 2003 to define interoperability guidelines [1, 2]. The underlying technology is Universal Plug and Play (UPnP) [3] for media management, discovery, and control. The UPnP, which includes a set of standard network protocols such as TCP/IP, HTTP, and Simple Object Access Protocol (SOAP), enables digital devices having networking capability to be connected.

DLNA compliant devices can seamlessly discover each other's presence on the same home network and share functional services or media content with each other. Four types of DLNA devices are defined in the standard: DMS, DMC, DMP,

---

T.-H. Yu · S.-C. Lo (✉)
Department of Computer Science and Information Engineering, National Dong Hwa Univeristy, Hualien 974, Taiwan, Republic of China
e-mail: sclo@mail.ndhu.edu.tw

J. J. (Jong Hyuk) Park et al. (eds.), *Multimedia and Ubiquitous Engineering*, Lecture Notes in Electrical Engineering 240, DOI: 10.1007/978-94-007-6738-6_41, © Springer Science+Business Media Dordrecht(Outside the USA) 2013

and DMR. A Digital Media Server (DMS) stores and provides media content to other types of devices. A Digital Media Controller (DMC) can discover media content and command a Digital Media Renderer (DMR) to play the content. A Digital Media Player (DMP) can discover and play media content directly, which can be considered as the combination of DMC and DMR.

The media sharing between DLNA supported devices contributes significantly to our comfortable life. By integrating sensor technology and automatic control to these home appliances, some research efforts [4–6] have been done for achieving a smart living environment. With the popularity of smart phones, using a mobile phone to monitor and control the living environment becomes a potential service model. For example, a mobile user when being out of home can retrieve and play media streaming from an indoor DMS by connecting the mobile phone to the home network using the Session Initial Protocol (SIP) [7].

In this paper, we provide a service platform using smart phones that can remotely access and control any home appliances. The home appliances include any DLNA or UPnP supported devices and traditional electronic devices such as air conditioners, TVs, and desk lamps. Our prototype system can remotely monitor and switch the power of a home device and share media content between an indoor device and a remote device.

Our developed platform is based on the Android system which provides open sources and many free libraries. To enable a smart phone to have extra control functions to other devices, we use the Arduino programmable platform [8]. Arduino is an open-source single-board microcontroller which can be easily programmed to control robots, lighting, etc. An Arduino board also provides the linkage to other external modules such as sensor modules and wireless communication modules.

The remainder of this paper is organized as follows. Section 2 explains some issues about our system design. Section 3 shows the design of each hardware/software component. Finally, the concluding remarks are given in Sect. 4.

2 System Design

We mainly extend smart living services from an indoor environment to an outdoor one. This platform provides the following services:

1. A user no matter in home or out of home can monitor and control home appliances and share digital content with home DLNA compliant devices through a mobile phone (For example, a remote device turns off the desk lamp and downloads a music file from an indoor DLNA device).
2. Two outdoor users can share digital contents as if they are in the same home network (For example, a remote device accesses image files from another remote device).

A Remote Control and Media Sharing System 331

3. An outdoor user can redirect digital contents from a home DMS to a local DMR (For example, a remote device commands a local DMR to retrieve and play a video from an indoor DLNA device).

There are two types of home devices in our system: DLNA devices and UPnP devices. DLNA devices provide or play media contents and UPnP devices provide remote control services. To enable traditional devices such as desk lamps and televisions to support for UPnP functions, we externally equip each of these devices with a control broad. Consequently, these traditional home devices become so called external UPnP devices (in contrast to internal UPnP devices with built-in UPnP functions). The control board works as a ZigBee End Device (ZED) that can be controlled by a ZigBee Coordinator (ZC) via the ZigBee wireless communication.

The core component in our system is the DroidHome which discovers and maintains DLNA and UPnP devices. Moreover, DroidHome plays the role of a proxy server for remote devices and informs these remote devices about the status of an indoor device by a message push mechanism. The message push mechanism is provided by an existing push service on the cloud called Cloud to Device Message (C2DM) [9, 10]. C2DM is released with the Android operation system by Google. Any Android device can be notified in background by the C2DM if the corresponding user has registered itself using a Google account.

2.1 Challenges and Solutions

Both in DLNA and UPnP networks, the service discovery and invocation follow the same method of Simple Service Discovery Protocol (SSDP). A control point or a DMC multicasts a discovery message (ssdp:discover) into the local home network. A device in the house responds its presence by sending back a replay message (ssdp:alive). A new joining DMS can also multicast this ssdp:alive message to inform other existing devices of its presence. This ssdp:alive message contains an Uniform Resource Locator (URL) to locate the file about the device profile. This device profile includes meta-information in XML format such as the device name, production factory, and a URL list to locate those service functions provided by the device. The retrieval of the device profile is based on the HTTP protocol. The service invocation is by sending a SOAP message to the device.

When we integrate any non-UPnP devices and remote devices into the above service platform, some problems are encountered. In the following, we discuss the new challenges and our proposed solutions.

1. **Remote DMS**: If we allow an indoor DLNA device to directly access digital contents from a remote device configured as a DMS, this remote DMS should join to the same home network. One simple way is to connect this remote device back to the home network using the Virtual Private Network's (VPN) Point-to-Point Tunneling Protocol (PPTP). However, this remote device cannot

show its presence by simply multicasting an ssdp:alive message into the home network due to a different setting of the subnet mask. The provided solution is that this remote device first unicasts the message to our DroidHome which in turns multicasts the message into the home network.

2. **Remote DMP**: If we configure a remote device as a DMP and connect it to the home network using the PPTP, this remote DMP can directly play the digital content of an indoor DMS. However, the drawback of this approach is that the remote DMP should stay connected to the home network for keeping the new status of an indoor DMS. Here, we use DroidHome to decouple this strong connection. DroidHome acts as a DMC and stores the service state of each indoor DLNA or UPnP device. A table to record the URLs to these device profiles is also maintained. A remote DMP simply contacts with DroidHome to retrieve this URL table for invoking a certain service. When DroidHome detects any state change (e.g., the joining or leaving of an indoor DMS), it will notify those remote devices using the C2DM push service.

3. **Private home network**: Almost all home networks are configured as private networks where all indoor devices use private IP addresses and an Internet Gateway Device (IGD) uses the only public IP address. This IGD provides the Network Address Translation (NAT) function to enable an indoor device to be reachable from the Internet. The core technique is by the port mapping. However, most NAT devices maintain dynamic port mapping, which makes an indoor device unreachable from the Internet. Here, we use an UPnP supported IGD which can be automatically discovered and remotely configured with a static port mapping. DroidHome would first register a mapping port to this IGD such that any other remote devices can connect to the DroidHome using the IGD's public IP address and the registered port number. Moreover, to enable a remote device to access the profile of an indoor device, DroidHome translates the private address in the URL table to the IGD's public address. Each address translation needs DroidHome to register a new mapping port to the IGD.

4. **External UPnP device**: In our system, an external UPnP device actually refers to a non-UPnP device. These devices naturally have no networking capability and service functions. We first equip each of these devices with a control board. This control board can switch the power or enable certain functions of the associated device by a programmable current relay unit or an infrared emitter. These control boards can communicate with our DroidHome using the ZigBee wireless communication. Second, we create a software device object for each external UPnP device in DroidHome. A device object stores the corresponding service profile and service functions about the physical device. Then, Droid-Home records the access path to this device object into the URL table. Consequently, a remote device can look up services provided by external UPnP devices and can invoke these functions through DroidHome.

3 Component Design

In the following, we explain the function and the design of each hardware component in our system.

3.1 Zed

The ZED is responsible for the control of the associated external UPnP device. Each ZED is implemented by an Arduino Uno platform and has the following types: infrared ZED and relay ZED. An infrared ZED can emit infrared signal to switch the power or the channel of the associated device. The relay ZED can switch the power of the associated device.

An infrared ZED contains the following hardware components inside. A current transformer (CT) sensor detects the current of a power line and is used to detect the power status of a device. An XBee module provides ZigBee communication and is configured as ZED mode. An IR module sends and receives infrared signal. A button module is used to set the infrared signal on the IR module with the same frequency as the remote controller of a home device. A relay ZED contains a CT sensor and an XBee module, and additionally a current relay module which can open or close the power. Beside these hardware components, each ZED maintains some data: power status (on or off) of the associated device, type of the ZED (infrared or relay), and identification (a unique sequence number) of the ZED.

3.2 DroidHome

This hardware component is composed of an Android device (a smart phone in our test system) and an Arduino Mega ADK platform through USB communication. The former subcomponent discovers and maintains indoor DLNA devices and internal UPnP devices in the home network (WiFi networking environment in our test system). The latter subcomponent discovers and controls indoor ZEDs using the ZigBee network.

The Arduino platform contains an XBee module which is configured with ZC mode. This ZC would periodically multicast a discovery message (performed by ZED discovery module) such that all surrounding ZEDs will respond this message with their types and ID data. These response data are recorded in the ZED table. The control command (e.g., power switch) to each ZED is issued from the ZED control module.

In the Android device, a registration table keeps the identification information for each authorized remote device, which includes an USN (Unique Service Name used to identified a device in our system), a C2DM token (a certification code given by the C2DM server), and a password. A DMC module is responsible for device discovery. The access path to the profile of a discovered device is stored in the URL table. This table keeps two versions of URLs (private URL and public URL) with the primitive and the translated IP addresses and port numbers, respectively, for each access path. An external device object table stores UPnP device objects for all discovered ZEDs. Each device object keeps the status and profile of the associated device. The request and response messages to and from the IGD for port mapping are handled by a port mapping module. A remote presence module is responsible to notify indoor devices about the joining or leaving of a remote DMS by multicasting messages into the home network. A server push module handles the contact to the C2DM server for pushing a notification to a remote device. An HTTP server module handles the request of device profile downloading and the request of service invocation.

3.3 Remote Device

This device is a smart phone installed with some software components. A register module processes the registration to the C2DM server and to the DroidHome. A DMP module can discover DMSs and render the output of service content. If this remote device is out of the home network, this device can discover indoor services by retrieving device profiles from the URL table of DroidHome. This task is performed by an import devices module. The notification of the state change of any indoor device is listened by a push receiver module. A DMS module is activated when the remote device would like to share digital contents. In this situation, the remote device waits for incoming requests to access its profile via a HTTP server module.

4 Conclusions

We have demonstrated a home service platform that integrates media sharing and power control to home appliances. Four distinguish features are highlighted. First, the restriction of media sharing among only indoor DLNA devices is broken. Mobile phones can be involved no matter in home or out of home. Second, traditional home appliances without automatic control and networking capability can be involved too in our platform by equipping with a simple external device. Third, our implementation is based on an open-source software stack, some existing equipments, and low-price hardware chips. Fourth, a variety of home services such as temperature sensing and video surveillance can be easily integrated into our service platform.

References

1. Digital Living Network Alliance. http://www.dlna.org/
2. Digital Living Network Alliance (2006) DLNA networked device interoperability guidelines v1.5, Mar 2006
3. UPnP. http://www.upnp.org/
4. Horng M-F, Chang B-C, Su B-H (2008) An intelligent intrusion detection system based on UPnP technology for smart living. In: Proceedings of 8th international conference on intelligent systems design and applications, Cairo. pp 14–18
5. Leu J-S, Lin W-H, Tzeng H-J (2009) Design and implementation of an OSGi-centric remote mobile surveillance system. In: Proceedings of IEEE international conference on systems, man and cybernetics, San Antonio, Oct. 2009. pp 2498–2502
6. Chen Y-S, Chen I-C, Chang W-H (2010) Context-aware services based on OSGi for smart homes. In: Proceedings of 3rd IEEE international conference on Ubi-media computing July 2010, China, pp 38–43
7. Oh Y-J, Lee H-K, Kim J-T, Paik E-H, Park K-R (2007) Design of an extended architecture for sharing DLNA compliant home media from outside the home. IEEE Trans Consum Electron 53(2):542–547
8. Arduino. http://www.arduino.cc/
9. C2DM. https://developers.google.com/android/c2dm/
10. Hansen J, Gronli T-M, Ghinea G (2012) Cloud to device push messaging on Android: a case study. In: Proceedings of 26th international conference on advanced information networking and applications workshops, Mar 2012, Japan, pp 1298–1303

A New Distributed Grid Structure for k-NN Query Processing Algorithm Based on Incremental Cell Expansion in LBSs

Seungtae Hong, Hyunjo Lee and Jaewoo Chang

Abstract To manage the frequent updates of moving objects' locations on road networks in an efficient way, we propose a new distributed grid scheme which utilizes node-based pre-computation technique to minimize the update cost of the moving objects' locations. Because our distributed grid scheme manages spatial network data separately from the POIs (Point of Interests) and moving objects, it can minimize the update cost of the POIs and moving objects. To process k-nearest neighbor (k-NN) query in our distributed grid scheme, we propose a k-NN query processing algorithm based on Incremental cell expansion which minimize the number of accesses to adjacent cells during POIs retrieval in a parallel way. Finally, we show from our performance analysis that our algorithm is better on retrieval performance than the k-NN algorithm of the existing work.

Keywords Distributed grid scheme · Query processing algorithm · Road network · Moving objects

1 Introduction

With the advancements on GPS and mobile device technologies, it is required to provide location-based services (LBS) to moving objects which move into spatial networks. Several types of location-dependent queries are significant in LBS, such

---

S. Hong · H. Lee · J. Chang (✉)
Department of Computer Engineering, Chonbuk National University, Chonju,
Chonbuk 561-756, South Korea
e-mail: jwchang@jbnu.ac.kr

S. Hong
e-mail: dantehst@jbnu.ac.kr

H. Lee
e-mail: o2near@jbnu.ac.kr

J. J. (Jong Hyuk) Park et al. (eds.), *Multimedia and Ubiquitous Engineering*,
Lecture Notes in Electrical Engineering 240, DOI: 10.1007/978-94-007-6738-6_42,
© Springer Science+Business Media Dordrecht(Outside the USA) 2013

as range queries [1], k-nearest neighbor (k-NN) queries [1–3], reverse nearest neighbor queries [4], and continuous queries [5]. Among them, the most basic and important queries are k-NN ones. The existing k-NN query processing algorithms use pre-computation techniques for improving performance [6–8]. However, when POIs need to be updated, they are inefficient because distances between new POIs and nodes should be re-computed. To solve it, S-GRID [9] divides a spatial network into two-dimensional grid cells and pre-compute distances between nodes which are hardly updated. However, S-GRID cannot handle a large number of moving objects which is common in real application scenario. As the number of moving objects increases, a lot of insertions and updates of location data are required due to continuous changes in the positions of moving objects. Because of this, a single server with limited resources shows low performance for handing a large number of moving objects. Therefore, we, in this paper, propose a new distributed grid scheme which manages the location information of a large number of moving objects in spatial networks. Based on our distributed grid scheme, we propose a new k-NN query processing algorithm based on incremental cell expansion which minimize the number of accesses to adjacent cells during POIs retrieval in a parallel way.

The rest of the paper is organized as follows. In Sect. 2, we present related works. In Sect. 3, we describe the details of our distributed grid scheme. Section 4 presents a new k-NN query processing algorithm based on our grid scheme. In Sect. 5, we provide the performance analysis of our k-NN query processing algorithm. Finally, we conclude this paper with future work in Sect. 6.

2 Related Work

In this section, we describe some related works on k-NN query processing in spatial networks. First, VN3 [6], PINE [7], and islands [8] were proposed to pre-compute the distance between POIs and nodes (or border points) in road networks. However, when POIs need to be updated, they are inefficient because distances between new POIs and nodes should be re-computed. To resolve the problem of the VN3, PINE and Island approaches, Huang et al. [9] proposed S-GRID (Scalable Grid) which represents a spatial network into two-dimensional grids and pre-computes the network distances between no1des and POIs within each grid cell. To process k-NN query, they adopt the INE algorithm [1] which consists of inner expansion and outer expansion. The inner expansion starts a network expansion from the cell where a given query point is located and continues processing until the shortest paths to all data points inside the cell have been discovered or the cell holds no data points. Whenever the inner expansion visits a border point, the outer expansion is performed from that point. The outer expansion finds all POIs in the cells sharing the border point. This process continues until k nearest POIs are found. In S-GRID, the updates of the pre-computation data are local and POI independent. However, S-GRID have a critical problem that it is not efficient in

A New Distributed Grid Structure

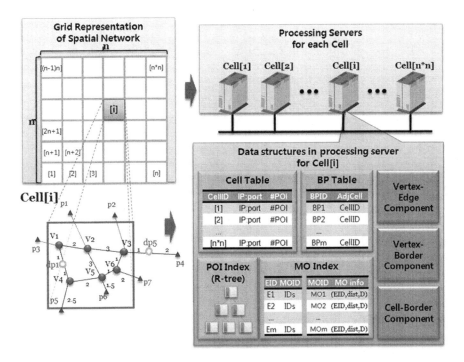

Fig. 1 Overall structure of distributed grid scheme

handling a large number of moving objects, which are common in real application scenario, because it focuses on a single server environment. That is, when the number of moving objects is great, a lot of insertions and updates of location data are required due to continuous changes in the positions of moving objects. Thus, a single server with limited resources shows bad performance for handing a large number of moving objects (Fig. 1).

3 Distributed Grid Scheme

To support a large number of moving objects, we propose a distributed grid scheme by extending S-GRID. Similar to S-GRID, our distributed grid scheme employs a two-dimensional grid structure for a spatial network and performs pre-computations on the network data, such as nodes and edges, inside each grid cell. In our distributed grid scheme, we assign a server to each cell for managing the network data, POIs and moving objects. Each server stores the pre-computed network data and manages cell-level two indices, one for POIs and the other for moving objects. Figure 2 shows an overall structure of our distributed grid scheme. We describe each component of our distributed grid scheme.

```
InnerExpansion Algorithm(q, k, Qv, Qdp, BPlist)
01. edge = findEdge(q)
02. for each POI∈findPOI(edge)          Qdp.update(POI, di st(q, POI))
03. for each v∈{edge.start_node, edge.end_node } Qv.update(v, dist(q, v))
04. for each bp∈Cell-Border Componenet  Qv.update(bp, dist(q, bp))
05. dMax=Qdp.dist(k)
06. do
07.     vx=Qv.deque, mark vx as visited
08.     if (vx is a vertex)
09.         for each adjacent vertex vy of vx in Vertex-Edge Component
10.             for each POI∈findPOI(ex,y)
11.                 Qdp.update(POI, dist(q,vx)+dist(vx,POI))
12.             Qv.update(vy, dist(q,vx)+dist(vx,vy))
13.             if (all POI in myCell is discovered or Qdp.maxdist()<dist(q,vx))
14.                 break
15.         else      BPList.update(vx, dist(q,vx))
16.         dMax=Qdb.dist(k)
17. while( d(q, vx) < dMax && Qv≠∅ )
```

Fig. 2 Inner expansion algorithm

4 K-NN query processing algorithm

In this paper, we propose a new k-NN query processing algorithm based on our distributed grid scheme, namely Incremental Cell Expansion (ICE) algorithm. First, our ICE algorithm finds all the border points and creates a list of cells containing the border points by doing the inner expansion. Next, a coordinate (server) sends the query to all the cells in the cell list. Secondly, servers receiving the query retrieve both POIs and other border points by doing outer expansion. Then, they send the retrieved POIs and border points to the coordinate from which query is originated. Thirdly, the algorithm checks whether or not there is a border point being nearer than the k-th POI. If true, the process is repeated until no border point is nearer than the k-th POI. To find k nearest neighbors, our ICE algorithm performs both inner expansion and outer expansion by using two priority queues, Qv and Qdp. Qv stores both relevant nodes and the distance between a query point and the nodes while Qdp stores both retrieved POIs and their distances from a query point. Thus, our ICE algorithm can improve retrieval performance by minimizing unnecessary visiting of adjacent cells. Figure 2 shows an inner expansion algorithm.

In addition, our ICE algorithm performs outer expansion in two cases; (i) the number of retrieved POIs is less than k and (ii) there remains a border point in the cell list. Figure 3 shows the outer expansion algorithm. First, the algorithm sends a query to the servers managing respective adjacent cells of the cell list. Then, by using a cell-border component, the servers insert both the border points of related cells and their distances from a query point into Qv. Secondly, by using a vertex-border component, the servers insert POIs within the related cells and their

A New Distributed Grid Structure 341

```
OuterExpansion Algorithm(q, k, BPlist)
Qdp=∅, Qv=∅
01. for each bpi∈BPlist
02.      for each bpj∈Cell-Border Component
03.          if (bpi bpj)    Qv.update(bpj, dist(q,bpi)+dist(bpi+bpj))
04.          for each POI∈myCell
05.              Qdp.update(POI, dist(q,bpi)+dist(bpi+POI))
06.      dMax=Qdp.dist(k)    bp=Qv.deque
07. return POIs in Qdp, bps in Qv
```

Fig. 3 Outer expansion algorithm

distances from a query point into Qdp At last, they return both the retrieved POIs and the border points to the coordinator where the query is originated.

5 Performance Analysis

We present performance analysis of k-NN query processing algorithm for our distributed grid scheme. We implement our grid under HP ML 150 G3 server with Intel Xeon 3.0 GHz dual CPU, 2 GB memory. In our experiments, we used multiple processes in a single server and each process manages a single cell. To provide an environment appropriate to a distributed grid scheme, we let each process use a different port number to communicate with other processes by using TCP/IP protocol. For spatial network data, we use San Francisco Bay map consisting of 220,000 edges and 170,000 nodes, and generate four sets of POIs (i.e., 2,200, 4,400, 11,000, 22,000) by using Brinkhoff algorithm [10]. These POIs are indexed by using R-trees. Moreover, we randomly select 100 nodes from San Francisco Bay map as query points. To measure the retrieval performance of k-NN queries, we average response times for all the 100 query points. Because the existing works VN3 [6], PINE [7], island [8] are very inefficient for the update of POIs due to their POI-based pre-computation techniques, they are not appropriate for dealing with a large number of mobile objects in spatial networks. Thus we compare our algorithm with S-GRID algorithm in terms of POI retrieval time.

Figure 4a first shows the performance of k-NN query processing with the different number of grid cells when k = 20 and POI density = 0.01. The performance of our algorithm is better than that of S-GRID when the number of grid cells is more than 10*10. This is because our algorithm performs outer expansion in a parallel way. Figure 4b shows the retrieval time of k-NN query with the varying value of k when the density of POI is 0.01 and the number of grid cells equals 20*20. We can say form the performance result that our ICE algorithm is better because it can minimize the number of accesses to adjacent cells during POIs retrieval.

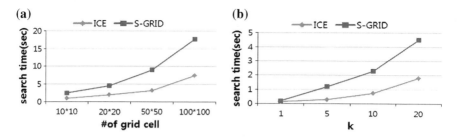

Fig. 4 Retrieval performance **a** with different number of grid cells **b** in terms of k

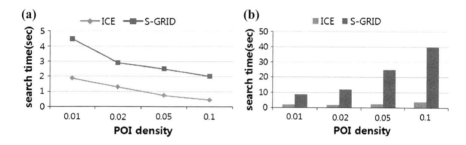

Fig. 5 Retrieval performance **a** in terms of the density of POIs **b** after updating POIs

Figure 5a shows the retrieval time of k-NN query with the varying density of POIs where the number of grid cells equals 20*20 and k = 20. As a result, we can reduce the cost of inner expansion within a cell and the number of adjacent cells to be visited. Figure 5b shows the retrieval time of k-NN query after updating POIs. For this experiment, we measure the search time of k-NN query when the 10 % of POIs is updated. In the case of S-GRID, the retrieval time is exponentially increased as the density of POIs increases. This is because S-GRID uses one R-tree to index all the POIs of the network and so the update of POIs in a cell affects the whole system. Whereas, because our grid scheme uses a separate R-tree per each grid cell to index POIs within it, the update of POIs in a cell does not affect all the grid cells globally. As a result, even though the number of updated POIs increases, the retrieval performance of our grid scheme is not dramatically increased.

6 Conclusion and Future Work

In this paper, we proposed a new distributed grid scheme to manage the location information of a large number of moving objects in spatial networks. Our distributed grid scheme makes use of a node-based pre-computation technique so that it can minimize the update cost of the moving objects' locations. Our distributed grid scheme splits a spatial network into two-dimensional grid cells so that it can update network data locally. Based on our grid scheme, we proposed a new k-NN query processing algorithm. Our ICE algorithm improves the retrieval performance of K-NN queries because it decreases the number of adjacent cells visited by transmitting a query to all the shared border points. Our experimental results show that our algorithm is better on retrieval performance than that of S-GRID. As a future work, we need to extend our grid scheme to handle a spatial network with dense and sparse regions in an efficient manner by using non-uniform grid cells.

Acknowledgment This research was supported by Basic Science Research Program through the National Research Foundation of Korea (NRF) funded by the Ministry of Education, Science and Technology (2010-0023800).

References

1. Papadias D, Zhang J, Mamoulis N, Tao Y (2003). Query processing in spatial network databases. In: Proceedings of VLDB, pp 802–813
2. Shahabi C, Kolahdouzan MR, Sharifzadeh M (2003) A road network embedding technique for K-nearest neighbor search in moving object databases. In: Proceedings of GeoInformatica, vol 7, no 3, pp 255–273
3. Jensen CS, Pedersen TB, Speicys L, Timko I (2003) Data modeling for mobile services in the real world. In: Proceedings of SSTD, pp 1–9
4. Benetis R, Jensen CS, Karčiauskas G, Šaltenis S (2006) Nearest and reverse nearest neighbor queries for moving objects. In: Proceedings of VLDB, pp 229–250
5. Huang YK, Chen C-C, Lee C (2009) Continuous K-Nearest neighbor query for moving objects with uncertain velocity. In: Proceedings of GeoInformatica, vol 13, no 1, pp 1–25
6. Kolahdouzan MR, Shahabi C (2004) Voronoi-based nearest neighbor search for spatial network databases. In: Proceedings of VLDB, pp 840–851
7. Safar M (2005) K Nearest Neighbor Search in Navigation Systems. Mobile Inf Syst 1(3):207–224
8. Huang X, Jensen CS, Saltenis S (2005) The islands approach to nearest neighbor querying in spatial networks. In: Proceedings of SSTD, LNCS 3633, pp 73–90
9. Huang X, Jensen CS, Lu H, Saltenis S (2007) S-GRID: a versatile approach to efficient query processing in spatial networks. In: Proceedings of SSTD, LNCS 4605, pp 93–111
10. Brinkhoff T (2002) A framework for generating network-based moving objects. In: Proceedings of GeoInformatica, pp 153–180

A New Grid-Based Cloaking Scheme for Continuous Queries in Centralized LBS Systems

Hyeong-Il Kim, Mi-Young Jang, Min Yoon and Jae-Woo Chang

Abstract Recent development in wireless communication technology and mobile equipment is making location-based services (LBSs) more popular day by day. However, because users continuously send queries to a server by using their exact locations in the LBSs, private information can be in danger. Therefore, a mechanism for users' privacy protection is required for the safe and comfortable use of LBSs. For this, we, in this paper, propose a grid-based cloaking area creation scheme in order to support continuous queries in LBSs. Our scheme creates a cloaking area rapidly by using grid-based cell expansion to efficiently support the continuous LBSs. In addition, to generate a cloaking area which lowers the exposure probability of a mobile user to a minimum level, our scheme computes a privacy protection degree by granting weights to the mobile users. Finally, we show from our performance analysis that our cloaking scheme shows better performance than the existing cloaking scheme.

Keywords Privacy protection · Continuous Queries · Cloaking scheme

---

H.-I. Kim · M.-Y. Jang · M. Yoon · J.-W. Chang (✉)
Department of Computer Engineering, Jeonbuk National University,
Jeonju, Jeonbuk, South Korea
e-mail: jwchang@jbnu.ac.kr

H.-I. Kim
e-mail: melipion@jbnu.ac.kr

M.-Y. Jang
e-mail: brilliant@jbnu.ac.kr

M. Yoon
e-mail: myoon@jbnu.ac.kr

J. J. (Jong Hyuk) Park et al. (eds.), *Multimedia and Ubiquitous Engineering,*
Lecture Notes in Electrical Engineering 240, DOI: 10.1007/978-94-007-6738-6_43,
© Springer Science+Business Media Dordrecht(Outside the USA) 2013

1 Introduction

Recent development in wireless communication technology and mobile equipment, location-based services (LBSs) become more popular. A location-based service is a service which is accessible by mobile devices through the communication network and utilizing the ability to make use of the geographical position of the mobile device. By using LBS, we can get various services such as finding the nearest Point of Interest (POI) like an ATM or a restaurant, and receiving the warning of traffic jam. However, we must send the exact location information to a LBS server when using these services. However, in this case, users' privacy may be leaked to unauthorized users and illegally used by them. For example, attackers can analyze users' leaked data and identify their life style. To solve these problems, a mechanism for users' privacy protection is required for the safe and comfortable use of LBSs.

There are many existing studies on the cloaking method of k-anonymity to protect users' privacy. The cloaking method makes a cloaking area, which includes a query issuer and $k - 1$ other users, while sending a query to LBS server. So, exact location information can be hidden with $1/k$ leaking probability. However, the existing cloaking methods have a problem when a user continuously request queries. While making a cloaking region for each time, the methods include different group of $k - 1$ users so that the unauthorized users are able to detect the query issuer by comparing the $k - 1$ users of successive time frames. To solve the problem, Xu et al. proposed Advanced KAA [1]. Advanced KAA calculates privacy degree of generated cloaking region. But it has two problems. At first, because it considers all candidate areas to generate minimal sized cloaking region, it takes much processing time. Secondly, because the Advanced KAA calculates privacy degree with random sampling of user data, it does not guarantee high similarity between $k - 1$ other users of current cloaking area and those of previous cloaking area. As a result, the privacy protection level may be reduced.

For this, we, in this paper, propose a grid-based cloaking area creation scheme in order to support continuous queries. Our scheme creates a cloaking area rapidly by using grid-based cell expansion to efficiently support the continuous queries. In addition, to generate a cloaking area which lowers the exposure probability of a mobile user to a minimum level, it computes a privacy protection degree by granting weights to mobile users. The rest of the paper is organized as follows. In Sect. 2, we introduce related works. In Sect. 3, we propose a grid-based cloaking area creation scheme supporting continuous queries. We present our performance analysis in Sect. 4. Finally, we conclude this paper with brief summary and future work in Sect. 5.

2 Related Work

The existing cloaking methods [2–5] don't protect user privacy to support continuous queries. While making cloaking region for each time, the methods include a different group of k − 1 users. Therefore, the unauthorized users are able to detect the query issuer by comparing the k − 1 users of successive time frames. There exists only work by Xu and Cai [1] to deal with this kind of problem. They proposed Advanced KAA (K-anonymity Area) method that calculates anonymity degree of newly added users, and finds minimal sized circle for generating the cloaking area with satisfying k-anonymity. Advanced KAA uses entropy for measuring privacy degree. The entropy is the amount of data that is needed to identify the query issuer in a cloaking area. If we assume that A is the cloaking area of the query issuer N, and there exist the group of m users, the entropy of A is measured by expression (1).

$$H(A) = -\sum_{i=1}^{m} p_i \log p_i \qquad (1)$$

Here, $H(A)$ is the value of entropy for A, p_i $(1 \le i \le m)$ is the detecting probability that the selected user is the query issuer N. After calculating the value of entropy for A, we can measure the Anonymity Degree of A by using expression (2).

$$D(A) = 2^{H(A)} \qquad (2)$$

After $T = 0$, we use following way to calculate the detecting probability. At first, we make the transition matrix M with α number of sample data which includes the information of users' movement. In M, the value in each cell means the number of sample data which is in both previous cloaking area and current cloaking area. The probability of users at $T = i$, p_i, is calculated by the product of the p_{i-1} and M.

But this method has a problem of computational overhead to find the minimum circle. With the many mobile objects, which mean the LBS users, the processing time of this method becomes larger with polynomial basis. Therefore, we propose a grid-based cloaking scheme to reduce the cloaking region processing time.

3 Grid-Based Cloaking Scheme for Continuous Queries

We use a centralized approach with the trusted third party called an anonymizer which creates the cloaking region. Figure 1 shows the system architecture consisting of three components, a mobile user, anonymizer, and LBS server. The query processing step is as follows. First, a mobile user sends a query to the anonymizer with user location information and he/she periodically updates the location data during service time. Secondly, the anonymizer generates a cloaking

Fig. 1 System architecture

area with k − 1 other users and sends the cloaking region to LBS server with a session ID. Thirdly, LBS server processes the query based on the cloaking region. Finally, the anonymizer filters out the query result based on the exact location of the query issuer and returns the exact result to the user.

On the other hand, we assume that the user location information is sent to LBS server by using GPS system. This is because, while sending the location data, an adversary can get the exact user location so that the privacy of query issuer can be in danger. Here, an adversary means an unauthorized user who leaks and uses the privacy data illegally. In serious cases, the service provider can be an adversary.

3.1 Generating the Initial Cloaking Region

Our algorithm first expands cells around the cell which the query issuer is located. During expansion, if the number of mobile users in expanded area (=k′) is greater than the value of user given k-anonymity, it sets the minimal boundary rectangle of the expanded cells as a temporary cloaking region. Here, the algorithm counts the number of cells in temporary area as the value of C.

Secondly, to generate minimal sized cloaking region, our algorithm sets the initial limitation number of cells to C and the initial value of k to k′. After that, it measures C and k′ in every cases while deleting some rows or columns for each directions. With this, the scheme can reduce the size of the temporary cloaking region. If the scheme finds an area which has same k′ and less number of cells, it sets the area to the temporary cloaking region and changes the current values of limitations. Or if the scheme finds an area which has same number of cells but has larger k′, it sets the area to the temporary cloaking region and changes the values of limitations. The algorithm will run until there is no cloaking region which satisfies the value of k-anonymity. By using this step, we can generate the minimal sized cloaking region and reduce the query processing time in LBS server. For example, Fig. 2 shows the example of this step. In Fig. 2a, by using width-based search information tree, the algorithm finds the minimal sized temporary cloaking region. Here, all nodes in the tree contain the C and k′ per each cases. Child nodes of each tree node include the information which can be deleted by one row or

A New Grid-Based Cloaking Scheme

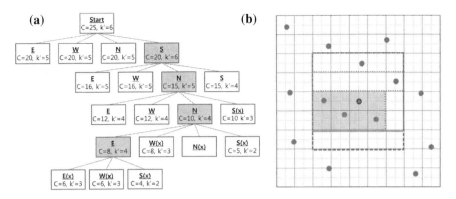

Fig. 2 Example of minimal sized cloaking region. **a** Information tree. **b** Minimal sized cloaking area

column for 4 directions. As a result, the algorithm finds the minimal sized cloaking area as shown in Fig. 2b (the colored area). Based on the initial cloaking area which was created in the step 1, the algorithm generates a cloaking area which can guarantee the privacy degree during the service time.

3.2 Generating a Cloaking Area with Guaranteeing Privacy Degree

In this step, the algorithm sets the weights for each user in the previous service time. During the service time (T), users in cloaking area can move to another places based on the networks. Therefore, at first, the algorithm generates temporal cloaking area (TC(i), i = current time), which contains all the users in the previous cloaking area. For example, Fig. 3a shows the cloaking region of T = i with k = 2. The region at T = i includes q and u4. Then at T = i + 1, the algorithm generates minimal boundary rectangle TC(i + 1). As shown in the Fig. 3b, TC(i + 1) includes q, u4 and u3.

Secondly, the algorithm calculates the entropy for guaranteeing privacy degree. For this, it makes the transition matrix (M). However, a transition matrix in the

Fig. 3 Example of setting temporal cloaking area. **a** T = i. **b** T = i + 1

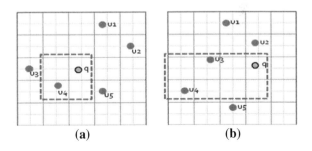

(a)

	q	u3	u4
q	12	3	5
u4	3	7	10

(b)

M.q = 0.5*12/20 + 0.5*3/20 = 0.375
M.u3 = 0.5*3/20 + 0.5*7/20 = 0.25
M.u4 = 0.5*5/20 + 0.5*10/20 = 0.375

Fig. 4 Example of calculating user probability. **a** Weight-based transition matrix. **b** User probability

existing Advanced KAA method has a problem that M does not reflect previous users' information. So in our work, we consider how many times each user in TC(i + 1) have been a member of previous cloaking regions (i.e. from T = 0 to T = i). Based on that, the algorithm creates α number of samples and makes M. For example, we assume that u4 and q in TC(i + 1) are members of the previous cloaking regions in Fig. 5, and the value of α equals to 20. Figure 4a shows the transition matrix. The probability of users at T = i, p_i, is calculated by the product of the p_{i-1} and M as shown in Fig. 4b. Here, we assume that the probability of q and u4 in T = i are 0.5.

By considering weight of each user, we can increase the probability of previous members. As a result, the privacy degree can be improved. Thirdly, based on the cell containing the query issuer, the algorithm expands the cells for generating a cloaking region when T = i + 1. Here, the algorithm calculates the entropy by using the formula (1), and then measures the privacy degree by the formula (2). If the measured privacy degree is larger than a given k-anonymity, the algorithm sets the minimal boundary rectangle to a cloaking area. For example, if the algorithm selects the u3 and q in Fig. 5b as members of candidate cloaking area, the entropy and the privacy degree can be calculated as follows.

$$H(A) = -(0.375 \times \log 0.375 + 0.25 \times \log 0.25) = 1.03064$$
$$D(A) = 2^{1.03064} = 2.04293$$

Here, the privacy degree is larger than k = 2, so the algorithm sets the minimal boundary rectangle including q and u3 to a cloaking region.

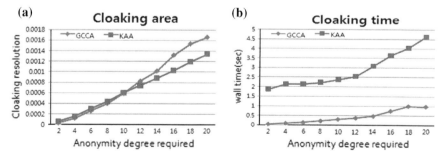

Fig. 5 Performance according to the value of k. **a** Size of cloaking area. **b** Cloaking time

4 Performance Evaluation

In this section, we show the performances of our Grid based Continuous Cloaking Algorithm (GCCA). We implemented GCCA by using MS Visual Studio.NET 2003 running on the Window XP system with 2.20 GHz Intel Core2 Duo CPU and 2 Gb memory. We generated moving objects data by using the network-based moving object generator [6]. We use real road network data of Oldenburg, Germany (15×15 km^2). For measuring cloaking area easily, we set the total size of the map as 1. We compare our GCCA with the existing advanced KAA (KAA) which was proposed by Xu [1]. We evaluated the performance by changing the anonymity level from 2 to 10. The value of session life time is set to 5, whereas the grid cell size is set to $1,000 \times 1,000$.

Figure 5a shows the size of cloaking area according to the value of k-anonymity. As the value of k is increased, the size of cloaking areas of both GCCA and KAA is increased. However, it is shown that GCCA generates 15 % larger area than KAA. This is because GCCA includes more users in cloaking region than KAA, by considering the users in the previous cloaking area for continuous query. Figure 5b shows the cloaking time according to the value of k. The average cloaking times of GCCA and KAA are 0.4273 and 2.8673, respectively. It is shown that GCCA achieves 50 % better performance than KAA. This is because GCCA can reduce the computational overhead by using grid cell expansion.

5 Conclusion

In this paper, we propose the grid-based cloaking area creation scheme to support continuous queries for LBSs. For reducing computational overhead while generating cloaking area, we design a grid cell expansion approach. In addition, to guarantee user privacy, we consider the weight-based privacy degree. From our performance analysis, it is shown that our GCCA algorithm is better than the existing scheme, in terms of cloaking time, service time and privacy degree. Meanwhile, both methods show similar performance in terms of the size of the cloaking area. As a future work, we need to enhance our GCCA algorithm to apply it to a distributed computing environment.

Acknowledgments This research was supported by Basic Science Research Program through the National Research Foundation of Korea (NRF) funded by the Ministry of Education, Science and Technology (2010-0023800).

References

1. Xu T, Cai Y (2007) Location anonymity in continuous location-based services. In: ACMGIS, pp 221–238
2. Mokbel M, Chow C, Aref W (2006) The new casper: query processing for location services without compromising privacy. In: Proceedings of the international conference on very large data bases, pp 763–774
3. Xu T, Cai Y (2009) Location anonymity in continuous location-based services. In: Proceedings of ACM conference on computer and communications security, pp 348–357
4. Yang L, Wei L, Shi H, Liu Q, Yang D (2011) Location cloaking algorithms based on regional characteristics. In: Proceedings of the international conference on computer science and automation engineering, pp 93–98
5. Jang M, Chang J (2012) New cloaking method based on weighted adjacency graph for preserving user location privacy in LBS. In: Proceedings of computer science and its applications, pp 129–138
6. Brinkhoff T (2002) A framework for generating network-based moving objects. GeoInformatica 6(2):153–180

New Database Mapping Schema for XML Document in Electronic Commerce

Eun-Young Kim and Se-Hak Chun

Abstract This paper considers a relational data system to store XML document efficiently. Also this paper proposes a data model to rewrite XML documents from data storage by representing data view and structure view at the same time and introduce a mapping schema to relational data base system from the data for electronic commerce.

Keywords XML document management · Relational database · Multi-format information retrieval · Electronic commerce

1 Introduction

Extensible Markup Language (XML) is a simple and flexible markup language. As XML has been used for data transaction in EDI from 1998, it supports all kinds of electronic commerce transaction [1]. There are two methods to map XML documents to RDBMS which are an approach for XML document only and an approach for definition of XML document structure with DTD. The element of DTD is been mapped as a relation of relational database system or as an attribute of relation according to element type, the number of repetition and whether an attribute is contained or not. The attribute of DTD is been mapped as the attribute of relation in the relational database system.

E.-Y. Kim
Department of Multimedia Contents, Sin Ansan University, 671 Chosi-dong, Ansan City, Kyunggi-do 425-792, Republic of Korea
e-mail: key@sau.ac.kr

S.-H. Chun (✉)
Department of Business Administration, Seoul National University of Science and Technology, Kongneung-gil 138, Nowon-gu, Seoul 139-743, Republic of Korea
e-mail: shchun@seoultech.ac.kr

J. J. (Jong Hyuk) Park et al. (eds.), *Multimedia and Ubiquitous Engineering*, Lecture Notes in Electrical Engineering 240, DOI: 10.1007/978-94-007-6738-6_44, © Springer Science+Business Media Dordrecht(Outside the USA) 2013

Fig. 1 Order element

```
<!ELEMENT order (product,quantity,price)>
<!ATTLIST order no ID #REQUIRED>
<!ELEMENT product (#PCDATA)>
<!ELEMENT quantity (#PCDATA)>
<!ELEMENT price (#PCDATA)>
```

Fig. 2 Order structure from DTD

```
<order no="120806">
<product>m234-2t</product>
<quantity>3</quantity>
<price>50000</price>
</order>
```

Figure 1 shows order elements in XML document, Fig. 2 shows a structure from DTD. Figure 3 shows RDMS which defines elements of no, product, quantity and price.

The existing method has difficulty to search according to attribute and element and to reproduce original XML document from stored RDBMS. There is a problem that changes a relational structure of RDBMS when attributes in element are added as normalization is needed when there are many duplicate attributes in relational DBMS.

This paper extends Kim and Chun's [2] model which considers a relational data system to store XML document efficiently. This paper proposes a data model to rewrite XML documents from data storage by representing data view and structure view at the same time and introduce a mapping schema to relational data base system from the data for electronic commerce. The paper is organized as follows. In Sect. 2, we describe the data model used in the study. In Sect. 3, we show the results of our analyses. Section 4 concludes this study.

2 The Extended Model Considering DTD

In the basic model, the XML document is represented as a graph that all nodes and edges are labeled and that is ordered among nodes and edges and that edges are directed on. When the XML document is represented as a graph $G = \{V,E,A\}$ where V is a vertex, i.e. set of nodes, E, a set of edges, and A, a set of attributes defined on start tag of element [2].

no	product	quantity	price
120806	m234-2t	3	50000

Fig. 3 Mapping results in RDBMS

New Database Mapping Schema

```
<?xml version="1.0" encoding="euc-kr"?>
<!DOCTYPE purchaseOrder [
    <!ELEMENT purchaseOrder(c_INFO,c_orderList,shipTo)>
    <!ATTLIST  purchaseOrder no ID #REQUIRED>
    <!ELEMENT customer ANY>
    <!ATTLIST customer id ID #REQUIRED>
     <!ELEMENT c_PHONE (#PCDATA)>
    <!ELEMENT orderList  (order)+>
    <!ELEMENT order (product,quantity,price)>
    <!ATTLIST  order no ID #REQUIRED>
    <!ELEMENT product (#PCDATA)>
    <!ELEMENT quantity (#PCDATA)>
    <!ELEMENT price (#PCDATA)>
     <!ELEMENT shipInfo (s_NAME,s_PHONE,s_ADDR1,s_ADDR2,s_MSG)>
    <!ELEMENT s_NAME (#PCDATA|EMPTY)>
    <!ATTLIST  s_NAME cID IDREF #IMPLIED>
    <!ELEMENT s_TEL  (#PCDATA)>
    <!ELEMENT s_PHONE (#PCDATA)>
    <!ELEMENT s_ADDR1 (#PCDATA)>
    <!ELEMENT s_ADDR2 (#PCDATA|EMPTY)>
    <!ELEMENT s_MSG (#PCDATA)>
]>
```

Fig. 4 An example of purchase order with IDFEF type

DTD means a mutual agreement on the XML document transmitted when XML documents are exchanged. Attributes referring elements in DTD are presented by defining the type of IDREF or IDREFS. Also the value of attributes of IDREF or IDREFS must be identical to the value of the specified property in ID type of

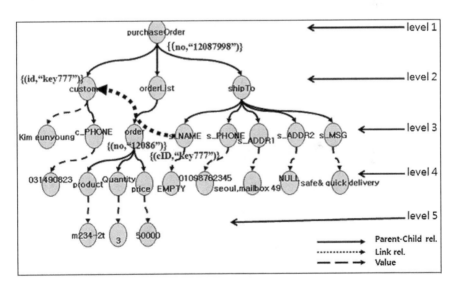

Fig. 5 A graph of the data model with reference information

different type. Figure 4 shows an example of purchase order with IDFEF type. Figure 5 shows the graph of the XML document with reference information between elements by using DTD.

When accompanied by DTD, the XML document can get the reference information between elements from the data of a DTD document. The XML document is extended like below if all the reference edge sets that are added on the graph of the XML document are $R(G)$.

$$G = (V, E, R, A)$$

3 The Mapping Schema to RDBMS

3.1 Four Objects of XML document

XML documents are represented as a graph. The graph consists of nodes, edges, attributes, reference edges, etc. and each component has its own inherent attributes as an object. When tuples are represented as attributes that contain V, E, A, R, each component object of a graph G, they are:

$$Vertex\,Object = (label,\,level)$$

$$Edge\,Object = (from,\,to,\,relation,\,order)$$

$$Attribute\,Object = (node,\,name,\,value,\,type)$$

$$Reference\,Edge\,Object = (refFrom,\,refTo,\,refAttr)$$

This tuple structure offers a unique structure regardless of the data and structure of the XML document.

3.2 Relational Structure

Figure 6 represents the graph G when VID(Vertex ID) is added in Vertex Object tuples, it is represented as follows:

$$Vertex\,Object = (VID,\,label,\,level)$$

$$Edge\,Object = (VID\,of\,from,\,VID\,of\,to,\,relation,\,order)$$

$$Attribute\,Object = (VID\,of\,node,\,name,\,value,\,type)$$

$$Reference\,Edge\,Object = (VID\,of\,refFrom,\,VID\,of\,refTo,\,refAttr)$$

New Database Mapping Schema 357

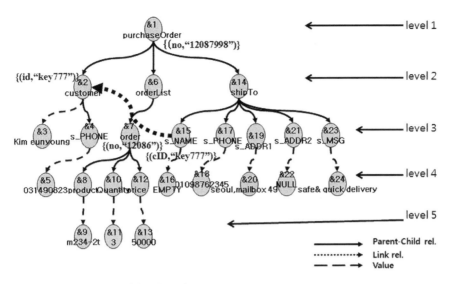

Fig. 6 The data model of the XML document

Figure 7 shows a mapping schema from the proposed data model to RDBMS. This proposed model is an applicable model in the case of designing a new database system for the XML document and in the case of using an existing database system like a relational database system. Because the data view and the structure view of the original XML document are lost in the data model for mapping to the existing relational database system [3–5] or the object-oriented system, not only can't XML generate again from the stored data, but also XML

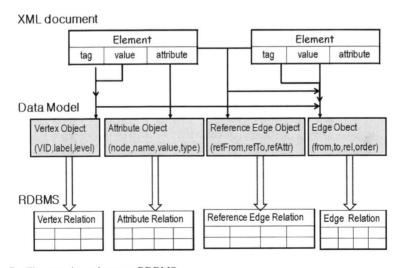

Fig. 7 The mapping schema to RDBMS

sub-graph corresponding to element as search result can't be returned [6]. In the proposed model, this problem is solved because the data view and structure view of the original XML document are stored.

4 Conclusion

In the proposed data model defines the data structure, depending on the specific properties because it does not attribute with a value of NULL data does not exist. The graph data model can be mapped naturally into existing RDBMS. Also the proposed model recreates the XML document from stored data when a graph data model is mapped to RDBMS all the data and structure view of the XML source document is stored. For a future study, we expect to do research on speed and accuracy valuation for RDBMS data stored in this way through keyword search queries.

References

1. Wills B (2009) The business case for environmental sustainability (Green). A 2009 HPS white paper. MicroSoft, "XML Scenarios". http://msdn.microsoft.can/xml/scenario/inro.asp
2. Kim E, Chun S-H (2012) New hybrid data model for XML document management in electronic commerce. Lect Notes Electr Eng 203:445–451
3. Du F, Sihen A-Y, Freire J (2004) ShreX: managing XML documents in relational databases. In: Proceedings of the 30th VLDB conference, Toronto, Canada
4. Kappel G, Kapsammer E, and Retschitzegger W (2000) X-Ray-towards integrating XML and relational database systems. Technical report, July 2000
5. Choi RH, Wong RK (2009) Efficient date structure for XML keyword search. DASFAA, pp 549–554
6. Atay M, Chebotko A, Liu D, Shiyong L, Fotouhi F (2007) Efficient schema based XML-to-relational data mapping. Inf Syst 32(3):458–476

A Study on the Location-Based Reservation Management Service Model Using a Smart Phone

Nam-Jin Bae, Seong Ryoung Park, Tae Hyung Kim, Myeong Bae Lee, Hong Gean Kim, Mi Ran Baek, Jang Woo Park, Chang-Sun Shin and Yong-Yun Cho

Abstract This paper suggests a location-based reservation management service model that can manage various reservation services and provide related information to users through their smart phones in real-time. The proposed service model is based on a smart phone to relieve inconvenience of existing reservation systems that have limited access effectiveness in space and time or long waiting time. To improve service satisfaction, the proposed service model includes a service server to manage highly qualified and reliable reservation schedules with user's location information from smart phones, and a client to support users to make their application plans through intuitive user interfaces. Therefore, the proposed model can

---

N.-J. Bae · S. R. Park · T. H. Kim · M. B. Lee · H. G. Kim · M. R. Baek · J. W. Park
C.-S. Shin · Y.-Y. Cho (✉)
Department of Information and Communication Engineering, Sunchon National University,
Suncheon, South Korea
e-mail: yycho@sunchon.ac.kr

N.-J. Bae
e-mail: bakkepo@sunchon.ac.kr

S. R. Park
e-mail: ghost214@sunchon.ac.kr

T. H. Kim
e-mail: mcteng@sunchon.ac.kr

M. B. Lee
e-mail: lmb@sunchon.ac.kr

H. G. Kim
e-mail: khg_david@sunchon.ac.kr

M. R. Baek
e-mail: tm904@sunchon.ac.kr

J. W. Park
e-mail: jwpark@sunchon.ac.kr

C.-S. Shin
e-mail: csshin@sunchon.ac.kr

J. J. (Jong Hyuk) Park et al. (eds.), *Multimedia and Ubiquitous Engineering*,
Lecture Notes in Electrical Engineering 240, DOI: 10.1007/978-94-007-6738-6_45,
© Crown Copyright 2013

help users to save a long waiting time in various service places, for example food stores, banks, and government office customer service centers.

Keywords LSB · Smart phone · Reservation management service

1 Introduction

Recently, a reservation service in a customer service center, restaurant and beauty shop becomes a common occurrence. However, because the reservation service is commonly provided through the Internet and a telephone, it is difficult to manage correctly a subscriber' demands and to predict correctly an awaiter's waiting time. Also, if the subscriber does not comply with reservation time or the awaiter comes out from the waiting line without previous notice, a problem in the reservation process may happen. Due to the higher distribution rate of a smart phone, through which location information can be used in real time, efficient location-based service and various reservation services can be provided.

The way to use real-time location information is by GPS and by base station of Smart Phone, which is provided for easy use in development tool of Smart Phone. This service model constructs database of ServiceProvider's location information, and compares user's location of Smart Phone by real-time judgment. Therefore, it is necessary to inscribe service ServiceProvider's location, and for user to install application to Smart Phone or to use QR-Code provided.

In order to make an efficient management of subscriber or awaiter, it is necessary for algorithm to deviate from service location, or to reflect real-time correction for reassignment its order of priority. The proposed service model makes a real-time management of location and correction to reset its order of priority. Therefore, it is necessary to receive real-time location information. And Server needs reservation algorithm for order of priority.

The proposed service model does not use another terminal. It is using location information and technology of Smart Phone with recent usage and distribution rate explosively increased. For this, it is possible to provide reservation service using ads with QR-Code attached to service location providing user's real reservation service, and to propose user's efficient reservation management service model that adjusts subscriber's priority on the basis of real-time location information.

2 Related Work

Recently, prevalence of smart phone has been increasing dramatically over the last few years. As a result, various applications (for example: guide traffic, directions etc.) using smart phone is being provided. Above all, one of the most popular services in provided services is QR-Code-based and location-based service.

2.1 QR-Code

Quick Response Code (QR code) is the trademark for a type of matrix barcode (or two-dimensional bar code) first designed for the automotive industry in Japan [1]. Bar codes are optical machine-readable labels attached to items that record information related to the item. Initially patented, its patent holder has chosen not to exercise those rights. Recently, the QR Code system has become popular to people due to its fast readability and greater storage capacity compared to standard UPC barcodes [1]. It recently use various places (for example: Business card, Flyers) printed by to publicity. Also, it use to induce access to web page of mobile [2, 3].

One of services QR-Code used is 'smart phone learning information code scan program' [4]. 'Smart phone learning information code scan program' is program that provides learning information by use camera of smart phone to recognize learning information code (or QR tag). It transfers automatically to screen of learning information code scan after this program activate learning information code scan program. The next can use learning information provide service by transfer to web service that smart phone learning information provided using link information of learning information code.

In this paper, QR-Code technology use to exchange first data between waiting service required place and user, it uses to receive reservation waiting management service (Fig. 1).

2.2 Location-Based Services

Location-based services are a general class of computer program-level services used to include specific controls for location and time data as control features in

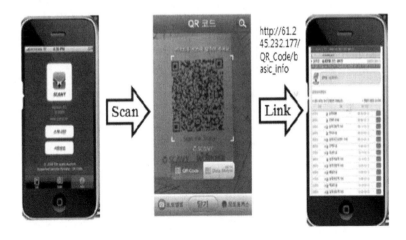

Fig. 1 'Daum Map' service

Fig. 2 Moving part of learning information code scan program using smart phone

computer programs. As such (LBS) is an information and has a number of uses in Social Networking today as an entertainment service, which is accessible with mobile devices through the mobile network and which uses information on the geographical position of the mobile device [5]. Recently, many people one of services used is 'Daum Map' [6]. When user inputs starting point and destination, 'Daum Map' service shows user's current location, destination and optimal path finding in user's screen. Also 'Daum Map' can choose means of transportation such as walk, subway, taxi, bus. Besides, it gives information such as the fare and the time required for each means of transportation (Fig. 2).

Interworking through geocoding is important, because provided service using location information use a map [7–9]. In this paper, LBS use for calculate required distance and arrival time to reset hard-wired reassignment system in server using location information of smart phone, and use to mark ServiceProvider location or arrival limit area.

3 Reservation Management Service Model Based Location Using Smart Phone

3.1 Service Model Construction

This service model is composed by server, service user (Smart Phone), and server has ServiceProvider's location and waiting management system and priority reassignment system. And service user (Smart Phone) would transfer location information regularly to server collected by geocoding and calculate reaching

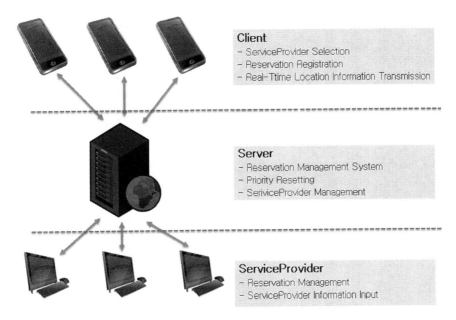

Fig. 3 The overall system configuration diagram

time. Deviation from reaching time range can move back or exclude client of waiting state in waiting lines in reassignment system. Figure 3 represents the whole system block diagram of user reservation management system based on location information using Smart Phone that is proposed by this paper. The proposed system is composed highly by ServiceProvider, server and client.

Client in Fig. 3 represents users available to service with Smart Phone. In this time, user selects his desired ServiceProvider from available ServiceProviders appeared on Smart Phone to user a specific service, and registers it in reservation service. The information of registered user and real-time location information is transferred to server. Server uses user's service register information received from client and real-time location information for reassignment and determining priority of waiting order, and uses the whole service subscribers and service reservation information registered in client to manage state of reservation. Also, Manages the ServiceProvider registration and the basic information. ServiceProvider means restaurant, bank and service center that provides reservation service. And, ServiceProvider can manage subscriber management and reservation service provided by connect to server.

3.2 Reservation Management Process Using Location Information

Reservation service model proposed in this paper progresses reservation management between user and a large number of users using real-time location information.

(1) Using geocoding gather smart phone location information, then transfers to the server.
(2) Compare entered ServiceProvider location information of database.
(3) Calculate distance, arrival time etc. in order of priority reset system then prepare to reset the order of priority.
(4) Transfers order of priority reset to the client.
(5) Approve User received order of priority reset.
(6) Reassign the order of priority, then transferred to awaiter management system to adjust the waiting lines.

Figure 4 represents module configuration diagram of each client and location information processing process for reservation management service. This figure includes location information, reservation service and basic information database related system modules.

Server of Fig. 4 is composed highly by waiting lines management system required for the reservation management, assignment and reassignment system for the reservation order of priority, subscriber management system and database to store information of ServiceProvider. Client is composed highly by location-information process system to process location information based geocoding and waiting lines registration module to register each service reservation of user.

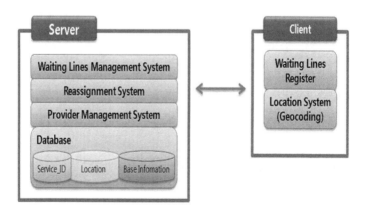

Fig. 4 Server, client configuration diagram

3.3 Client

The user install application or using separately provided QR-Code by Service-Provider in the client after service start by connect to reservation service. The client provides reservation information including waiting time and a number of awaiter to user. Also, User may receive various benefits including coupon and event information.

3.4 Waiting Management

User by use information of ServiceProvider and transmitted location in the client judges whether user arrives to the destination or not. If user then arrives to destination, user is bookable. However, if the user is unable to arrive on time and are limited reservation Also, user a high ratio of canceled is expressed to screen of ServiceProvider.

If you need to reset after distinguishing users in reassignment system, then reduce waiting time and waiting lines in reservation management system.

3.5 Order of Priority Assignment and Reassignment

If User deviates feasible arrival distance on time using location information real-time transmitted from smart phone, then except from awaiter's waiting lines, and then transmit subscriber information to the reservation management system. If user postpones reservation time or cancels reservation, then alike transmit subscriber information to the reservation management system after judge suitable order of priority. Also, If user have a high rate of reservation cancelation or don't input authorization code, then rejudge order of priority.

- Deviated client from arrival limit distance
- Cancel a reservation by client
- Postpone a reservation by client
- Client, a high rate of canceled
- Client Adjusted by ServiceProvider

4 Conclusion

This paper proposed a reservation management service model based on Smart Phone location information technology and can determine the ranking of providing service based on reservation service user's location information and context information and give an efficient support of waiting management. The proposed service model uses individual Smart Phone and QR-Code technology that has excellent use and access of reservation management service. Also, various applications of reservation service with the proposed service model applied are expected to make an efficient reservation management based on user's location and context information and to raise satisfaction of service user's service quality. The future study will be progressed apt for designing and realizing real reservation management service system according to the proposed service model.

Acknowledgments This work was supported by the National Research Foundation of Korea (NRF) grant funded by the Korea government. (MEST) (No. 2012-0003026).

References

1. Wikipedia http://www.doopedia.co.kr/doopedia/master/master.do?_method=view&MAS_IDX=101230001172183
2. Lee K (2011) Characteristics of QR code ad and its effects on usage satisfaction and consumers' behavior as a commercial communication tool. Korean J Advert 22(3):103–124
3. Park J (2012) A research on expansion of library service by using QR code. Korean library. Inf Sci Soc 43(3):1–27
4. Jung W (2010) A design of U-learning study support system. In: 2010 Korea multimedia society autumn conference, vol 13, no 2
5. Wikipedia http://en.wikipedia.org/wiki/Location-based_service
6. 'Daum Map' http://map.daum.net/
7. Choi DS (2010) Google API-based expression and utilization plan. The Korean institute of maritime information and communication sciences 2010 autumn conference, pp 672–674
8. Jin K (2008) A study of government policy plans on ubiquitous based lbs services revitalization. Thesis
9. Google, Google Geocoding API,. https://developers.google.com/maps/documentation/geocoding/?hl=ko-KR
10. Barkhuus L, Dey A (2003) Location-based services for mobile telephony: a study of users' privacy concerns. In: Proceedings of INTERACT 2003, 9th IFIP TC13 international conference on human-computer interaction
11. Gruteser M, Grunwald D (2002) Anonymous usage of location-based services through spatial and emporal cloaking. In: Proceedings of the first international conference on MobiSys

A Real-time Object Detection System Using Selected Principal Components

Jong-Ho Kim, Byoung-Doo Kang, Sang-Ho Ahn, Heung-Shik Kim and Sang-Kyoon Kim

Abstract The detection of moving objects is a basic and necessary preprocessing step in many applications such as object recognition, context awareness, and intelligent visual surveillance. Among these applications, object detection for context awareness impacts the efficiency of the entire system and it requires rapid detection of accurate shape information, a challenge specially when a complicated background or a background change occurs. In this paper, we propose a method for detecting a moving object rapidly and accurately in real time when changes in the background and lighting occur. First, training data collected from a background image are linearly transformed using principal component analysis (PCA). Second, an eigen-background is organized from selected principal components with excellent ability to discriminate between object and background. Finally, an object is detected by convoluting the eigenvector organized in the previous step with an input image, the result of which is the input value used on an EM algorithm. An image sequence that includes various moving objects at the same time is organized and used as training data to realize a system that can adapt to changes in

---

J.-H. Kim · H.-S. Kim · S.-K. Kim (✉)
Department of Computer Engineering, Inje University, Gimhae,
Gyeongsangnam-do 621-749, Republic of Korea
e-mail: skkim@inje.ac.kr

J.-H. Kim
e-mail: luckykjh@daum.net

H.-S. Kim
e-mail: kimhs@inje.ac.kr

B.-D. Kang
Researcher, STAR Team, Korea Electronics Technology Institute,
Bucheon-si, Gyeonggi-do 420-734, Republic of Korea
e-mail: deweyman@gmail.com

S.-H. Ahn
Department of Electronic Engineering, Inje University, Gimhae,
Gyeongsangnam-do 621-749, Republic of Korea
e-mail: elecash@inje.ac.kr

J. J. (Jong Hyuk) Park et al. (eds.), *Multimedia and Ubiquitous Engineering*,
Lecture Notes in Electrical Engineering 240, DOI: 10.1007/978-94-007-6738-6_46,
© Springer Science+Business Media Dordrecht(Outside the USA) 2013

lighting and background. Test results show that the proposed method is robust to these changes, as well as to the partial movement of objects.

Keywords Object detection · Principal Components Analysis (PCA) · Eigen-background · Mixture of Gaussian

1 Introduction

Computer vision technology, which was originally developed for human computer interaction, has been applied to a variety of fields such as user interface designs, robot learning, and intelligent surveillance systems. Object detection, which accurately and effectively separates an object from its background, is an essential technology. Without proper foreground/background separation, it would be difficult to detect objects or analyze gestures in the next step of the system such as that required in vision-based robotic manipulation, augmented reality, and gesture recognition.

Therefore, a number of researchers have been studying how to separate an object from its background. Representative methods utilize the difference between a previous frame or a previously saved background and the current frame [1], compressed video information [2], object movement [3], or visual attention [4].

Although methods that utilize difference of images between frames are easy to realize, they are extremely sensitive to illumination changes and are unable to detect objects when there are no moving images or when only a part of an image moves [5]. Methods that use compressed video information do not require a previously saved background and is fast. However, these methods have a downside in that they use different detection methods depending on the technique of compression used [2].

A method based on motion, such as structure from motion, has a fast detection speed, but it also has difficulty in determining the accurate shapes of objects and it is also sensitive to changes in illumination. Finally, methods that use a visual attention model present fast detection speed but they have difficulty in extracting meaningful object contours as well as accurate shape information.

Eigenspace is a technique that can overcome most of these problems. The basic idea behind this method is to analyze the background information obtained from training images and then to construct an eigen-background to recall the expected background. This eigen-background can be used to separate an object from its background, allowing the detection of non-moving objects. The improved eigenspace models presented in [6, 7] adapt the background after the formation of the initial eigen-background. That is, these models continue to learn the eigen-background while inputting new images as training data. However, methods using an eigen-background usually separate a foreground object using the eigenvalues and corresponding eigenvectors with a high explanation rate. Therefore, using these methods lead to background noise to be detected as an object. In [6], on the other hand, this adaptation is carried out using a synthetic background obtained from the current image and eigen-background after the removal of any foreground object. Also in [6], the dimension of the eigenspace—i.e. the number of eigenvalues and corresponding eigenvectors selected—is kept to an optimum number by applying a clustering algorithm that classifies images based on their illumination conditions. Unfortunately, the criterion to select eigenvalues employed by the method in [6] is still based on the largest eigenvalues.

In this paper, we propose a robust object detection system that addresses most of these problems. In the heart of the proposed method is the construction of an advanced eigenspace that is able to capture more effective information about the background, especially under different lighting conditions and background changes. This step is achieved by employing a clustering algorithm that selects the eigenvalues based on their "power of explanation", rather than their numerical value.

First, we construct training data using images, including moving people, chairs, etc. We can then use these images to create eigen-backgrounds that are accommodative to background changes. Next, eigen-backgrounds are constructed after the background information has been analyzed through PCA [8]. Next, an eigen-backgrounds that distinguishes an object clearly from the background by using a clustering algorithm is selected. Next, an object is detected by convoluting the eigenvector organized in the previous step with an input image, the result of which is the input value of an EM algorithm.

An EM clustering algorithm, which is an unsupervised learning method (i.e., it does not require a user's artificial goal value), is used for the detection method. A mixture of Gaussians (MOG) and fuzzy c-means (FCM) [9, 10] are used as an EM algorithm. An object detection system that is robust to changes in illumination is realized using a combination of the EM algorithm and the eigenspace.

2 Object Detection System

2.1 Overview

Figure 1 shows the main structure of our object detection system. First, the system uses various background images as PCA training data. It analyzes the training-data-set information using PCA and then selects the eigenvalue that properly distinguishes moving objects from their backgrounds. The system then constructs an eigen-background using a selected eigenvalue and uses this background as input data for clustering. The system deducts the resultant value achieved by multiplying the input image and the eigen-background based on the pixel unit. Finally, the system detects moving objects from images clustering based on the resultant value obtained using MOG.

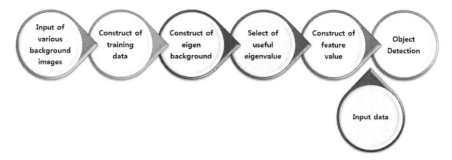

Fig. 1 Main structure for object detection system

2.2 Eigen-Background

In this paper, we make an eigen-background using PCA to detect moving objects from the background in an image.

First, the system acquires background samples that are used as training data to construct an eigen-background. Second, the system analyses the training data using PCA and generates an eigen-background after extracting principal components of the background.

The system then changes 2D images into 1D line vector in order to use the training data as PCA input images.

$$S = [I_1, I_2 \cdots I_M] = \begin{bmatrix} x_{11} & x_{12} & \cdots & x_{1M} \\ x_{21} & x_{22} & \cdots & x_{2M} \\ \vdots & \vdots & \ddots & \vdots \\ x_{N1} & x_{N2} & \cdots & x_{NM} \end{bmatrix}, \quad I_J = \begin{bmatrix} x_{1j} \\ x_{2j} \\ \vdots \\ x_{Nj} \end{bmatrix}, \quad (1 \leq j \leq M) \quad (1)$$

S is an training data images, j is the index of an image existing within the entire set. N is the amount of training data, and N is the number of features.

In this paper, the number of features N is 76,800 because we use 240×320 images. The average of training data is calculated using the following Eq. (2):

$$\Psi_i = \frac{1}{M} \sum_{j=1}^{M} x_{ij}, \quad (1 \leq i \leq N, 1 \leq j \leq M)$$

$$\Psi = [\Psi_1, \Psi_2, \Psi_3, \ldots, \Psi_N]^T$$

(2)

where x_{ij} is an i order pixel value in I_j. We use 100 training images in this paper. The deviation is presented as shown in Eq. (4) through determinant (3).

$$\Phi_j = I_j - \Psi \quad (3)$$

A Real-time Object Detection System Using Selected Principal Components

$$A = \begin{bmatrix} \Phi_{11} & \Phi_{12} & \cdots & \Phi_{1M} \\ \Phi_{21} & \Phi_{22} & \cdots & \Phi_{2M} \\ \vdots & \vdots & \ddots & \vdots \\ \Phi_{N1} & \Phi_{N2} & \cdots & \Phi_{NM} \end{bmatrix} = [\Phi_1, \Phi_2, \ldots, \Phi_M] \tag{4}$$

We create a covariance matrix to ascertain the eigenvalue and eigenvector. Covariance matrix C is obtained using the following equation:

$$C = \frac{1}{M} \sum_{j=1}^{M} \Phi_j \Phi_j^T \tag{5}$$
$$= AA^T$$

C is an $N \times N$ dimension matrix. The number of particular dimensions of this matrix N is 76,800 for images. It is difficult to calculate 76,800 eigenvalue S and their corresponding eigenvectors uses a $76{,}800 \times 76{,}800$ dimension covariance matrix. Therefore, the system calculates the eigenvalue and eigenvector in $M \times M$ dimensions considering an eigenvector on $A^T A$.

$$A A^T v_j = \lambda_j v_j \tag{6}$$

When both sides are multiplied by A^T

$$A^T A A^T v_j = \lambda_j (A^T v_j) \tag{7}$$

$A^T v_j$ becomes eigenvector v'_j for $A^T A$. This matrix has $M \times M$ dimensions.

$$A^T v_j = v'_j, v_j = A v'_j \tag{8}$$

The eigenvector v_j, of an $M \times N$ matrix is calculated using Eq. (8).

An M numbered eigenvalue λ_j and N numbered eigenvector corresponding to each eigenvalue are determined using the a covariance matrix C.

$$B_j = \frac{1}{M} \sum_{k=1}^{M} v_j \Phi_k^T \quad (j = 1, 2, \ldots, M) \tag{9}$$

B_j is an eigen-background that has eigenvalue order j, and v_j is an eigenvector that has a j order eigenvalue. B_j is determined as eigenvector v_j that has an eigenvalue order j and a deviation of Φ^T.

A general eigen-background method uses an eigenvalue that is arranged in descending order and satisfies Eq. (10).

$$\frac{\sum_{j=1}^{m} \lambda_j}{\sum_{j=1}^{T} \lambda_j} \geq th\,(=0.9) \tag{10}$$

T is the number of eigenvalues, creates an eigen-background using m-numbered eigenvalues that set the threshold to th (0.9), and reflects an explanation rate of up to 90 %. The eigenvalue that has the highest account rate is not suitable for separating an object from its background and has many noisy elements. Although *th* (threshold) is set above 0.9, the separation of the object from its background is not clear because the first eigenvalue is included.

$$B = \frac{1}{m}\sum_{j=1}^{m} B_{lj}, \quad (1 \le l \le m, \; 1 \le l \le N) \tag{11}$$

$$D_j = \left| I_j \times B \right| \succ t \tag{12}$$

If an Input image has a higher value than threshold *th*, it is an object; if not, it is the background. To construct an eigen-background, the system analyzes the components of the training data.

2.3 Object Detection Using EM Algorithm

In this paper, we use a clustering algorithm to determine an eigenvalue that is useful for detecting objects from their backgrounds. We use mixture of Gaussians among clustering algorithms.

2.3.1 MOG (Mixture of Gaussian)

An MOG is a density estimation method used for improving the method of modeling the distribution density of a sample data set as one probability density function. It models the distribution of data using a number of Gaussian probability density functions.

A complete Gaussian probability density function defined as a linear combination of K number of Gaussians is represented as Eq. (13).

$$f(D_t = u) = \sum_{i=1}^{K} \omega_{i,t} \cdot \eta \left(u; \; \mu_{i,t}, \sigma_{i,t} \right) \tag{13}$$

where $\eta \left(u; \; \mu_{i,t}, \sigma_{i,t} \right)$ is the *i*th Gaussian component with intensity mean $\mu_{i,t}$ and standard deviation $\sigma_{i,t}$ and $\omega_{i,t}$ is the weight based on the *i*th component. Typically, K ranges from three to five depending on the available storage, and here, we use three as the value for K.

Equation (14) calculates the weight $\left(\omega_{i,t} \right)$, average $\left(\mu_{i,t} \right)$, and standard deviation $\left(\sigma_{i,t} \right)$ initiated when an image is input.

$$\left| D_{i,t} - u_{i,t-1} \right| \le T \cdot \sigma_{i,t-1} \tag{14}$$

In the above equation, if a deviation in the absolute value is lower than or the same as the standard deviation, T, which controls the threshold standard deviation rate, becomes 2.5. The next values are renewed if they are lower than threshold T times the standard deviation. In this paper, 2.5 is used as the threshold T value.

$$\omega_{i,t} = (1 - \alpha)\,\omega_{i,t-1} + \alpha \quad \mu_{i,t} = (1 - \rho)\mu_{i,t-1} + \rho D_t$$
$$\sigma_{i,t}^2 = (1 - \rho)\,\sigma_{i,t-1}^2 + \rho \left(D_t - \mu_{i,t} \right)^2 \tag{15}$$

In the above equation, ρ is the training rate and is calculated using Eq. (15).

$$\rho \approx \frac{\alpha}{\omega_{i,t}} \tag{16}$$

Here, $0 \leq \alpha \leq 1$ the user's defining training rate.

If the value in Eq. (14) is higher than the standard deviation which is the modulus deviation T times the threshold, the next expression is renewed.

$$\omega_{i,t} = (1 - \alpha)\,\omega_{i,t-1} \tag{17}$$

Finally, an object is separated from its background using the following expression.

$$\sum_{k=i_1}^{i_M} \omega_{k,t} \geq \Gamma \tag{18}$$

When the total weight of a pixel is more than Γ, it is identified not as an object but an background.

3 Experimental Results

The environment for our experiment was Visual C++ on a 3.4 GHz Intel Pentium with Dual CPU, 2 GB RAM and Windows operating system. The proposed method is experimental to image sequences that were captured from IJUData (indoor and outdoor environment with various changes in constituents and illumination, using an HVR-2030 webcam).

3.1 Method Based on Difference Images and Comparison Test

Figure 2 shows the results of a comparison test between a method using difference images and a general eigen-background method and the method proposed in this paper when frames such as those in Fig. 2a are input. Figure 2b show the result of

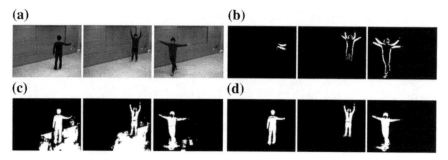

Fig. 2 **a** Input images, **b** difference image, **c** result of general eigen-background methods and **d** images obtained using proposed method

using a difference image that is different with the previous frame. As you can see in the figure, a body part with no movement is not detected, and only a moving hand is detected. Figure 2c are the results of detection using a general eigen-background. As you can see in the figure, because the principal components have a high explanation rate with lots of noise and shadows, the noise and shadows are detected from the background. Figure 2d show a method using an eigen-background that is improved through the use of the clustering proposed in this paper. No moving body parts are detected accurately, as can be seen in the figure. Because the proposed method in this paper selects a useful eigenvalue that separates a background and object, it does not leave after images. It also removes noise and shadows, and detects an object robustly.

3.2 Experiment on Change in Number of Objects Depending on Various Backgrounds and Lighting Conditions

To analyze the efficiency of the proposed system, we experimented on changing the number of objects according to various backgrounds and lighting conditions. Figure 3a show the results of detecting two passing pedestrians passing under natural outdoor lighting. Figure 3b are the results of an experiment in a hallway under mixed natural and indoor lighting conditions. Figure 3c show the results of an experiment done under indoor lighting with a complicated background. Figure 3a shows that an object is detected well under natural light but is recognized incorrectly as a shadow under robust light. Figure 3b shows that under a mixture of natural and indoor lighting, the system detects objects without shadows. Figure 3c shows the result of detecting an object from a large number of other objects with a complicated background. Although many parts of the correct object in Fig. 3a are similar to the background region, the object region is correctly detected.

Figure 3d shows the object detection results for a situation in which the colors of the cars and roads are not distinguishable due to a bright natural light. According to our results, the improved eigen-background proposed in this paper

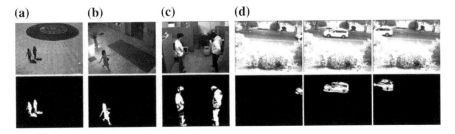

Fig. 3 Experimental results obtained using various backgrounds and natural lighting

separates light and noise from a background efficiently based on the environment, shadow conditions under different lighting, and a non-moving object. The method proposed in this paper solves the problem of an after image, and the inability to detect non-moving objects. It also does not require artificial background initialization and detects objects from noise and shadows more strongly than a method using a general eigen-background.

4 Conclusion

In this paper, we constructed an improved eigen-background to detect an object from its background. First, we constructed various background images as training data to detect an object adaptively under changes in the environment. We then analyzed the training data using PCA and selected the principal components used in analyzing an object from its background using a clustering algorithm.

We solved the problems in which a complicated background is recognized incorrectly as an object using a principal component, which analyzes a background and an object well without depending on the explanation rate and is the traditional method used to analyze the main elements in object detection. The main selected component is used as an eigen-background and is input to the MOG. As a result of convoluting an eigenvector and an input image used as an input value, the existing object detection system using MOG responds sensitively to changes in light. This reduces errors such as noise generated in an image or when an object is recognized as a background when it has little or no movement.

References

1. Yyilmaz A, Javed O, Shah M (2006) Object tracking: a survey. ACM J Comput Surv 38:1–45
2. Stein AN, Herbert M (2009) Local detection of occlusion boundaries in video. Image Vis Comput 27:514–522
3. Horn BKP, Schunck NG (1981) Determining optical flow. Artif Intell 17:185–203

4. Itti L, Koch C, Niebur E (1998) A model of saliency-based visual attention for rapid scene analysis. IEEE Trans PAMI 20:1254–1259
5. MCHugh JM, Konrad J, Saligrama V, Jodoin PM (2009) Foreground-adaptive background subtraction. IEEE Signal Process Lett 16:390–393
6. Rymel J, Renno J, Greenhill D, Orwell J, Jones GA (2004) Adaptive eigen-backgrounds for object detection. In: International conference on image processing, vol 3, pp 1847–1850
7. Zhang J, Zhuang Y (2007) Adaptive weight selection for incremental eigen-background modeling. In: International conference on multimedia and expo, pp 851–854
8. Turk MA, Pentland AP (1991) Face recognition using eigenfaces. In: IEEE conference on computer vision and pattern recognition. pp 586–591
9. Cheung SCS, Kamath C (2005) Robust techniques for background subtraction in urban traffic video. J Appl Signal Process 1:2330–340
10. Yang MS, Wu KL, Hsieh JN, Yu J (1999) Alpha-cut implemented fuzzy clustering algorithms and switching regressions. IEEE Trans Syst 18:1117–1128

Trajectory Calculation Based on Position and Speed for Effective Air Traffic Flow Management

Yong-Kyun Kim, Deok Gyu Lee and Jong Wook Han

Abstract Trajectory modeling is basic work for 4D-Route modeling, conflict detection and air traffic flow management. This paper proposes a novel algorithm based on coordinate prediction for trajectory calculation. We demonstrated through simulations with flight position and ground speed, and experimental results show that our-trajectory calculation exhibits much better performance in accuracy.

1 Introduction

The air traffic management (ATM) system improves safety and efficiency of air traffic by preventing collisions against other aircraft, obstacles and managing aircraft's navigation status. To achieve these purpose the air traffic control system identifies aircraft and displays its location, displays and distributes flight plan data, provides flight safety alerts, and processes controller's requests.

Despites technological advances in air navigation, communication, computation and control, the ATM system is still, to a large extent, built around a rigidly structured airspace and centralized, mostly human-operated system architecture.

The accuracy route calculation in En-route airspace impacts ATM route predictions and Estimated Times of Arrival (ETA) to control fix points. For the airspace controllers, inaccurate trajectory calculation may results in less-than-optimal maneuver advisors in response to a given traffic management problem [1].

Y.-K. Kim (✉) · D. G. Lee · J. W. Han
Electronics and Telecommunications Research Institute, 161 Gajeong-dong,
Yuseong-gu, Daejeon, Korea
e-mail: ykkim1@etri.re.kr

D. G. Lee
e-mail: deokgyulee@etri.re.kr

J. J. (Jong Hyuk) Park et al. (eds.), *Multimedia and Ubiquitous Engineering*,
Lecture Notes in Electrical Engineering 240, DOI: 10.1007/978-94-007-6738-6_47,
© Springer Science+Business Media Dordrecht(Outside the USA) 2013

There has been significant research in the fields of air traffic flow management and basic trajectory modeling. With the rapid development of air traffic management technology, more and more aircraft would fly in the sky simultaneously [2].

One key factor in route modeling is the uncertainty in present and future estimations of the velocity and position vectors of aircraft.

Many route calculation algorithm account for this uncertainty [3]. On the other hand, only limited research has been done on stochastic route calculation. Given the limited literature on route calculation under uncertainty, some studies concerned with developing probabilistic route calculation model s conclude by stating that there is a need to better understand and utilize route probability estimations in route calculation algorithm [4, 5].

In this paper we propose efficient En-route trajectory calculation algorithm. The remainder of this paper is structured as follows. In the next section, trajectory calculation techniques and theoretical background about trajectory calculation is presented. We present some experimental results of our proposed scheme in Sect. 3, and finally conclude with the conclusion in Sect. 4.

These uncertainties may be due to sensor noise or due to unpredictable disturbances such as wind.

2 Trajectory Calculation Techniques

2.1 Trajectory Calculation Theory

First, consider a fairly simplified model for trajectory calculation problem. A flow is defined as a set of flights between a departure airdrome and an arrival airdrome. The following simplifications are made (Fig. 1).

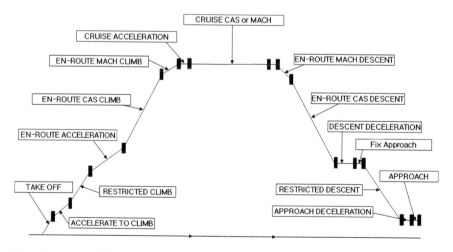

Fig. 1 Typical model of a trajectory calculation modeling

Trajectory Calculation Based on Position and Speed

The airspace is considered as an Euclidean space, where all airdrome are at altitude 0. Latitudes and longitudes on the ellipsoid earth surface are converted into (x, y) coordinates by a stereographic projection, and the altitude in feet shall be our z coordinate [4].

2.2 Parameters for Trajectory Calculation

For trajectory calculation, we must consider about concept of speed and conversion between other speeds, speed variation by altitude changing and wind parameter.

Fir, airspeed is the speed of an trajectory calculation relative to the air. Among the common conventions for qualifying are: Indicated Airspeed (IAS), Calibrated Airspeed (CAS), True Airspeed (TAS) and Ground speed (GS).

IAS is the airspeed indicator reading uncorrected for instrument, position, and other errors. From current European Aviation Safety Agency (EASA) definitions: Indicated airspeed means the speed of an aircraft as shown on its pitot static airspeed indicator calibrated to reflect standard atmosphere adiabatic compressible flow at sea level uncorrected for airspeed system errors.

Most airspeed indicators show the speed in knots (i.e. nautical miles per hour). Some light aircraft have airspeed indicators showing speed in miles per hour.

CAS is indicated airspeed corrected for instrument errors, position error and installation errors.

CAS is CAS values less than the speed of sound at standard sea level (661.4788 kn) are calculated as follows:

$$V_C = A_0 \sqrt{5 \left[\left(\frac{Q_C}{P_0} + 1 \right)^{\frac{2}{7}} - 1 \right]} \tag{1}$$

where
V_C is the CAS.
Q_C is the impact pressure sensed by the pitot tube.
p_0 is 29.92126 inches Hg at standard sea level.
A_0 is 661.4788 kn; speed of sound speed at standard sea level.

This expression is based on the form of Bernoulli's equation applicable to a perfect, compressible gas. The values P0 and A0 are consistent with the International Standard Atmosphere (ISA).

TAS is the physical speed of the aircraft relative to the air surrounding the aircraft. The true airspeed is a vector quantity. The relationship between the true airspeed (V_t) and the speed with respect to the ground (V_g) is

$$V_t = V_g - V_w \tag{2}$$

where
V_w is wind speed vector.

Aircraft flight instruments, however, don't compute true airspeed as a function of groundspeed and wind speed. They use impact and static pressures as well as a temperature input. Basically, true airspeed is calibrated airspeed that is corrected for pressure altitude and temperature. The result is the true physical speed of the aircraft plus or minus the wind component. True Airspeed is equal to calibrated airspeed at standard sea level conditions.

The simplest way to compute true airspeed is using a function of Mach number

$$V_t = A_0 \bullet M \sqrt{\frac{\tau}{\tau_0}} \tag{3}$$

where M is Mach number, τ is Temperature (kelvins) and τ_0 is Standard sea level temperature (288.15 K)

Second, speed variation by altitude changing means that when aircraft are climb or descent.

The rate of climb (RoC) is the speed at which an aircraft increases its altitude. This is most often expressed in feet per minute and can be abbreviated as ft/min. Else where, it is commonly expressed in meters per second, abbreviated as m/s. The rate of climb in an aircraft is measured with a vertical speed indicator (VSI) or instantaneous vertical speed indicator (IVSI). The rate of decrease in altitude is referred to as the rate of descent or sink rate. A decrease in altitude corresponds with a negative rate of climb.

There are two airspeeds relating to optimum rates of ascent, referred to as Vx and Vy.

Vx is the indicated airspeed for best angle of climb. Vy is the indicated airspeed for best rate of climb. Vx is slower than Vy.

Climbing at Vx allows pilots to maximize the altitude gain per unit ground distance. That is, Vx allows pilots to maximize their climb while sacrificing the least amount of ground distance. This occurs at the speed for which the difference between thrust and drag is the greatest (maximum excess thrust). In a jet airplane, this is approximately minimum drag speed, or the bottom of the drag vs. speed curve. Climb angle is proportional to excess thrust.

Climbing at Vy allows pilots to maximize the altitude gain per unit time. That is, Vy, allows pilots to maximize their climb while sacrificing the least amount of time. This occurs at the speed for which the difference between engine power and the power required to overcome the aircraft's drag is the greatest (maximum excess power). Climb rate is proportional to excess power.

Vx increases with altitude and Vy decreases with altitude. Vx = Vy at the airplane's absolute ceiling, the altitude above which it cannot climb using just its own lift.

Last, we consider about wind parameters. Wind parameter can divide two components (weather fonts and thermal wind) on a large scale.

Weather fronts are boundaries between two masses of air of different densities, or different temperature and moisture properties, which normally are convergence zones in the wind field and are the principal cause of significant weather. Within surface weather analyses, they are depicted using various colored lines and symbols.

The air masses usually differ in temperature and may also differ in humidity. Wind shear in the horizontal occurs near these boundaries. Cold fronts feature narrow bands of thunderstorms and severe weather, and may be preceded by squall lines and dry lines.

Cold fronts are sharper surface boundaries with more significant horizontal wind shear than warm fronts. When a front becomes stationary, it can degenerate into a line which separates regions of differing wind speed, known as a shear line, though the wind direction across the feature normally remains constant. Directional and speed shear can occur across the axis of stronger tropical waves, as northerly winds precede the wave axis and southeast winds are seen behind the wave axis.

Horizontal wind shear can also occur along local land breeze and sea breeze boundaries.

Thermal wind is a meteorological term not referring to an actual wind, but a difference in the geostrophic wind between two pressure levels p_1 and p_0, with $p_1 < p_0$; in essence, wind shear. It is only present in an atmosphere with horizontal changes in temperature.

In a barotropic atmosphere, where temperature is uniform, the geostrophic wind is independent of height. The name stems from the fact that this wind flows around areas of low (and high) temperature in the same manner as the geostrophic wind flows around areas of low (and high) pressure.

$$f_{VT} = K \times \nabla(\phi_1 - \phi_0) \qquad (4)$$

where the ϕ_x are geopotential height fields with $(\phi_1 > \phi_0)$, f is the Coriolis parameter, and K is the upward-pointing unit vector in the vertical direction. The thermal wind equation does not determine the wind in the tropics. Since f is small or zero, such as near the equator, the equation reduces to stating that $\nabla(\phi_1 - \phi_0)$ is small. This equation basically describes the existence of the jet stream, a westerly current of air with maximum wind speeds close to the tropopause which is (even though other factors are also important) the result of the temperature contrast between equator and pole.

3 Experimental Results of Proposed Scheme

This section describes the method for computing the various parameters used to compute the position, speed and our route calculation algorithm.

First of all, we need aircraft's position and its speed. Aircraft's position consists of latitude and longitude.

For calculating aircraft's position at specified time, we need airspeed (TAS or GS), wind speed and wind direction.

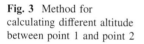

Fig. 2 Result of position prediction algorithm simulation

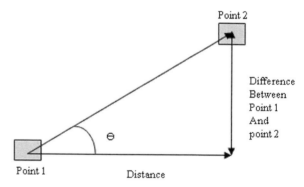

Fig. 3 Method for calculating different altitude between point 1 and point 2

Aircraft's position can calculate using vincenty's formula at specified time.

For using position prediction algorithm, predict aircraft's position is shown in the Fig. 2.

And now we consider about altitude for increase accuracy (Fig. 3).

By computing climb rate, we can calculate aircraft's position in terms of other altitude.

Trajectory Calculation Based on Position and Speed

```
D:\Route\Route>Climb
Input Start Fix Name : SEL
SEL 's Altitude <ft>: 4000

Fix : SEL
1265542E  372449N
Altitude : 4000.000000 <ft>

Input Stop Fix Name : BELMI
BELMI 's Altitude <ft>: 6000

Fix : BELMI
1265929E  371249N
Altitude : 6000.000000 <ft>

WindSpeed<Knot> : 37
Wind Direction<WindFrom, Degree> : 157
True Air Speed<Knot> : 700

Two Fix Point Distance is 22.886766 Km ( 75087.814865 ft )
Tracking Angle is 2.895073 Degree

Ground Speed is 1237.226827 ( ft/sec )
Estimate Time is 60.690419 ( sec )


   0 Sec     372447N 1265540E
  15 Sec     372149N 1265637E
  30 Sec     371851N 1265733E
  45 Sec     371553N 1265829E
  60 Sec     371255N 1265925E
  60 Sec     371247N 1265927E
```

Fig. 4 Result of climb late compute algorithm simulation

Fig. 5 Result of trajectory calculation using position and speed

```
Route Information...
BIGOB -> GOTLO -> BULLS -> KAKSO -> SEL
Time : 250

Aircraft Position 250.000000 Sec
BIGOB -> GOTLO -> BULLS ->  351640N 1283610E
```

Climb rate is calculated as below:

$$\text{Climb rate} = TAS \times \sin\theta \tag{5}$$

Using climb rate, trajectory calculation results in shown in Fig. 4.

With the position and speed, the trajectory of the aircraft can be derived, which is shown in Fig. 5.

4 Conclusion

In this paper, we propose trajectory prediction to accurately predict and calculate trajectory a maneuvering aircraft. For increase estimation, we consider wind speed, wind direction and altitude.

From now on, it is further suggested that the proposed algorithm may be extended to the trajectory modeling, which may further improve 4-D trajectory prediction.

Acknowledgments This research was supported by a grant (07항공-항행-03) from Air Transportation Advancement Program funded by Ministry of Land, Transport and Maritime affairs of Korean government.

References

1. Banavar S, Grabbe SR, Mukherjee A (2008) Modeling, optimization in traffic flow management. Proc IEEE 96(12):2060–2080
2. James KK, Lee CY (2000) A review of conflict detection and resolution modeling methods. IEEE Trans Intell Transp Syst 1(4):179–189
3. Adan EV, Erwan S, Senay S (2009) A two-stage stochastic optimization model for air traffic conflict resolution under wind certainty. In: 28th digital avionics systems conference 2009, pp 2.E.5-1–2.E.5-13
4. Lin X, Zhang J, Zhu Y, Liu W (2008) Simulation study of algorithm for aircraft trajectory prediction based on ADS-B technology. In: Asia simulation conference—7th international conference on systems simulation and scientific computing, pp 322–327
5. Terence SA (2007) A trajectory algorithm to support en route and terminal area self-spacing concepts, NASA/CR-2007-214899

Part VII
Multimedia Entertainment

Design and Implementation of a Geometric Origami Edutainment Application

ByeongSu Kim, TaeHun Kim and JongHoon Kim

Abstract Edutainment lies at the intersection of games and learning and includes the benefits of both. To increase students' interest in mathematics and improve their problem-solving abilities, we have designed and implemented a geometric origami edutainment application. This application is based on three heuristic axioms and simple data structures. Students can create any angle or polygon by folding/unfolding the 2D paper displayed on the screen of the Android based system. This application has the benefit of an intuitive interface, and requires a higher level of thinking from the students than that required by traditional origami. We created the application to entertain the students, but also to assist in improving their mathematical problem-solving abilities.

Keywords Origami · Geometry · Problem-solving · Edutainment · Android

1 Introduction

The digital age in which we currently reside is changing rapidly, and such change will only get faster and more unpredictable as time goes on. In the education field, digital tools are being used to a greater extent in learning activities. In particular, games have had such a large influence on students that they have changed the way these students learn today. Teachers utilize games to enhance the motivation and engagement of students' learning [1].

B. Kim · T. Kim · J. Kim (✉)
Department of Computer Education, Teachers College,
Jeju National University, Jeju, Korea
e-mail: jkim0858@jejunu.ac.kr

B. Kim
e-mail: pigpotato79@naver.com

T. Kim
e-mail: gtranu@naver.com

J. J. (Jong Hyuk) Park et al. (eds.), *Multimedia and Ubiquitous Engineering*,
Lecture Notes in Electrical Engineering 240, DOI: 10.1007/978-94-007-6738-6_48,
© Springer Science+Business Media Dordrecht(Outside the USA) 2013

Recently, smartphones and tablet PCs have been identified as emerging educational materials. Students carry their own mobile devices, so that they can access a variety of different multimedia sources anywhere and anytime. For example, if a digital game that a student enjoys playing is related to learning English and it can improve his/her English speech, it is not only a game, but also useful learning material. We refer to this kind of game as edutainment.

Edutainment is a hybrid genre that relies heavily on visuals and narratives or game-like formats, but also incorporates some type of learning objective [2]. In this study, we developed a geometric origami edutainment application to enhance students' interest in learning geometry and to improve their problem-solving abilities.

Origami is the Japanese name for the centuries-old art of folding paper into representations of birds, insects, animals, plants, human figures, inanimate objects, and abstract shapes [3]. Most origami designs are attempted by a combination of trial and error and/or heuristic techniques based on the folder's intuition. However, as the designs include various patterns, they could be presented as algorithms based on a set of mathematical conditions [4]. There are many other options for describing shapes other than origami, such as a set of polygon layers. The most attractive feature of origami, however, is that a wide variety of complex shapes can be constructed using a few axioms, simple fixed initial conditions, and one mechanical operation (folding) [5].

2 Application Design and Development

2.1 Learning Content

Having extracted learning content related to geometric features from the school curriculum of the Republic of Korea [6], we created problems based on the geometric content and reformatted these for implementation on the Android system. The problems are displayed on the screen at each level, from easiest to most difficult, in the same order as the learning content:

- Plane figures (right angle, right triangle, square, 45° angle);
- Triangles (isosceles triangle, equilateral triangle, acute triangle, obtuse triangle);
- Quadrilaterals and polygons (trapezoid, parallelogram, rhombus, rectangle, equilateral polygon).

2.2 Design of Android Application

An activity is the basic unit from which the user interface of the Android application is constructed. In other words, an activity is considered to be the screen that a user observes and deals with when responding to system events. In this

Design and Implementation of a Geometric Origami Edutainment Application   389

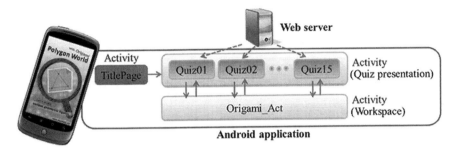

Fig. 1 Structure of activities and interactions

application, there are three main activities: the title page, quiz presentation, and workspace (see Fig. 1).

When a user enters the title page and then starts a quiz presentation, the application attempts to download the data for the first quiz from the Web server. The user can then read the requirements of the quiz and work out a solution in the workspace by manipulating the origami to create a geometric figure for submission as the answer of the current quiz. For example, if the quiz requires a right angle to be constructed, the user would make several folds, thereby creating the angle and then submit the answer by touching three intersection points made by the folded lines and pressing the "submit" button. Figure 2 depicts an activity diagram for Origami_Act, which shows the flow and logic of the computations and presentations by the system when responding to a user action.

Implementation of application. Figure 3 shows screen capture images of a quiz presentation and the workspace. Although many people can fold origami in a variety of different ways to create animals, flowers, insects, or other shapes, it is difficult and complex for programmers to design a 3D application that can be manipulated in the same way as traditional origami. In this study, we imposed certain conditions when running the origami application to allow intuitive manipulation on Android devices.

- The paper represents a 2D environment in which six operations can be performed: restart, pick, fold, unfold, cancel, and submit.
- The user can fold the paper inwards and unfold it outwards.
- The user can fold the paper from 'point to point' (see Fig. 4). However, the user cannot fold the one side over to the other side.
- The user cannot make two consecutive folds on the paper; in other words, the folded paper must be unfolded first, before the user can fold it again.

When a user folds the paper, the shape of the folded paper is shown as one of three different types (see Fig. 5).

When the user presses the 'PICK' button, four open vertexes appear at the four corners of the paper, and the user can then touch one of these. Let the first vertex selected by the user be $V1$. The user then selects a second point, $V2$, within or on the edge of the paper. Three different types of folded paper are formed according to

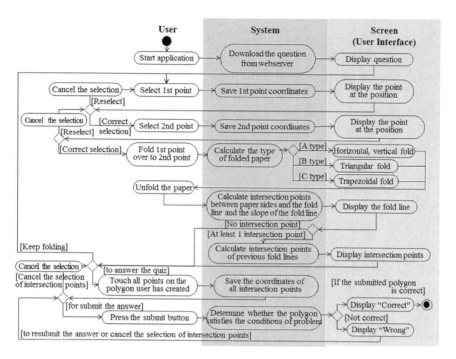

Fig. 2 Activity diagram for Origami_Act

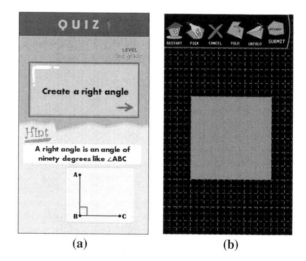

Fig. 3 Screen capture images of a quiz presentation and the user's workspace. **a** Quiz presentation. **b** Workspace

the positions of V1 and V2. Let the neighboring vertex of V1 in a clockwise direction be R, and that in an anticlockwise direction be L. Further, let Q(R) be the quadrant of the circle with center R and Q(L) be the quadrant of the circle with

Design and Implementation of a Geometric Origami Edutainment Application 391

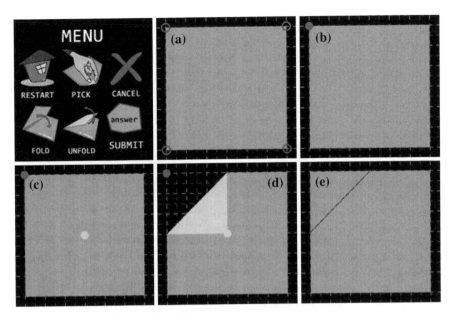

Fig. 4 Operation menu (*top left*) and the shape of the origami based on the user's action. **a** State when the user presses the 'PICK' button. **b** State when the user touches an open point. **c** State when the user touches an inner point. **d** State when the user presses the 'FOLD' button. **e** State when the user presses the 'UNFOLD' button

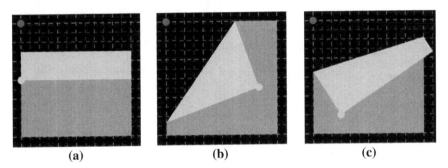

Fig. 5 Three variations of folded paper. **a** Type A: Rectangular type. **b** Type B: Triangular type. **c** Type C: Trapzoid type

center L. We defined the following simple axioms with respect to the patterns of the folded paper.

$$\text{Type A: } V2 \in \overline{V1R} \cup \overline{V1L} \tag{1}$$

$$\text{Type B: } V2 \in Q(R) \cap Q(L) \tag{2}$$

$$\text{Type C: } V2 \in (Q(R) \cup Q(L)) - (Q(R) \cap Q(L)) - (\overline{V1R} \cup \overline{V1L}) \tag{3}$$

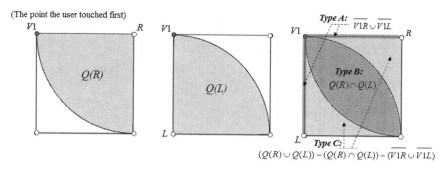

Fig. 6 Origami axioms with respect to the type of folded paper

If point V2 is on line V1R or line V1L described by Axiom (1), the paper is folded as Type A (rectangle type). If point V2 is on the intersection of quadrant Q(R) and Q(L) described by Axiom (2), then the paper is folded as Type B (triangle type). If point V2 does not lie on line V1R or line V1L, and is on the outside of the intersection of quadrant Q(R) and Q(L) as described by Axiom (3), the paper is folded as Type C (trapezoid type). Figure 6 illustrates these axioms.

The axioms defined above can be proved mathematically [7]. However, it was more convenient and simple to design and implement the application.

Example of playing the game. When users attempt to find the answer to a quiz, they should consider the following: How many times should the paper be folded? How can a fold line be made at a particular position? How can a line be created with the same length or angle as a previously created line? Using this process during the learning process stimulates the user's thinking. If the quiz requires a parallelogram to be constructed, the user can find a solution by carrying out several fold/unfold steps and then touching the vertexes to outline the parallelogram (see Fig. 7). While creating this shape, the user can also calculate the lengths of line segments using the grid in the background.

Once the user has created the parallelogram as the solution, he/she can submit it by pressing the 'SUBMIT' button. When the system receives the coordinates of the polygon vertexes, it calculates the distances and angles and checks that the submitted polygon meets the requirements of the quiz.

Data structure. When the user selects a point on the paper in the application, key information about the point is saved in a data structure. Assume a tuple P, with four different variables representing the point; that is, the key index, x-coordinate, y-coordinate, and array L containing the index numbers of lines on which P is located. This information is used to create the path of the polygon (see Fig. 8).

If the user folds the paper four times as in Fig. 8b, four points and four lines are created. The points are defined by the data depicted below:

P1 (1, 100, 50, L[1, 2])	P2 (2, 100, 150, L[2, 3])
P3 (3, 225, 50, L[1, 4])	P4 (4, 225, 150, L[3, 4])

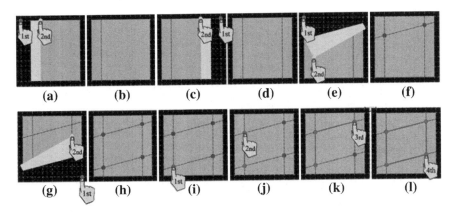

Fig. 7 Procedure for creating a parallelogram (PICK, FOLD, and UNFOLD mean pressing the corresponding menu buttons, while 'touch' implies the user's action). **a, c, e, g** PICK-touch-FOLD. **b, d, f, h** UNFOLD. **i** Touch ist point. **j** Touch 2nd point. **k** Touch 3rd point. **l** Touch 4th point

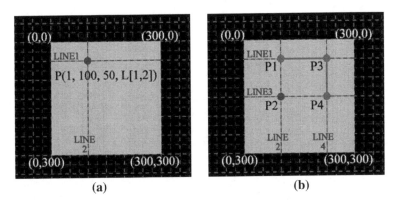

Fig. 8 Data structure and use of points. **a** Data saved for a point. **b** Creating a path when the user touches P1, P3 and P4 in order

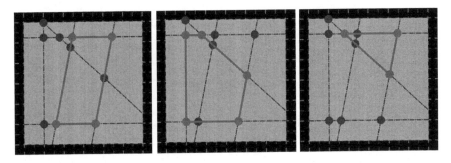

Fig. 9 Some *polygons* on the paper using the same points

When the user touches P1 and P3, the application checks array L of each point. If both arrays have the same index numbers of lines, they are connected with a red line. When the user touches another point, the application operates in the same way. In this algorithm, users can design any polygon using the points they have created by folding the paper and connecting the points (see Fig. 9).

3 Conclusion

In the digital age, the boundaries between games and learning applications have disappeared. In the geometric origami edutainment application we developed, users can have extraordinary experiences creating angles and polygons with digital paper. From an education perspective, drawing a polygon on real paper requires only one dimension of thinking, whereas creating a polygon using this application requires a higher dimension of thinking. We expect that this application can assist students by improving their geometric problem-solving abilities. In addition, it could be used as educational material in ICT based mathematics learning.

References

1. Garris R, Ahlers R, Driskell JE (2002) Games, motivation, and learning: a research and practice model. Simul Gaming 33(4):441–467
2. Okan Z (2003) Edutainment: is learning at risk? Brit J Educ Technol 24(3):255–264
3. Kasahara K, Misaki I (1967) Creative origami. Japan Publications, Tokyo
4. Lang RJ (1996) A computational algorithm for origami design. In: 12th annual ACM symposium on computational geometry. SCG, Pennsylvania, pp 98–105
5. Nagpal R (2001) Programmable self-assembly: constructing global shape using biologically-inspired local interactions and origami mathematics. PhD thesis, Department of Electrical Engineering and Computer Science, Massachusetts Institute of Technology
6. Ministry of Education, Science, and Technology (2008) The school curriculum of the Republic of Korea, Ministry of Education, Science, and Technology, Seoul
7. Alperin RC (2000) A mathematical theory of origami constructions and numbers. N Y J Math 6:119–133

Gamification Literacy: Emerging Needs for Identifying Bad Gamification

Toshihiko Yamakami

Abstract Gamification is a collection of game-design-origin know-how that facilitates service engagement. It is used in a wide range of applications, marketing, enterprise management, and education. The positive side of gamification provides useful tools for improving engagement. At the same time, there are negative sides of gamification when it is used excessively or maliciously. The author proposes the concept of gamification literacy.

1 Introduction

The market size of mobile social games in Japan has demonstrated a radical growth in the past couple of years. The investigation of best practices in mobile social games shows that the design is based on micro-management of goal-achievement cycles with intensive feedback systems. This intensive feedback system can be utilized as gamification in order to improve engagement in a wide range of services. However, excessive game techniques of mobile social games lead to some social problems.

This provides an important lesson teaching us that we have to master a new kind of literacy in order to manage gamification in the virtual world of consumer services and enterprise systems. The author proposes the concept of gamification literacy. The author examines some anomalies of mobile social games and discusses the components of gamification literacy.

The aim of this research is to identify a new type of literacy in the emergence of gamification in the virtual world.

T. Yamakami (✉)
ACCESS, Software Solution, 1-10-2 Nakase, Mihama-ku, Chiba-shi 261-0023, Japan
e-mail: Toshihiko.Yamakami@access-company.com
URL: http://www.access-company.com/

Literacy refers to the ability to read for knowledge, write coherently, and think critically about the written word (Wikipedia).

Game design has attracted a wider scope of audience because of its non-game applications. McGonigal discussed how game design techniques and mechanisms can solve real-world problems [1]. She explained how the combination of an artificial goal, a set of artificial rules, and feedback with voluntary participation creates challenges and fun for the user.

Literacy that deals with gamification has not been covered well in the past literature.

The originality of this paper lies in its examination of a new kind of literacy in service design using gamification.

2 Observation

2.1 Industry Landscape of Mobile Social Games

The market size of mobile social games in Japan grew on the order of 100 billion Japanese yen since 2010. The amount in 2012 is estimated by Mitsubishi-UFJ-Morgan-Stanley securities to be 400 billion Japanese yen. Early mobile social games that were designed for feature-phones were relatively simple. However, the rapidity of market growth provided rich cash flows and boosted the profits of early winners. The rich cash flows enabled rich graphic representations and intensive data mining that produced a completely different landscape with an intensive selection process.

2.2 Social Problems

The overheating market situation has brought about several social problems depicted in Table 1. The industry started to depend on heavy users that pay tens of thousands Japanese yen per month. In the case of GREE, 3.5 % of heavy users provide approximately 70 % of its revenue.

Table 1 Social problems

Aspect	Summary
Addicted users	Combination of higher goals and routine simple mini-games promote addiction with a daily routine repetition
Speculative systems	Gambling-like achievement using a very small probability of success with extreme, rare rewards promote speculative attitudes in users
Unfair probability adjustment	Completion gatcha, a gambling system, uses arbitrary probability adjustments with misleading expectation of completion

The top-ranking Japanese mobile social game vendors have enjoyed high profitability, approximately 50 %. This high profitability is driven by the heavy users that pays more than five thousand Japanese yen per month. The focus on charging heavy users is an easy way to create efficient revenue-generating engines. Therefore, the top-ranking game vendors have increased their dependence on charging heavy users in the past couple of years. The most popular technique is the completion gatcha. Completion gatcha is a method of awarding rare in-game items in mobile games only when the player has bought a full set of other in-game items. It is a kind of lottery game played on a mobile phone.

In May 2012, the Consumer Affairs Agency of Japan announced that completion gatcha had led legal issues. The Consumer Affairs Agency decided to ban "kompu gatcha" (completion gatcha) online games played on mobile phones starting from July 2012 under the Law against Unjustifiable Premiums and Misleading Representations.

3 Gamification Literacy

3.1 Gamification

The framework of a game defined by J. McGonigal for a game is depicted in Fig. 1.

It should be noted that this framework can be applied to a wide range of services including social services. When an appropriate goal and a set of rules can be designed, many services can be leveraged by the applications of game-origin techniques to increase user engagement.

The positive attitudes of human beings explored by positive psychology are depicted in Fig. 2.

This indicates that a human beings are an extraordinarily positive creatures with emphasis on internal rewards, hard work, an eagerness to help others, high engagement of promises, and so on.

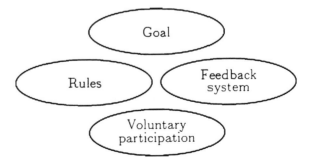

Fig. 1 Framework of a game (by J. McGonigal)

Fig. 2 Positive attitudes of human beings

All these attributes contribute to constructing positive societies. The engineering of hope with the expectation of achieving rewards drives people in the virtual world.

These positive attitudes represent the possibility that artificial goals and rules can accommodate great enthusiasm from users when managed appropriately. No fantasy is required when the appropriate feedback mechanism is ensured, because internal achievement provides high-level user motivation. This is a core factor that feature-phone based mobile social games in Japan demonstrated through radical user-base growth and revenue growth in the past couple of years, even though they provide simple multimedia representations that are inferior to those of smartphones.

3.2 Definition of Gamification Literacy

The components of gamification literacy are depicted in Table 2.

This shows how people have awareness of virtual-world persuasion. First, people love to compete. The virtual world amplifies this tendency through a wide range of visualizations. Also, the virtual world amplifies this through 24 h a day, 365 days a year of competition. Competition can be enhanced with globalization and anonymity.

Second, the virtual world can manipulate probability in selection and rewards. People easily build up false expectations in the artificial setting of the virtual world.

Third, people are susceptible to time. Time is something we have to measure in order to identify. People have weak points in the management of asynchronous events. This attribute can be easily exploited by virtual-world service providers. For example, we are sensitive to time-limited offers, one of the classic persuasion techniques.

Fourth, people are susceptible to a speculative mindset and gambling. The real-world has many constraints on gambling. The virtual world has very relaxed

Gamification Literacy: Emerging Needs

Table 2 Components of gamification literacy

Component	Description
Competition literacy	Awareness of control techniques using social competition
Probability literacy	Awareness of control techniques using probability, including misleading probability impression
Time management-based persuasion literacy	Awareness of in-game persuasion using time limitation and clock-based visualization techniques
Speculation literacy	Tolerance of speculative game mechanisms
Addiction literacy	Awareness of combination of low-level and high level game machismo that drives addiction to a game
Rareness-based persuasion literacy	Awareness of rare item-based persuasion
Social invitation literacy	Awareness of background persuasion techniques of friend invitation
Completion literacy	Awareness of in-game persuasion techniques using completing a set

regulations on gambling and speculation. Therefore, we have to raise our own awareness of gambling and the speculative mindset in the virtual world.

Fifth, people are susceptible to a routine-task-based stimulus. This can be easily constructed when a repetitive routine task is combined with a higher goal. Many mobile social games use routine-task techniques to leverage the addiction to a game. We need to be aware of this addiction-promoting process.

Sixth, the virtual world can coin a wide range of rarity. Rareness is another classic example of persuasion. In the virtual world, rareness is easily manipulated compared to in the real world. People need to increase their awareness of this type of persuasion in the virtual world.

Seventh, many service providers use the technique of social invitation. As Facebook reached one billion active users, social invitation became recognized as a powerful marketing tool. As people increase their encounters with social invitation in the virtual world, it increases the necessity for the awareness of social invitation. In the technique of social invitation, there are two different roles, inviters and invitees. Both sides need a new kind of awareness of the consequences of social invitation. Also, people need to master a kind of ability to refuse or ignore social invitations for a healthy social life in the virtual world. Social invitation can invoke social problems such as Ponzi schemes even in the real world. In the era of the socially connected Internet, it is more important to leverage social invitation literacy for a healthy Internet life.

Eighth, there is the mechanism of completion in the virtual world. As a user approaches the completion of a task, there comes a stronger desire to complete that task. This psychological process was exploited in mobile social games in Japan in 2011, and that led to the ban of completion gatcha in July 2012. Completion gatcha is considered to be a driving factor that enabled a multi-billion-dollar business for mobile social games in Japan. Generally, completion can be used to lure users to elicit actions to complete a task or collection. In order to protect users from paying extremely high amounts like hundreds of dollars per month, people need to become aware of this kind of persuasion technique in the virtual world.

The lessons that we learn from the rise of the mobile social game business in Japan include the power of data-mining capabilities for game-tuning. The today's data mining infrastructure provides a real-time analysis capability to collect behavior logs from millions of users and parse them for clues in order to tune games by the hour. In order to cope with this massive data-empowered infrastructure, end users need to have knowledge of behind-the-scenes engineering principles and to have the will-power to resist the temptations leveraged by these design principles. Gamification literacy provides them with the basic skills.

The list shown above is not exhaustive. As people's lives are merged with the virtual world, service providers in the virtual world continue to invent different patterns of persuasion. It is necessary to raise awareness of this increasing persuasion in the virtual world. It is also necessary to constantly maintain the set of gamification literacy in order to protect end users and stability of the virtual-world economy.

Not only consumers, but also workers in the workplace may experience more time staying in the virtual world. This raises an awareness of virtual-world workplace literacy. Gamification can leverage the performance of the virtual-world workplace. However, it is beyond the scope of this paper.

3.3 Gamification Classifier Framework

Examples of bad gamification in the mobile social game industry are depicted in Table 3.

In order to systematically deal with gamification literacy, the author provides a view model of the gamification classifier framework, as depicted in Fig. 3.

This multi-stage framework provides a skeleton that accommodates the flow-chart of gamification identification.

3.4 Education of Gamification Literacy

When the technology to impact end users on a massive scale emerged, we had to coin a new type of literacy in order to deal with it. Communication technology created radio and television technologies that provide real-time massive distribution of information. Such a type of real-time information broadcasting was impossible before the invention of that technology. The convenience of technologies also brought concerns such as brain-washing and biased broadcasting. These concerns built up the so-called media literacy, the literacy to read the behind-the-scene intention and provide unbiased judgments of broadcast information.

Gamification Literacy: Emerging Needs

Table 3 Examples of bad gamification

Item	Description
Untrustworthy manipulation of gambling	Setting up rewards as a gambling system. As a direct award, the user is entitled with the right to enter a lottery. The probability of winning the lottery may be controlled in an arbitrary way, such as controlling the winning probability to be extremely small when a user gets closer to completion of a target set
Untrustworthy tuning of a game	Changing game parameters in an untrustworthy way without notices
Intensive competition	Putting an extremely rare item for daily top ranking, weekly top ranking, or monthly top ranking users
Intensive time-limited offer	Setting a bonus time such as gaining double points or triple points for a limited time span, such as five minutes or ten minutes to heat-up the competition
Arbitrary management	Creating arbitrary parameters, levels, enemies. Arbitrarily raising the upper limit to lure consumers such as by giving the impression of a final limit of 100, but then raising it to over 100 after a certain time frame
False impression	Setting up non-human characters to be listed in the top rank. Arbitrary management of the "rareness" of items
Extravagant advertisement	Misleading advertisement of the rareness of gift items, time-limited offers, and so on

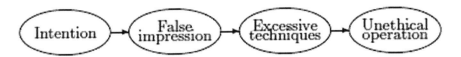

Fig. 3 Gamification classifier framework

When Internet technology started to flourish, many people experienced new types of conflicts such as flaming on bulletin boards. In addition, new areas of concern about privacy protection on the Internet emerged. These concerns built up the so-called Internet literacy, the literacy to detect privacy issues and avoid unnecessary interpersonal conflicts in the social interactions leveraged by Internet technology.

As social service engineering emerges and the stay time on social services becomes longer, it is necessary to deal with behind-the-scenes social service techniques. The author calls this gamification literacy, with emphasis on the back-end technology that enhances user engagement in social services.

A detailed discussion of gamification literacy is beyond the scope of this paper. The author hopes that coining this new terminology will raise awareness of the impacts, and risks of social service engineering that has come to prevail on this planet in the past couple of years. The author feels that this is a starting point to build up a gamification literacy that will lead to part of modern literacy education in the coming years.

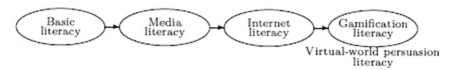

Fig. 4 Transition of literacy

The transition of literacy as the methods of massive real-time proliferation of information raised is depicted in Fig. 4.

4 Discussion

4.1 Advantages of the Proposed Approach

The mass media industry has been challenged by emerging Internet services. Not only for information retrieval, but also social interaction is now empowered by Internet services including social network services such as Facebook, Twitter, and so on.

The stay time in such virtual world social lives continues to increase. There is no foreseeable obstacle to prevent its growth.

Additionally, many teenagers are becoming digital citizens who begin their primary social lives in the virtual world.

These changes have brought about concerns about how we can develop appropriate manners and techniques to deal with a new style of social life.

The increase stay time in social lives in the virtual world leverages a strong demand for social service engineering. New growth in social service engineering has both positive and negative side effects.

For the positive side, the services utilizes advanced computer and communication technologies with anytime and anyplace capabilities to improve the social quality of life. For the negative side, the immaturity of social service engineering may damage social lives and the virtual world economy with inappropriate use of technology. Some negative aspects are already visible in the heating-up of the mobile social game industry in Japan.

The author coins the concept, gamification literacy, in order to highlight the importance of awareness of new types of literacy we need for the new type of social life that exists today.

Gamification literacy helps people:

- increase their willpower to resist temptations in online services with awareness of backend mechanisms,
- prevent themselves from paying extremely high bills through awareness of operational manipulation,
- prevent themselves from becoming network service addicts through awareness of backend persuasion methodologies.

4.2 Limitations

This research is a descriptive study. The quantitative measures for verifying multiple aspects of gamification literacy discussed in this paper remain for further study.

Concrete education design methodology of gamification literacy is beyond the scope of this paper.

5 Conclusion

The game industry needs to cope with this problem through their self-regulation and ethics. At the same time, considering the huge social graph and rapid propagation of word-of-mouth mechanisms on social network services, it is important to raise awareness on the consumer side. Not only for games, but also a wide range of social services can utilize the general principles of gamification.

The author coins the concept, gamification literacy, to highlight the needs for such a new awareness and techniques for daily life.

Reference

1. McGonigal J (2011) Reality is broken: why games make us better and how they can change the world. The Penguin Group, New York

Automatic Fixing of Foot Skating of Human Motions from Depth Sensor

Mankyu Sung

Abstract This paper presents a real time algorithm for solving foot skating problem of a character whose motions are controlled in on-line puppetry manner through motion depth sensor such as Kinect. The IR (Infrared Light) projection-based motion sensor is very sensitive to occlusion and light condition. One significant artifact from these characteristics is foot skating, which means that feet are floating over the ground even when the character is standing still. Our algorithm is working as a post processing that applies the orientation data from sensors to a character first, and then keeps monitoring the current character status and applying foot skating solver next whenever the problem occurs. For this, we propose an adaptive threshold method that make positional and velocity threshold change automatically. Inter-frame smoothing is then followed to make sure there is no discontinuity on the motions.

1 Introduction

The emergence of new motion sensor technology provides users a new experience of a natural user interface for interactive contents. A primal example is Microsoft's *Kinect* [1]. Ever since it was introduced to video game community, it has drawn a lot of attention for its effectiveness in capturing movement. Natural interaction with contents using full body increases immersive feeling to the contents, which is a critical point for attracting people. The current skeleton tracking of Kinect sensor is based on the *depth image* [2, 3]. The depth image represents the distance map from the sensor to the environment including users. Through the machine learning algorithm or by solving numerical kinematic configuration of users, most of the sensors are able to obtain the joint positions and orientation at real time rate.

M. Sung (✉)
Keimyung University, 1895 Dalgubeol-Daero, Daegu 704-701, Korea
e-mail: mksung@kmu.ac.kr

J. J. (Jong Hyuk) Park et al. (eds.), *Multimedia and Ubiquitous Engineering*,
Lecture Notes in Electrical Engineering 240, DOI: 10.1007/978-94-007-6738-6_50,
© Springer Science+Business Media Dordrecht(Outside the USA) 2013

Current motion sensors, however, do have some limitations. Because they are not a motion capture device—which use more than eight optical cameras to compensate the occlusion problem—the data from sensor has a jittering and significant artifacts such as foot skating [4]. The foot skating means that one or both feet are floating over the ground even when the character is standing still. This is the most disturbing artifact in character animation. This artifact comes from the several reasons. First, estimated depth data from the sensor might have some errors. This is inevitable because the sensor, which uses IR (Infrared sensor) projection, is very sensitive to light condition and occlusions. Second, the skeleton tracking algorithm might have some bugs. Although current depth sensors provide SDK for building applications that need skeleton tracking data, the orientation and position data often have low confidence values [5].

In this paper, we address this foot skating problem. Our approach is not targeting on improving the existing depth camera or accuracy of tracking algorithm. Instead, we try to solve this problem as a post processing job. That is, our algorithm is working right after applying original data onto a character. We then keep checking the character status after applying original data. If any artifact such as foot skating is happening on the character, we trigger the foot skating solver to fix it. Figure 1 shows the result of our algorithm. Although jerkiness of motion data can be smoothed out through a filtering technique, the foot skating cannot be solved with ease. First, we need to figure out when foot skating happens. Second, if foot skating occurs, then we need to determine the exact foot position. Third, we have to change the leg configuration in order to locate the foot to the position exactly while keeping original leg configuration as much as possible. This might have a lot of answers, but we need to choose right one based on our criteria. At the final step, we need to minimize the pop-out effect that might be caused due to the change of configuration.

In our algorithm, we define the *constrained frames* as frames that one or both feet touch the ground. We recognize the constrained frame by analyzing the foot velocity and distance from the ground. Under the condition that in general the foot plants continue for multiple frames, once the constrained frames start, then we use the very first foot position as the target foot position. Then, we move the root joint to the target when the target is located too far away. In this process, we apply simple moving averaging (SMA) technique to minimize discontinuity. Next, we apply the special IK solver to locate the foot to the goal position. The IK solver

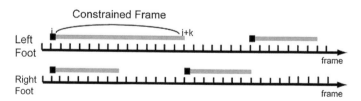

Fig. 1 In general, the constrained frames continue for multiple frames. We choose the foot position of first frame of each constrained frame blob as a target

Automatic Fixing of Foot Skating

should consider the skin mesh collapsing during computation so that the skin mesh does not twist. Finally, we apply rotation interpolation with neighboring frames for cleanup any pop-out.

2 Related Work

For obtaining depth image, several different approaches have been used for the camera. Stereo, Time-of-flight (TOF) and IR-structured-light method are three major technologies for implementing the depth camera today [6, 7]. In the case of Kinect, it uses the continuously projected infrared structured light to estimate 3D structure of objects [8, 9]. Currently, it is a unique consumer hardware that has interactive rate with full range of body types and sizes.

Foot skating has been a nagging problem in character animation community for a long time, especially for animators using motion capture data. Kovar et al. proposed an explicit solver for the problem [10]. They use an analytic IK solver which adjusts leg length to keep the foot to an exact position on the ground for the motion capture data. In our approach, we use the similar fast analytic IK solver to obtain the hinge orientation of leg. However, one fundamental difference is that our foot skating is working in on-line manner. Therefore, our algorithm does not have any information about next frames whereas motion capture data have full information about previous and future data.

3 Algorithm

3.1 Adaptive Constrained Frame Detection

Our foot skating cleanup algorithm consists of five steps. First step is to detect constrained frame. Essentially, this step checks whether the current character's feet are contacting the ground or not. We have two different ways for doing it. First, we can do all checking on the sensor domain. From the user's foot position received from sensor, we can figure out if its foot is touching the ground or not. Second, we can do that on the character domain. After applying data on the character first, we check the current character's foot position in the virtual environment. The sensor domain computation is highly risky because it is very sensitive to the sensor location and user position. So, we choose the character domain approach in our algorithm.

In particular, we determine the foot's status through two parameters. These parameters are foot position and velocity. Let's say that current foot position of P_f and the foot position at previous time frame is $\acute{P}_f$, then the velocity V_f is computed as following:

$$V_f = r\left(\frac{P_f - \dot{P}_f}{\Delta t}, n\right)$$

$$r(v, n) = \text{roundXL}(v, n)$$

where Δt is the inter-frame duration time and $r\,(v, n)$ returns the rounded value at nth digit of v.

Given y coordinate of P_f, which is the up vector coordinate that represents the distance from the ground, and its velocity V_f, we check whether these two values are under the threshold value $P_{threshold}$ and $V_{threshold}$ at the same time. For relaxing the sensitivity, we apply rounded values instead of the original value. If two conditions are satisfied, we mark that frame is constrained. One point to note is that the whole character's global position is closely related with the root (pelvis) joint position. Therefore, initial root position of the character is very important. In our algorithm, we calculate the length between foot (end-effector) and root joint, and lift up the character as far away from the ground as possible so that the character stands up above the ground. One downside is that setting constant threshold values might be too risky because these values should be changed depending on applications. Finding threshold value is a time consuming job that requires a lot of try-and-errors. In our approach, we take *adaptive* approach for finding these values. Basically, the user set the rough initial threshold value at the beginning, then the algorithm incrementally adjusting these values automatically. One important cue that makes this technique possible is that for human motion, one of the feet should be on the ground all the time. Therefore, while we applying the initial threshold values to detect constrained frames, if both feet are turned out to be not-constrained, then the threshold values need to be changed to make one or both feet be constrained. Let's say $P_f(r)$ is the right foot position, $P_f(l)$ be the left foot position, $V_f(r)$ and $V_f(l)$ are their velocity respectively. If both feet are non-constrained, then we check which values are above the threshold values on both feet at the same time. For example, if $P_f(r)$ and $P_f(l)$ are above the threshold value $P_{threshold}$, then new $P_{threshold}$ are set as $MIN\left(P_f(r), P_f(l)\right) + \delta$. Likewise, if $V_f(r)$ and $V_f(l)$ are above the threshold value $V_{threshold}$, then the new $V_{threshold}$ is set as $V_{threshold} = MIN\left(V_f(r), V_f(l)\right)$. In our approach, we set the $\delta = 0.1$.

3.2 Determining Foot Position

Once we find out that the current frame is constrained, then we need to fix the foot on the ground during the time being constrained. The most important job for this is to decide the exact foot position. The new foot position, P_t, is then fed into IK solver as an argument for obtaining proper leg configuration. In general, the constrained frames continue for a fair of amount of time. During that time, we need only a single foot position for fixing the foot. If we are working on off-line processing, where we have all previous and future data, we can easily find a single

Automatic Fixing of Foot Skating

foot position that minimizes deviation from original data over constrained frames [12]. However, because we are working on on-line processing, we don't know how long this constrained frame period continues. Therefore, we use the foot position of the first occurred constrained frame as target foot position.

3.3 Adjusting Root Joint Position

The target foot position may be too far to be reached even when the character is fully stretching its leg. This problem can be solved by adjusting the root joint position to the target foot position. Let o be the offset vector from the hip to the root joint and P_r be the root joint position. Also, let P_t the target foot position and l be the full limb length. Then, the target foot joint, P_t, is reachable only if $\|P_t - (P_r - o)\| \leq l$. In other words, when we think of it as a sphere that centered at $P_t - o$ with radius l, the P_r should be inside the sphere in order to be reachable. Because both feet can be constrained at the same time, we have to consider maximum two spheres at the same time. In this case, the root should be projected onto surface of two spheres' intersection region [10, 11]. Because of relocation of root joint, there might be small discontinuity on the character. To minimize the discontinuity, we apply the simple moving average (SMA) on the root joint position stream. Suppose that newly computed root position from this projection technique is $\acute{P}_r$ and original position is P_r, then SMA performs $P_r = \frac{(\acute{P}_r + P_r)}{2}$ to get the final position. This computation is iterating during the constrained frames.

3.4 Applying IK Solver

Once we fix the root position, next step is to adjust the leg to place the foot in the target position. Specifically, we need to find the hip and knee angle. This is accomplished by two-link analytic IK algorithm. We use the similar method that Lucas et al. used for their paper [12, 13]. The IK solver is composed of two phases. The first phase is to rotate the knee angle so that the length between hip and target foot position $(\overrightarrow{P_h P_t})$ matches the length between hip and current foot position $(\overrightarrow{P_h P_f})$. The knee angle θ_k can be obtained as follows:

$$\theta_k = \arccos\left(\frac{l_1^2 + l_2^2 + 2\sqrt{l_1^2 - \acute{l}_1^2}\sqrt{l_2^2 - \acute{l}_1^2} - \|P_h - P_f\|^2}{2\acute{l}_1 \acute{l}_2}\right)$$

where l_1 and l_2 denote the length of thigh and knee joint respectively and $\acute{l}_1$ and $\acute{l}_2$ are the length of projected vector onto rotation axis. The rotation axis is obtained by cross product between $\overrightarrow{P_h P_k}$ and $\overrightarrow{P_f P_k}$ where P_k is the knee joint position.

One significant problem happens when the θ_k is close to 180 degrees, which corresponds to a fully straight leg. As addressed in[4], this causes an unnatural extension and contraction of knee angle for small change of target position. To prevent it, we limit the knee ration to the maximum knee angle, say θ_{max} (170°, in our algorithm), and if the angle is bigger than θ_{max}, we put a damping on the angle. Suppose that original knee angle before phase 1 is $\acute{\theta}_k$, and the difference between $\acute{\theta}_k$ and θ_k becomes $\Delta\theta_k$. Then, the damped knee angle $\overline{\theta_k}$ is computed as follows:

$$\overline{\theta_k} = \acute{\theta}_k + \int_{\acute{\theta}_k}^{\theta_k} I\left(\frac{\theta - \theta_{max}}{\pi - \theta_{max}}\right) d\theta$$

$$I(\theta) = 2\theta^3 - 3\theta^2 + 1$$

$\frac{\theta - \theta_{max}}{\pi - \theta_{max}}$ is the normalization operation with θ_{max}. $I(\theta)$ is the cubic polynomial guaranteeing C1 continuity because $I(1) = 0$, $I(0) = 1$ and $\frac{dt}{d\theta}(1) = \frac{dt}{d\theta}(0) = 0$.

One problem is that all additional hip rotations q about the axis $\frac{P_h P_t}{P_h P_t}$ satisfy the target constraint. Among all possible rotations, we choose the orientation that makes the hip and knee angle closer to the original orientations relative to their parent coordinate systems.

When we adjust the joint angle, we have to consider the skin mesh twisting problem as well. Most of skinning is built with linear blending skinning technique where each vertex of skin mesh is assigned weight values for small set of joints. the left and the right character have an exactly same skeleton pose but the left character's skin is twisted. To prevent this problem, given old joint orientation q and newly determined joint angle $\acute{q}$, we first computes angle ϕ about the twisted axis, say X axis, then multiplied $\acute{q}$ with the corresponding rotation $\Delta\theta_\phi$ to compensate the twist.

4 Experiments

We perform an experiment to validate our algorithm with Microsoft Kinect sensor. The NITE middle ware from PrimeSense is used for retrieving position and orientation data form the sensor. Also, Open source game engine (OGRE) is used for rendering characters. All experiments were carried out on an Intel Xeon 3.20 GHz processor PC with a graphics acceleration card. Figure 2 shows our experiment setup. We use the Kinect sensor as the motion sensor and perform actions 2–3 m in front of the sensor. The skeleton has 15 joints. But, currently the NITE middle ware supports only 11 of them.

Figure 2 is screen shots that compare two cases when we do not use our algorithm and when we use the foot skating clean up algorithm. Note that yellow balls on the heel appear when the frame is identified constrained. Red balls indicate the foot positions when foot IK solving finished. When the yellow balls

Automatic Fixing of Foot Skating 411

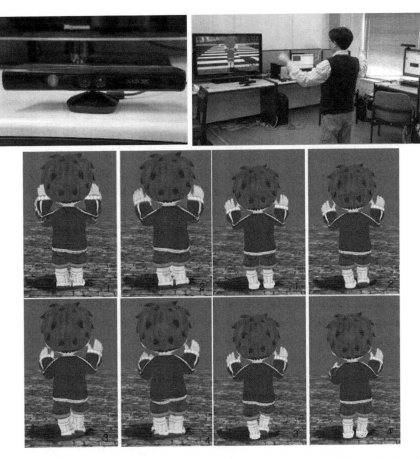

Fig. 2 *Top left*: Kinect sensor for experiments. *Top right*: An 3D character is controlled by the user in on-line puppetry manner. *Bottom left*: Screen shots before applying foot skating cleanup. *Bottom right*: Screen shots after applying foot skating cleanup. *Yellow ball* indicates whether this frame is constrained or not

match red balls totally, it means that our algorithm fixed foot on the ground. For evaluating performance, we display the frame rate information on the top left of the screen at run time, which shows more than 30 frame/s.

5 Discussion

In this paper, we have introduced a simple and practical foot skating cleanup algorithm for a character whose motions are controlled in the on-line puppetry manner through the depth sensor. One advantage of our algorithm is that even when character's pose is collapsed because of some errors in the sensor, our algorithm can keep reasonable pose all the time as long as the frame is identified constrained. Also, not only for legs, but we can also easily apply our algorithm for arms. One limitation of our approach is that our algorithm depends on the middle ware providing joint angle data through skeleton pose estimation algorithm from depth image. As a result of this limitation, we are not currently supporting multi-user foot skating cleanup, although the sensor itself has the capability for identifying multiple users.

Acknowledgments This research was supported by the Bisa Research Grant of Keimyung University in 2012.

References

1. Tsunoda K, Fitzgibbon A (2010) Kinect for Xbox 360—the innovation journey (June 2010). Microsoft Research Faculty Summit
2. Shotton J, Fitzgibbon A, Cook M, Sharp T, Finocchio M, Moore R, Kipman A, Blake A (2011) Real-time human pose recognition in parts from single depth images. In: CVPR'2011: Proceedings of the 2011 IEEE Computer Society conference on computer vision and pattern recognition. IEEE Computer Society
3. Zhu Y, Fujimura K (2010) A bayesian framework for human body pose tracking from depth image sequences. Sensors 10(5):5280–5293
4. PrimeSense: PrimeSense™ NITE 1.3 Algorithms notes (2010) PrimeSense
5. PrimeSense: PrimeSense™ NITE 1.3 Control programmer's guide (2010) PrimeSense
6. Moeslund TB, Hilton A, Krüger V (2006) A survey of advances in vision-based human motion capture and analysis. Comput Vis Image Underst 104:90–126
7. Ganapathi V, Plagemann C, Koller D, Thrun S (2010) Real time motion capture using a single time-of-flight camera. In: CVPR'2010: Proceedings of the 2010 IEEE Computer Society conference on computer vision and pattern recognition. IEEE Computer Society
8. Microsoft Kinect Teardown: Microsoft Kinect Teardown, http://www.ifixit.com/Teardown/Microsoft-Kinect-Teardown/4066/
9. Carmody T, How motion detection works in Xbox kinect, http://www.wired.com/gadgetlab/2010/11/tonights-release-xbox-kinect-how-does-it-work/
10. Kovar L, Schreiner J, Gleicher M (2002) Footskate cleanup for motion capture editing. In: SCA'2002: Proceedings of the 2002 ACM SIGGRAPH/eurographics symposium on computer animation, New York, NY, USA. ACM, pp 97–104
11. Shin HJ, Lee J, Shin SY, Gleicher M (2001) Computer puppetry: an importance-based approach. ACM Trans Graph 20:67–94

Part VIII
IT and Multimedia Applications

A Study on the Development and Application of Programming Language Education for Creativity Enhancement: Based on LOGO and Scratch

YoungHoon Yang, DongLim Hyun, EunGil Kim, JongJin Kim and JongHoon Kim

Abstract Students meet with many problems during programming education in computer education. Programming education have been several research that are published that helps develop students' creativity in the process of exploring solutions through flexible. Especially in elementary school students, receive training in programming for the first time feel constraint to lean programming language. But the use of EPL reduces difficulties. So Students can focus on the thinking. Therefore, in this paper, a draft was produced and applied to observe the creative elements increasing of elementary school students. And modify and supplement the draft through improvements to a draft obtained by applying.

Keywords Educational programming language · Computer education · Creativity

---

Y. Yang · D. Hyun · E. Kim · J. Kim (✉)
Department of Computer Education, Teachers College, Jeju National University,
Jeju, Korea
e-mail: jkim0858@jejunu.ac.kr

Y. Yang
e-mail: atriple1981@naver.com

D. Hyun
e-mail: gody5@naver.com

E. Kim
e-mail: eunjgmail@naver.com

J. Kim
Polytechnic Seoul Kangseo College, Seoul, Korea
e-mail: jjkim70@kopo.ac.kr

J. J. (Jong Hyuk) Park et al. (eds.), *Multimedia and Ubiquitous Engineering*,
Lecture Notes in Electrical Engineering 240, DOI: 10.1007/978-94-007-6738-6_51,
© Springer Science+Business Media Dordrecht(Outside the USA) 2013

1 Introduction

Programming education is very suitable education for the direction of education. Program planning, designing, implementing and revising procedure is very similar to solving social problems in the reality. Students will, therefore, experience problem solving indirectly and enhance their creative problem-solving skills through programming in various methods on given problems [1].

EPL ensures to design one's thoughts easily through the use of the intuitive command and block-script based coding. This will reduce repulsion of programming education and initiate interests. Students may also visually identify their program movement process thereby check and modify errors.

The study prepared draft teaching materials to examine the effect of EPL education, and executed TORRANCE TTCT (Drawing) Type-A inspection, a creative thinking test, before and after implementing the education, respectively [2]. Draft teaching materials are crafted based on 40 min classes in the amount of eight sessions suitable for circumstances of schools; education is implemented on elementary students in fourth grade by an elementary school teacher who pursues the master and doctor's courses.

Teaching materials are revised and supplemented considering education details, such as the cognitive level of students and conditions of schools based on the inspection results of before and after education and experience of using draft teaching materials.

2 Preceding Research

There are many researches reporting that programming education via LOGO enhances creativity. With regard to thinking skills, Clement surveyed the affect of LOGO programming through numerous researches and published the results that LOGO affects mathematical knowledge and enhances thinking skills (problem-solving ability and creativity) [3]. In addition Clement (1991) divided 73 students in the 8 years into three groups; a LOGO programming group, a group using creativity enhancement programs, not LOGO, and a comparison group, and examined the affect of LOGO experience in creativity. The experiment result provided affirmation that there was Figural Creativity learning transfer after learning LOGO and also uncovered that LOGO can contribute in promoting verbal creativity [1].

The preceding research shows a positive role of LOGO in the enhancement of creativity. However, as there are no comparative analyses with various new EPLs, it is challenging to describe the pros and cons of LOGO only when compared with EPL. Hence, there is a limitation in selecting ELP that considered the characteristics of students.

One of the most popular ELP that has introduced recently is Scratch. Today, vibrant research is being conducted on Scratch, and here are a few examples as follows.

As a result of applying the learning content configured base on creative problem solving model (CPS) to improve complex cognitive skills and strategy to stimulate learners' intrinsic motivation during programming in sixth grades discretionary activity time, the study confirmed that Scratch programming learning is effective in improving learners' intrinsic motivation and problem solving skills [4].

The programming education using Scratch is a positive influence in effectiveness and satisfaction with respect to the cognitive area of learning. Especially, the performance of Scratch use is statically higher to visual-inclined learners. Therefore, we can presume that the programming process of elementary students has a great impact in learning effect.

However, Scratch also demonstrated independent influence only and there is a lack of research on differentiating the benefit of Scratch only through comparison with other ELPs.

Hence, the study carried out comparative analysis between Scratch and LOGO, which are known to provide help in improving creativity through preceding research and examined the enhancement of detailed elements of creativity by developing draft teaching materials, and revised problems suggested from field application of draft teaching materials in order to provide materials that can be utilize in education and EPL selection suitable for education purpose as well as student's characteristics.

3 Development of Draft Materials

3.1 Education Details and Configuration System

Most of all, in order to teach programming, somewhat esoteric topic to elementary students, we need to continuously induce their interests. If we suggest visual stimulus which lead to a result, students will feel a sense of accomplishment alongside improved interest on learning as well as reduce repulsion on programming. In addition, the content should be configured in education details that could improve their creativity. The study developed teaching materials that ensure programming and demonstrating creativity as well as expanding the scope of their thinking, rather than a mere suggestion and analysis of new programs.

While LOGO and Scratch have features that they can easily approach as educational programs, they have a difference in programming environment. Suppose LOGO writes program when command is input, Scratch uses command that uses a method to drag and stack blocks expressed in graphic.

Hence, we come to determined that the approach methods should be different in teaching the two educational languages. We used LOGO to advance outputs through stepwise education of command while Scratch is used to increase the difficulty of making content by subject.

Details of Teaching Material Development

It is a transitional stage towards the formal operational period from concrete operational period, and the study selected as follows considering the programming language features and the level of elementary school students in fourth grade who encounter programming for the first time [4].

LOGO

We let students to introduce various diagrams and draw them using each different command by session, and introduced variables as well as control command and incorporated a variety of features of LOGO more in detail.

Scratch

The study abstracted educational elements which can be used in Scratch in the elementary school level based on programming concept. We abstracted educational elements ensuring to learn and practice the basic functions and operating methods of Scratch whereby configured in education data development.

Configuration System of Teaching Materials

The steps to derive creative outputs were initially introduced in the creative research of Wallas in 1926 [5]. Wallas divided the courses of obtaining creative outputs in four steps as Fig. 1.

Learners will become familiar with the knowledge of details relevant to individuals in the Preparation Step, analyze and understand acquired information in the Incubation Step, a solution will be provided in the Illumination Step and the solution will be verified in the Verification Step.

Base on these, the study first configured steps to apply LOGO and Scratch programs and developed educational materials accordingly. The study configured Fig. 2 by considering the features of each program language for program configuration of single session amount, and divided into four steps to ensure manifestation of self-directive creative thinking and encourage divergent thinking by associating with learning objectives.

Step 1 (Concept understanding and finding principles).

Understand the details of today's study as well as related concept and principles and execute simple programs firsthand. Concept and principles should be provided with figures and also include details studied in the previous sessions.

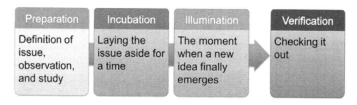

Fig. 1 The Wallas model for the process of creativity

A Study on the Development and Application

Fig. 2 Material configuration system

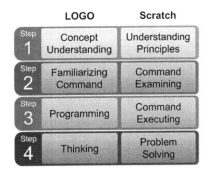

Step 2 (Familiarizing and examining command).

In the steps, learns examine a concept related to problems to study today and explore rules and instructions based on completed output. The discovered rules and command will be used at the programming.

Step 3 (Programming and command executing).

This step will ensure programming using the rules and command discovered in the previous step.

Step 4 (Contemplating and problem solving).

We designed this step to make new programs using a concept or express differently by modifying programs related to details already learned.

4 Improvement Measures Study Through Field Application

4.1 Field Application and Analysis

Based on the developed draft worksheet, we organized four groups, which divided into men and women, respectively, targeting each class of fourth grade of Shinjeju Elementary School (Scratch) and Gangjung Elementary School LOGO) Jeju Special Self-governing Island, and implemented pre-post inspections through TORRANCE TTCT (diagram) creativity test A-type [2]. We also configured the homogeneous group by abstracting samples from each group as well as comparative analysis on detailed elements creativity including pre-post analyses.

As a result of comparative analysis on the impact of each language on the creativity area of students, both languages as shown in Table 1 were helpful in improving creativity. Especially, whereas LOGO causes positive influence in the fluency area among creativity area, Scratch affects on resistance and abstractness. These results implies that the use of various functions via the command of LOGO are suitable to the features of fluency, and that the process of program implementation using the graphic of Scratch can be checked visually suits the characteristics of abstractness and resistance, which result in improving the area of each creativity.

Table 1 Analysis of pre-post creativity enhancement

EPL	Gender	Creative element	(Average)	t	Freedom degree	Significance probability (both side)
LOGO	Male	Fluency	−17.38	−4.950	7	0.002**
		Originality	−9.63	−1.925	7	0.096
		Abstractness	−8.38	−0.706	7	0.503
		Sophistication	−5.38	−2.641	7	0.033*
		Resistance	−5.63	−0.802	7	0.449
	Female	Fluency	−25.10	−6.138	9	0.000**
		Originality	1.80	0.596	9	0.566
		Abstractness	−7.30	−0.959	9	0.362
		Sophistication	−1.70	−0.749	9	0.473
		Resistance	−1.50	−0.271	9	0.793
Scratch	Male	Fluency	−6.67	−2.082	8	0.071
		Originality	−5.44	−1.135	8	0.289
		Abstractness	−31.56	−3.914	8	0.004**
		Sophistication	−2.33	−1.228	8	0.254
		Resistance	−14.11	−2.615	8	0.031*
	Female	Fluency	−11.78	−2.075	8	0.072
		Originality	−4.22	−0.907	8	0.391
		Abstractness	−25.22	−2.889	8	0.020*
		Sophistication	−3.22	−1.442	8	0.187
		Resistance	−13.11	−5.063	8	0.001**

$*\ p < 0.05$, $**\ p < 0.01$

It may seem that the features of each ELP played positive roles in the enhancement of creativity, but a programming process using a command in the case of LOGO suffers a downside in that it is difficult to detect errors—if they occur—in the execution of the program process, while a programming process using Scratch block entails a restriction and unnecessary long process of programming in using advanced function. Hence, we should solve a challenge to establish detailed and systematic education process suitable for the development steps of students when teaching the programs considering these limitations of the programs.

4.2 Improvement Measures for Draft Teaching Materials

Education Details and Configuration

In a programming process, it is necessary to configure a process that emphasizes problem solving based on learning simple programming techniques. Therefore, it is needed to provide additional details of teaching and learning plans as well as learning materials produced in the existing eight sessions. In particular, trails should be partitioned to ensure students to fully understand the concept in the step that introduce (variables and logical operation) the upper mathematical concept.

Education Details

Programming courses that require systematic design, namely the function call and recursive call, need adequate thinking and the time to practice by adjusting difficulties, and various methods should be introduce to students to ensure to solve problems by adding the basic algorithm details. The focus should be on the use of effective combination, not mere basic function learning. And the length of code should be reduced as well. Even if an overall length of the code become lengthens, it should be bundled up by feature in order to reduce fatigability of students. This problem was displayed especially with Scratch's code but the lengthier of a code that uses a block, the more the visual fatigability.

Therefore, functions should be properly divided by feature whereby approach using a calling method by the need of the function.

5 Improved Teaching Material Development

5.1 Teaching Material Improvement Details

The study believed that there is no problem for elementary school students to use EPL. It was, however, challenging for fourth graders to understand the mathematical concept. Therefore, we extended the learning materials that are prepared in eight sessions of the draft teaching materials into 15 sessions and add more sessions in the process of concept introduction to ensure seamless application for students to apply in class. With respect to the basis of the 15 sessions, our suggestion is based on the circumstance of schools in which compiles about 16 h of one semester among 64 h of the total annual training hours of discretionary activity as information education for students in fourth, fifth, and sixth grades. In addition, in order to improve thinking skills using the acquainted features, the study adjusted the configuration of sessions and provided additional relation/logical operation as well as algorithm topics.

In the case of LOGO, we stressed the process that solve problems through the recursive call of functions, and as for Scratch, we divided code preparation by feature so as to call when needed in order to reduce visual fatigability resulting from lengthy code and emphasized by displaying the flow of thinking in a diagram form.

6 Conclusion

The study produced and input draft teaching materials on LOGO and Scratch, the most used EPL and conducted cross-tab analyses on the influence of the detailed areas of creativity between the two languages, which are unprecedented in preceding research in addition to the improvement of creativity often observed in the existing research. Moreover, we identified problems in design and the course of draft teaching development through on-site inspection and confirmed the matters that required attention when the materials are used in the actual field whereby developed suitable teaching materials for the characteristics of each language in addition to increase the vast use of the materials.

To foster creative citizens who meet the demand of the age, it is vital to continue stepwise education through the establishment of elaborated education courses, instead of mere one-off education. With respect to computer education, we believe it is possible to establish efficient education courses if the instructional designed used in the study is fragmented and incorporated to various circumstances.

There are many educational languages besides LOGO and Scratch used in the study and a variety of languages are used abroad in education courses. Hence, Korea should also develop research on a variety of ELPs as soon as possible.

References

1. Clements DH (1991) Enhancement of creativity in computer environments. Am Educ Res J 28(1):173–187
2. Torrance EP (1999) Torrance test of creative thinking: norms and technical manual. Scholastic Testing Services, Bensenville
3. Clements DH, Battista MT (1989) Learning of geometric concepts in a logo environment. J Res Math Educ 20(5):450–467
4. Jeongbeom S, Soenghwan C, Taewuk L (2008) The effect of learning scratch programming on students' motivation and problem solving ability. J Korean Assoc Inf Educ 12(3):323–332
5. Wallas G (1926) The art of thought. Harcourt, San Diego

Design and Implementation of Learning Content Authoring Framework for Android-Based Three-Dimensional Shape

EunGil Kim, DongLim Hyun and JongHoon Kim

Abstract This study was conducted to create a more tangible educational environment by allowing learners to directly control three-dimensional learning contents through the touch interface of smart devices. Furthermore, because there are limitations to the acquisition and provision of three-dimensional learning contents due to difficulties in producing them, the proposed framework was designed to allow teachers and learners to directly produce and share contents. The proposed framework is based on intuitive XML language and the application was built to enable playback and authoring in Android-installed devices. In addition, a server environment was constructed for contents sharing. The feasibility of the proposed framework was verified through expert evaluation and its potential for utilization of new learning contents was positively evaluated.

Keywords Android · 3D learning contents · m-Learning

1 Introduction

Today's smart devices have expanded the scope of educational contents because have much more advanced hardware compared to PDAs in the past and are equipped with various sensors such as GPS, acceleration, and compass.

E. Kim · D. Hyun · J. Kim (✉)
Department of Computer Education, Teachers College, Jeju National University,
Jeju, Korea
e-mail: jkim0858@jejunu.ac.kr

E. Kim
e-mail: computing@korea.kr

D. Hyun
e-mail: gody5@naver.com

J. J. (Jong Hyuk) Park et al. (eds.), *Multimedia and Ubiquitous Engineering*,
Lecture Notes in Electrical Engineering 240, DOI: 10.1007/978-94-007-6738-6_52,
© Springer Science+Business Media Dordrecht(Outside the USA) 2013

Fig. 1 Framework for bidirectional contents authoring and playing

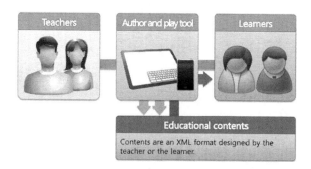

Educational media using smart devices can provide infinite learning contents through the Internet and enable the provision of three-dimensional contents with visual, sensory expressions.

Based on these advantages of smart devices, this study investigated a method of creating learning contents that can be directly used by learners using touch sensors. Furthermore, three-dimensional learning contents that can express even the sense of texture, etc. to attract learners' interest were produced. However, for effective education, measures to widely supply 3D learning contents were required. To create 3D learning contents, programming abilities as well much time, effort, and costs are required to implement the development environment.

Thus, in order to achieve quantity and quality of 3D learning contents, this study proposed a contents production framework in which anyone can easily produce and use learning contents. This implies that from the educational perspective, teachers and students with different educational levels can produce a wide variety of learning contents as shown in Fig. 1.

Android provides frequent updates for continuous improvement and many new hardware control APIs [1]. This study defined XML format for 3D learning contents based on Android and designed a framework for analyzing and showing them to learners in a controllable form.

2 Theoretical Background

2.1 Android

The core of the Android platform is the Linux kernel which plays the role of an operating system that manages device drivers, resources, and so on. Above the kernel, there are OpenGL for 3D graphic, etc. [2]. As part of the Android project, Google developed the Dalvik virtual machine for optimal design of low-power mobile devices after a lot of research. It can reduce the capacity of applications by combining various Java class files into .dex and reusing duplicate information. Most applications are developed in Java and access the kernel and libraries through the Dalvik virtual machine [3]. The application framework contains telephone communication, position tracking, and content provider, and applications are developed using APIs [4].

2.2 OpenGL

OpenGL is managed by the Khronos Group Consortium founded by many global companies.

In the OpenGL pipeline as shown in Fig. 2, geometric data such as vertices, lines, and polygons go through Evaluator and Per-Vertex Operation, but pixel data (pixels, textures, etc.) go through different paths, gather at Rasterization and are recorded in the Frame Buffer.

Display List stores OpenGL commands, geometric and pixel information for later execution and it can improve performance because cache can be applied. Evaluator transforms the information inputted through a polynomial function into coordinates to obtain vertices used in curves and surface shapes. One point in a 3D space is projected into a vertex on a screen through Per-Vertex Operation. Furthermore, if advanced functions such as lighting are activated, lighting effect-related operation is performed for the color information of vertices. The main feature of Primitive Assembly is clipping. After this step, a geometric primitive with a perfect shape consisting of transformed or clipped vertices is created for color, depth, texture coordinates, and rasterization steps.

Pixel data are processed along a different path from the geometric data in the OpenGL rendering pipeline, and the pixels saved in an arrangement of a specific format in the system memory are read first. To these data, operations such as scaling and basis application are applied. This process is called Pixel Transfer Operation, and the result can be saved in the texture memory or directly rasterized.

In the Rasterization step, all the geometric data and pixel data are transformed into fragments. Each fragment is an array that includes colors, depth, line thickness, dot thickness, and anti-aliasing. Each fragment corresponds to a pixel in the

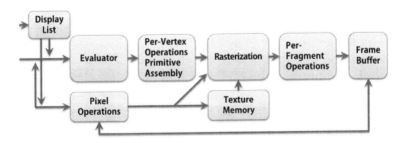

Fig. 2 OpenGL pipeline [5]

Frame Buffer. These fragments are changed or removed through Per-Fragment Operations before they are stored in the Frame Buffer.

3 Construction of Framework

The overall flow of the framework proposed in this study for producing and playing learning contents for 3D object is shown in Fig. 3.

When the application and learning contents are started, the Main Activity communicates in XML for update information with the server through Thread at the background. Thus, learners can receive continuously updated learning contents. Not only the update information, but also the learning contents are defined in XML which allows communication in reduced volumes. The XML information of the learning contents is analyzed by XML Parser in the DOM method. The analysis results are created in a tree form which allows easy interaction with learners for insertion and deletion of 3D objects [6]. The space coordinates of the required vertices and the texture space coordinates for expression of texture are calculated through Vertex Operation. The calculation results are processed by OpenGL through the Renderer interface of GLSurfaceView and displayed as 3D object on the screen [7].

3.1 Contents Authoring and XML Script

The contents authoring part also works in the Android-embedded device. Users input 3D object by touching them to the 3D projection guideline and pile them up to gradually create the total shape. For the positions of the inputted shapes, the row, column and layer data are created in XML which is parsed during playback and the space coordinates of each 3D object are calculated. User can select a surface quality of the 3D object and the resource ID of the image representing the selected quality is saved in XML. Then it is applied to the surface when the contents are played back together with the space coordinates of the 3D object.

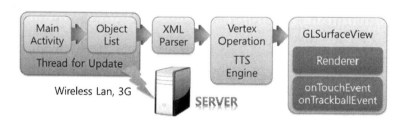

Fig. 3 Total flowchart for learning contents

A learning scenario consists of text in XML. The scenario text includes explanation about learning method and is converted to speech in multilanguages and told to learners using the TTS engine of Android [8]. Learning scenarios output the viewpoint of 3D objects in line with the sequence through interaction with users.

3.2 Contents Playback and Control

The contents playback part largely consists of XML Parser that analyzes XML information and saves it by element, Renderer that shows the 3D object on the screen, and the TTS engine that transforms text to speech. The interaction between contents and user is based on touch and trackball control.

The XML information is analyzed in the DOM method, and the resource elements related to the screen display of 3D object are divided by delimiters and sent to Vertex Operator. The Vertex Operator calculates the space vertex coordinates of each face to express the location of the 3D object together with their surface texture. The length of one side of the 3D object is defined by calculating the value of the longest axis so that the prospect of the OpenGL virtual camera is not exceeded, and the center of the entire 3D object is specified as the origin of the virtual space so that it will not go out of the screen during control [9, 10]. The space coordinates of the vertex calculated by the Vertex Operator and the surface texture information are sent to the Renderer. In the renderer, the 3D object is rasterized through the OpenGL ES API provided by the Android album, saved in the Frame Buffer and displayed on the screen.

The scenario elements of XML are analyzed by Parser and saved in the multi-dimensional array for each angle element. When user requests the saved information by entering the sequence by touch input, the text element values are sent to the TTS engine and outputted as speech and the external resource information is displayed as Android View.

4 Implementation of Authoring and Playback Framework

To verify the feasibility of the proposed framework for authoring and playing 3D object learning contents, it was implemented in the development environment based on Android 2.3 Gingerbread.

Users can freely rotate the 3D object to the desired direction via the Android touch and trackball interface and the figures were instantly drawn by the Frame Buffer. The rotation information, text saved in the XML scenario can be provided as buttons for user operation as needed. In addition to the default learning contents, users can also produce and play scenarios.

The learning contents that are created by user through the authoring part can be provided to other users through sharing. When shared, the XML file of the learning contents is sent to the server and the server updates the learning contents list and version. Other users who run the application can compare their version with the server version and download the XML file of the updated contents. Then the updated list of contents is displayed through the Android ListView.

This framework application was installed and tested in a smart phone with the hardware specifications of CPU QSD8250 1 GHz, RAM 512 MB, and WVGA display. It took less than 1 s for parsing the XML script, preparing the TTS engine, and displaying the content on the screen. However, considering the various factors of the mobile environment, external resource loading was not added to the XML.

5 Expert Evaluation

The usability and possible improvements of the prototype application implementing the framework were evaluated through experts who were 20 regular licensed teachers of class 1 who had at least 10 years of experience as elementary school teacher.

The expert evaluation method was to answer a questionnaire after demonstration and use of the application. The experts checked one of the answers according to 5-level Likert scale in 2.5 point intervals, or selected an answer and stated their opinions depending on the questions.

The prototype application was largely divided into playback and control part and authoring part for this survey. The answers to the questions about the playback and control part were quantified by the 5-level Likert scale. The prototype application received 8.875 points for the reflection of learner controls in the expression of learning contends, and 9 points for realistic expressions. Furthermore, the touch interface of smart devices was positively evaluated at 8.25 for user interaction with contents. Thus, the use of touch and trackball interface of smart devices for control of learning contents seems to be effective. The authoring part using touch interface received 8.375 points for easiness of creating learning contents and 7.75 points for provision of features required for creating learning contents.

The demand for provision of learning contents about various basic 3D object in addition to the cube shape that was developed for the prototype was the highest at 30 %, followed by the demand for easier and simpler authoring part, the demand for manual for authoring and use of contents at 20 %, and the demand for the development of applications that can be run on other platforms at 15 %.

The analysis results for the requirements for activation of 3D learning contents using smart devices, the requirement for a sufficient volume of default learning contents was the highest at 60 %, followed by the requirement for interaction among learners using various sensors which is the advantage of smart devices.

This result suggests that users demand learning methods using various sensors (gravity, acceleration, GPS, etc.).

6 Conclusions

This study implemented a prototype of learning contents for 3D object that can be used in education on the basis of highly portable smart devices equipped with various sensors. The learning contents of our study allow teachers and learners to directly produce learning contents through touch interface and to share contents with one another as they are written in XML. Furthermore, scenarios can be defined to control the flow of learning contents and it is also easy to interact with the contents through the touch interface. From the aspect of communication environment of mobile devices, they are also positive for the new learning environment using smart devices considering that small size XML learning contents can be quickly shared.

The feasibility of the application produced on the proposed framework was verified through expert evaluation by teachers and the future improvements were discussed. The evaluation result was positive for the new educational environment, and the use of various sensors of smart devices for interaction with learning contents was required. Future studies will investigate learning methods using various sensors such as gravity, acceleration, and GPS in addition to touch, and develop learning contents that can be directly produced and controlled by learners based on these methods.

References

1. Murphy ML (2010) Beginning Android2. Apress, New York
2. Hashimi S, Komatineni S, MacLean D (2010) Pro Android2. Apress, New York
3. Dalvik virtual machine internals. http://developer.android.com/videos/index.html#v=ptjed OZEXPM
4. Google. http://developer.android.com/guide/basics/what-is-android.html
5. Munshi A, Ginsburg D, Shreiner D (2008) OpenGL ES 2.0 programming guide. Addison-Wesley Professional, Boston
6. w3schools. http://www.w3schools.com/Dom/dom_parser.asp
7. Google. http://developer.android.com/guide/topics/graphics/opengl.html
8. Text-To-Speech and Eyes-Free Project. http://developer.android.com/videos/index.html#v=xS-ju61vOQw
9. Pulli K, Aarnio T, Miettinen V, Roimela K, Vaarala J (2007) Mobile 3D graphics. Morgan Kaufmann, San Francisco
10. Khronos. http://www.khronos.org/opengles/sdk/docs/man

A Study on GUI Development of Memo Function for the E-Book: A Comparative Study Using iBooks

Jeong Ah Kim and Jun Kyo Kim

Abstract Currently used electronic books (hereafter referred as "e-book") do not reflect people's memo-taking behavior patterns, as an intuitive Graphic User Interface (hereafter referred as "GUI") is not used in e-books. A study was carried out to suggest a GUI prototype that applies users' memo-taking behavior patterns to the memo function of iBooks to enable greater usability. The prototype herein suggested is a memo GUI prototype that can apply people's real-life memo-taking behavior to the memo function based on the iBooks interface. For the next step, a usability test was conducted on the suggested prototype and the iBook's memo interface through four environmental factors and eighteen evaluation factors. Five research subjects participated in the usability test on two types of interface, and a questionnaire was analyzed using a paired t-test (T-test). Analysis of the questionnaire showed that users were highly satisfied with the usability of the newly suggested prototype, compared to iBooks.

Keywords E-book application · GUI · Usability test

1 Introduction

1.1 Background and Purpose

Technology development today brings out new media such as smartphones, tablet PCs, and e-books providing a viewer function. These remove the inconvenience of carrying books around and offer the convenience of selecting books through the

J. A. Kim
Design Department, Chung-Ang University, Seoul, Korea
e-mail: sam2496kr@hotmail.com

J. K. Kim (✉)
Visual Design Department, Chung-Ang University, Seoul, Korea
e-mail: kjk3134@korea.com

J. J. (Jong Hyuk) Park et al. (eds.), *Multimedia and Ubiquitous Engineering*, 431
Lecture Notes in Electrical Engineering 240, DOI: 10.1007/978-94-007-6738-6_53,
© Crown Copyright 2013

Internet and applications instead of visiting a bookstore in person. Although e-book applications have diversified in accordance with this rapidly changing market, little research has been performed on the GUI environment of e-book applications from the perspective of usability. Specifically, research is now required on a memo function for users who would like to 'take notes', rather than on information provided in e-books. To achieve this, the study aims to research individuals' memo-taking behavior in an analogue environment to suggest a memo GUI prototype that can apply their memo-taking behavior to e-books.

1.2 Range and Methods of Research

For the study, two types of memo-taking behavior patterns were derived from 10 study participants' memo-taking behavior, and they were applied to the iBooks memo interface to suggest a GUI prototype. Then a usability test was conducted on the iBook's current memo interface and the newly suggested interface. Usability test factors were derived through 5 experts' verification, and a survey questionnaire was prepared for the usability test. It was conducted on five research subjects for three days, and analyzed using a T test. Furthermore, the study range was restricted to the iBooks interface environment, and the survey was conducted on a prototype, not a developed version of the suggested interface.

2 Memo

2.1 Questionnaire Survey of Memo-Taking Behavior

A questionnaire survey of 150 people was conducted from March to April, 2012. The questionnaire was conducted on individuals who used analogue books to identify what their behavior was when taking memos, and on the usability of the currently-used iBook memo function. The result showed that auxiliary tools (e.g. post-it) used for taking memos account for 38.8 % of memo-taking when reading books, followed by hardly taking memos, taking memos on the book, and others, which accounted for 30.1 %, 20.8 %, and 10.3 % respectively. Furthermore, when using auxiliary tools (e.g. Post-It notes) to take memos, the results showed that underlining parts related to Post-It memos accounts for 12 % while randomly sticking those memos accounts for 10 %; taking memos in a blank space does not decrease the readability of a book. According to an analysis of the results, memo-taking behavior can be classified into the two following patterns: first, people take memos in blank spaces in the book, and secondly, people use auxiliary tools to take memos.

2.2 Memo Functions in E-Book Reader Applications

2.2.1 A Comparison of Memo Functions in e-book Reader Applications

Currently, around the world, the Amazon Kindle and iBooks are the most well-known e-book readers and applications. The Jungle Kindle is an e-book reader, and iBooks and similar e-book applications are being developed around the world for the Android and iOS platforms. These e-book platforms offer readers different things in terms of size, GUI, and so on. When comparing the Amazon Kindle and iBooks' memo function, the following becomes apparent: the Amazon Kindle was developed for use on the iPad and Android, and the Amazon Kindle application and Aldiko book reader application can be used on the Android, but not all platforms offer the memo function (Fig. 1).

2.2.2 iBooks and Usability

The iBook's memo function is simple to use. Drag and select a part to take a memo, and then press a memo selection button. Then a memo pop-up window appears, and users can take memos. A memo can then be typed. After the memo is done, tap any space other than the memo pop-up window, and the pop-up window disappears and a memo icon appears on the end at the right. A simple question-naire survey was conducted concerning the usability of the iBook's memo function. The results showed low satisfaction in that only 25 % of the respondents were satisfied with the usability of the iBook's memo function. On the question of whether or not the function is convenient to use, just 33 % of the respondents responded positively, which indicates poor usability for users. A short essay question about what inconveniences related to the current memo function occur produced many different opinions, including: difficult typing, hard to retrace

type	iBooks	Amazon Kindle	type	iBooks	Amazon Kindle
Memo GUI			Memo Icon	too. It can disguise othing more serious ust try to catch the cannot reach a safe i pound has all the	ie as when Rumpus
	– Large form for entering notes	– Context shouldn't change		– Shows date note entered	– Very small icon indicate note is present
	– Similar design with Post–it	– User enters Note on 6 lines of text disappears.		– Tapping note brings up form on left	– Tapping note brings up form on left

Fig. 1 A comparison of memo functions

memos, hard to take several memos in one line, the memo pop-up window blocks one's view of one's writing, memos are non-interactive and hard to use, use of writing or pictures is highly recommended, etc.

2.3 Prototype Memo GUI

A new prototype memo GUI was suggested, adjusted for GUI composition factors, four environmental factors, and inconveniences derived from the survey: difficult typing, hard to retrace memos, hard to take several memos in one line, memo pop-up windows blocking one's view of one's writing, memos are non-interactive and hard to use, use of writing or pictures was highly recommended, etc. The characteristics of the prototype are five-fold, as follows: First, many memo icons can be put into one line. Second, a sketch can be offered. Third, location of memo icons can be changed by users. Fourth, windows are slightly transparent to show the text below. And, fifth, font size and color can be modified. This prototype memo GUI is described in Fig. 2 in detail.

3 Usability Test of iBook Memo and Prototype Memo

3.1 Usability Test Method

A comparative study was conducted between the prototype memo GUI, to which new functions have been added, and conventional iBook memo functions. Prerequisites for testing usability were the extraction of evaluation factors and the composition of four environmental factors into a matrix.

3.2 Collected Factor Data for Usability Test

This study conducted a usability evaluation using eighteen usability evaluation factors of e-book applications. By creating a matrix with the collected factors, five experts distinguished overlapping parts and colors of unnecessary parts to show the importance status as a number, and extracted factors from collected data. After this, factors were derived from the experts' evaluation. Table 1 shows these derived factors to test usability, which can be used as a basis for designing and deriving a questionnaire for a usability test. The following eighteen factors can be suggested as a guideline for an evaluation framework to be a principle for usability tests of tablet PCs. A usability test can be conducted using the following factors:

New prototype GUI		New prototype GUI	
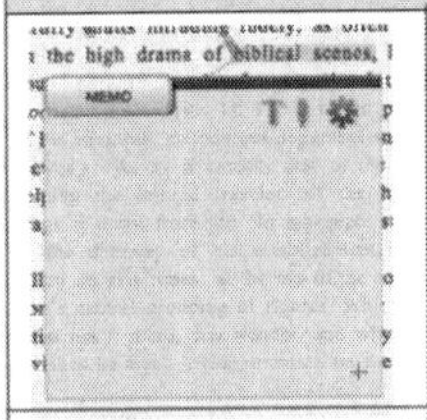	−New design for memo, and new functions added. −Sketch function is available −Transparency is added	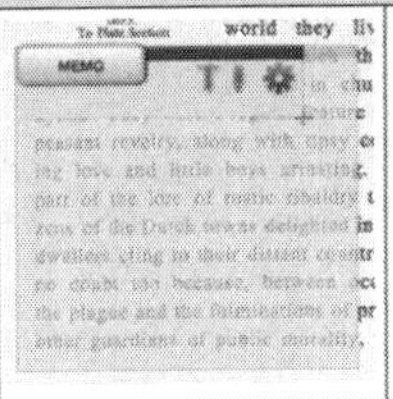	−Memo can be eddited its size by user by drag and grop
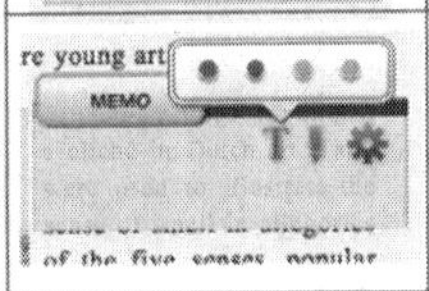	−Memo can use many colors of Fonts	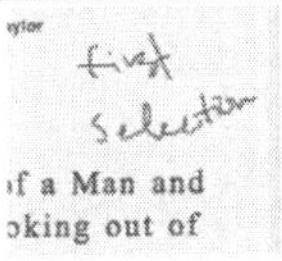	−It is available to write without background
	− Memo icons can be moved & edited by users freely		

Fig. 2 Prototype for iBook memo function

3.3 Selection of Research Subjects

The matrix was composed by using the above usability evaluation factors and the four environmental elements of e-book applications extracted above, and an evaluation was performed to give scores based on each evaluation factor. Prior examination of each five users showed that all of them had used an iPad for more than six months, and two of them had previously used an e-book while the other three had not. Therefore, before the experiment, a time of 30 min was set in which users with no e-book experience could use an e-book; this was to reduce the difference between those who had used an e-book and those who had not.

3.4 Results and Analysis on Usability Test

The usability test questionnaire was analyzed using T test. In other words, an average score for each question item showed the degree of research subjects'

Table 1 Evaluation of e-book application usability

Usability test factors of e-book applications
Consistency, Efficiency, Readability, Simplicity, Aesthetics, Alert, Operability, Accessibility, Effort, Intuition, Frequency of Mistakes, Clarity, Learning Ability, Satisfaction, Personalization, Help, Memorization, Understanding of Users, Feedback

satisfaction. If the significant probability (p value) is $0.05 \leq \alpha$, a null hypothesis is rejected and the analysis shows a significant difference. If the significant probability (p value) is $0.05 > \alpha$, a null hypothesis is not rejected and the analysis shows no significant difference. Only evaluation results showing a significant difference were evaluated, and usability was differentiated on items showing 3.0 (normal), a significant difference. In other words, a difference of 3.0 or above is normal or above and can be manipulated while a difference below 3.0 causes inconvenience to manipulate. Figure 3 shows is a questionnaire to compare usability tests.

4-5-1 Questionnaire Analysis

Items showing a significant probability (p-value) of less than 0.05 were classified. The criteria for this were the eighteen types of usability test items, and GUI visual factors of e-books such as color, layout, typography, multimedia, graphics, and navigation. Multimedia was deleted in that it is not applicable to the criteria. According to analysis results, a new prototype showed a significant probability of 0.05 or below in the following twelve evaluation items of color: Consistency, Efficiency, Readability, Simplicity, Aesthetics, Accessibility, Effort, Clarity, Learning Ability, Satisfaction, Personalization, and Memorization. The new

	Usability Factors	N	iBooks		New Prototype		F value	P value
			average	standard deviation	average	standard deviation		
Consistency	Graphic	10	3.60	.966	3.80	.789	1.353	.257
	Navigation	10	3.60	.516				
	layout	10	3.70	.949	4.00	.943		
	Multimedia	10	3.20	.632	4.10	.876		
	Color	10	3.70	.675	3.80	.789		
	Typographic	10	3.00	.943	4.60	.516		
	Total	60	3.47	.812	3.10	.994		
Efficiency	(Graphic, Navigation, layout, Multimedia, Color, Typographic) Total	60	3.35	.899	4.20	.632	1.482	.211
Readability	Total	60	2.93	1.056	3.70	.823	.327	.894
Simplicity	Total	60	3.23	1.047	3.00	.816	2.147	.074
Aesthetics	Total	60	3.28	.922	3.60	.516	.421	.832
Accessibility	Total	60	2.83	.924	3.60	1.075	9.636	.000
Effort	Total	60	2.75	.856	3.70	.675	2.284	.059
Intuition	Total	60	3.23	.851	3.90	.738	1.949	.101
Frequency of mistake	Total	60	2.63	.843	3.70	.675	1.507	.203
Clarity	Total	60	3.07	.756	3.60	1.075	1.763	.136
Learning ability	Total	60	3.32	.676	3.50	1.179	1.292	.281
Satisfaction	Total	60	3.00	.759	3.37	.920	.894	.492
Personalization	Total	60	2.33	1.115	3.40	.699	1.200	.322
help	Total	60	2.05	.928	3.40	.516	.860	.514
Memorization	Total	60	1.77	.945	4.20	.675	2.262	.061
Feedback	Total	60	1.67	.729	2.70	.949	.869	.508

Fig. 3 T-test results

prototype showed a significant probability of 0.05 in ten evaluation items in layout, in color, in typography, in graphic, and in navigation. As such, compared to the conventional iBook's memo function, the newly suggested prototype's memo function resulted in higher satisfaction in that it showed a significant probability of 0.05 or below, and it showed a very significant difference in evaluation items. Also, the new prototype showed a significant probability of 0.05 in ten evaluation items in accessibility as well.

4 Conclusion

The study herein was carried out to suggest a GUI prototype where usability based on memo-taking behavior patterns can be applied to the iBook's memo function as an e-book application. A memo GUI prototype that can apply individuals' memo-taking behavior to memo functions based on the iBook's interface environment was suggested. For the next step, a usability test was conducted on the suggested prototype and the iBook's memo interface using four environmental factors and eighteen heuristic evaluation factors verified by five UI experts. Five research subjects participated in the usability test on two types of interface, and a questionnaire was analyzed using T test. Through the questionnaire analysis, users were more satisfied with the usability of the newly suggested prototype interface, compared to the iBook interface. As a result, a new prototype contains typographic improvements to the current inconvenient environment, the addition of a memo icon and a new editing interface, various GUI improvements including the ability to adjust size automatically, and improvements to the overall design. A test comparing the usefulness of this newly improved environment using the most commonly used e-book platform, iBooks, was done and proved that usefulness had greatly increased.

Acknowledgments This work has been supported by the Chung-Ang University research fund.

References

1. Gong J, Tarasewich P (2004) Guidelines for handheld mobile device interface design. In: Proceedings of decision sciences, Institute annual meeting
2. Bahr GS, Nelson MM (2007) Development of a multiple heuristics evaluation table (MHET) to support software development and usability analysis. In: Universal access in human computer interaction: coping with diversity. Springer, Berlin
3. Nielsen J (2000) www.useit.com/alertbox/20000319.html
4. Gardiner E, Ronald GM (2010) The electronic book. In: Suarez MF, Woudhuysen HR (eds) The Oxford companion to the book. Oxford University Press, Oxford, pp 164
5. Kim JA, Kim JK (2012) Methods of portable PC GUI usability evaluation. J Digit Des 12(1):289–298, Korea Digital Design Society

6. Jeong Ah Kim, Jun Kyo Kim, Study on investigation and analysis of UI design trend of e-book applications, J Korea Soc Des Trend 36:253–264
7. Kim SH (2011) A study on the graphic user interface design for improving usability. Seoul National University of Science and Technology, Seoul
8. Chang W, Ji YG (2011) Usability evaluation for smart phone augmented reality application user interface. Soc E-bus Stud 35–47
9. e-book Oxford Dictionaries (2010) Oxford university press experimental study on usability evaluation of e-Book terminal of Seungjin Kwak and Kyoungjin Bae
10. Mehrabian A (1968) Communication without words. Psychol Today 56(4):53–56
11. Conati C, Gertner A, VanLehn K (2002) Using Bayesian networks to manage uncertainty in student modeling. User Model User-Adap Inter 12:371–417
12. Dadgostar F, Ryu H, Sarrafzadeh A, Overmyer S (2005) Making sense of student use of nonverbal cues for intelligent tutoring systems. In: Proceedings international conference of ACM SIGCHI, vol 122, pp 1–4
13. Wentzel K (1997) Student motivation in middle school: the role of perceived pedagogical caring. J Educ Psychol 89(3):411–419
14. Lehman B, Matthews M, D'Mello S, Person N (2008) What are you feeling investigating student affective states during expert human tutoring sessions. In: ITS
15. Ekman P (1989) The argument and evidence about universals in facial expressions of emotions. Wiley, New York
16. Meijer M (1989) The contribution of general features of body movement to the attribution of emotions. J Nonverbal Behav 13(4):247–268
17. Pavlovic V, Sharma R, Huang T (1997) Interpretation of hand gestures for human-computer interaction: a review. IEEE Trans Pattern Anal Mach Intell 19(7):677–695

Relaxed Stability Technology Approach in Organization Management: Implications from Configured-Control Vehicle Technology

Toshihiko Yamakami

Abstract The requirement for quickly solving complicated problems poses a fundamental challenge to the modern organization. In order to cope with this challenge, an organization needs to tune its management methodology to reduce time and costs and to increase its organizational efficiency. The author proposes a theory called a Relaxed Stability Organization (RSO) framework from implications learned from Relaxed Stability Technology (RST) in Configured-Control Vehicle (CCV) technology. The author discusses the overall organizational challenges in general. Then, the author discusses how the viewpoint of RST technology can be applied to organizational management. The author presents the framework of RSO Theory.

1 Introduction

Gamification is a paradigm that utilizes techniques originating from game theory to improve user engagement with services. This framework can be applied to services, marketing, and education. This technique also promises to be useful for improving enterprise management. The rapidly changing industrial landscape increases demands for agility in the decision making process and execution process in the enterprises. In order to cope with these demands, the author proposes the Relaxed Stability Organization (RSO) framework as an analogy of Configured-Control Vehicle concept in military aviation technology. In this paper, the author outlines the concept of an RSO framework and its implications for computer-supported cooperative work.

T. Yamakami (✉)
ACCESS, Software Solution, 1-10-2 Nakase, Mihama-ku, Chiba-shi 261-0023, Japan
e-mail: Toshihiko.Yamakami@access-company.com
URL: www.access-company.com

J. J. (Jong Hyuk) Park et al. (eds.), *Multimedia and Ubiquitous Engineering*,
Lecture Notes in Electrical Engineering 240, DOI: 10.1007/978-94-007-6738-6_54,
© Springer Science+Business Media Dordrecht(Outside the USA) 2013

2 Background

The aim of this research is to identify a framework that can cope with the agility of decision and execution in the fast-changing industrial landscape.

Organization is an open system. Grudin presented eight challenges for groupware from social dynamics [1]. Past research presented a range of different approaches, business process reengineering, open innovation, agile process, and so on.

The originality of this paper lies in its examination of relaxed stability in the context of organizational management.

3 Lessons from Relaxed Stability Technology

3.1 What is the Aim of RST?

Fly-by-wire (FBW) is a system that replaces the conventional manual flight controls of an aircraft with an electronic interface (Wikipedia). The movements of flight controls are converted to electronic signals that are transmitted by wires. The fly-by-wire system also allows automatic signals sent by the aircraft's computers to perform functions without the pilot's input, as in systems that automatically help stabilize the aircraft.

In aviation, relaxed stability is the tendency of an aircraft to change its attitude and angle of bank of its own accord (Wikipedia).

Fly-by-wire technology facilitated the Control-Configured Vehicle (CCV), which is an aircraft that utilizes fly-by-wire flight controls. In military and naval applications, it is now possible to fly military aircraft that have relaxed stability. Relaxed Stability Technology escapes from the restrictions induced by aviation design that focuses on stability. Although this increases the risks of aircraft crashes, it also increases flexibility in air-combat aviation. The increased flexibility in aircraft attitude control leads to an increased kill ratio.

3.2 Why Don't CCVs Have Extra Hardware?

Increased flexibility in aviation was the target of research in Control-Configured Vehicle technology in the 1970s. The basic design for aircraft was based on stability. The demand for stability restricted the flexibility of aviation, which lead to a drawback in air-combat capabilities. Relaxed Stability Technology was introduced in order to increase the flexibility of air-combat capabilities. In order to bring relaxed stability, the initial test aircraft was equipped with special hardware (e.g. special wings). This special hardware facilitated an increased flexibility in aircraft attitude. During development, the sensor and control technology was

Fig. 1 Softwarization of special hardware in advancement of RST technology

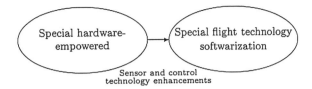

Table 1 Softwarization of special hardware

Aspect	Description
Advantages	Improved indefectibility. Flexibility of added features
Disadvantages	Increased risk of instability. Increased demands of control processing power

enhanced. The final CCV achieved relaxed stability without the initial special hardware. The transition is depicted in Fig. 1.

The implications of this softwarization of special hardware are summarized in Table 1.

The most significant advantage is the improved indefectibility. The lack of physical characteristics makes the radar-detection of the relaxed stability-empowered aircraft difficult. This brings a critical advantage in the air-combat situations.

3.3 Implications for Organizational Management from RST

Advances in information technology increase the rapid speed of industry changes.

The needs to cope with these changes have increased on a global scale. Market changes are fast and radical, and today's organization have to cope with these challenges using fast decision making and execution capabilities.

In order to cope with the challenges of market dynamism, organizations need to use a fusion use of different aspects of organization structures, as depicted in Fig. 2.

Market demands require increased flexibility of organizational capabilities, with softwarization of organization management, as depicted in Fig. 3.

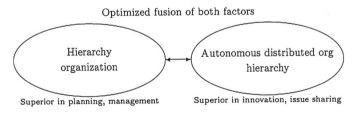

Fig. 2 Utilization of multiple aspects of organization structures

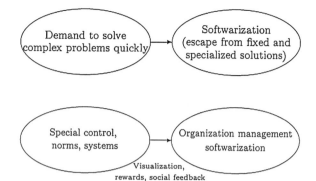

Fig. 3 Softwarization of organization management

Fig. 4 Examples of the softwarization of organization management

In order to facilitate the improvement of agility in organizations, it is necessary to relax the existing constraints. These constraints were created to maintain the stability of organization. Relaxation requires additional spontaneous stabilization mechanisms to deal with decreased stability.

Examples of softwarization of organization management are depicted in Fig. 4.

There are several approaches to relaxed organizations. For example, it is feasible to construct a special team with specially-trained members. Another example is software control of an organization using visualization, rewards, and social feedback. The other is the softwarization of organization management, using software-controlled feedback based on ubiquitous real-time monitoring systems.

4 RSO Theory

4.1 Definition

RSO is defined as an organization that is managed by a highly coordinated visualization and feedback system in order to improve agility and flexibility with the intentional elimination of norms and regulations in existing organizations. In this context, "existing organizations" refers to organizations with legacy norms, regulations and fixed structures, such as hierarchy organizations.

4.2 What are Sensors and Controllers in RSO?

The lessons from mobile social game design for gamification of organization management are depicted in Table 2.

It is interesting to note that these techniques are deployed by skillful managers with conscious or unconscious manners in a real world landscape. Frankly speaking, these techniques are universal metrics that deal with the creation of

Table 2 Lessons from mobile social game design

Aspect	Summary
Visualization	Visualization of achievement and the next target
Real-time human rewards	Rewards from human beings in a real-time manner
Sense of honor	Constant awareness of honor

human hopes. Hope represents a certain type of expectation that combines rewards with a confidence of achievement. Engineering that deals with the creation of human hopes has been widely recognized in the game design of mobile social games.

4.3 RSO Framework

The transition to a flexible organization is depicted in Fig. 5.

In order to pursue a flexible organization, it is necessary to remove fixed norms. The fixed norms are replaced with dynamic management with monitoring and rewards. Monitoring and rewards contribute to building a positive expectation with a flexible goal-achievement cycle.

A framework that facilitates RSO is depicted in Fig. 6.

Flexible organizations are open to external resources. Spontaneous collaboration is facilitated through various sharing and socialization mechanisms such as lunch socials. Teams are virtual, therefore, the human capital is sought on demand. Sometimes, massive external resources are utilized using cloud-sourcing where a large-scale problem is split into small pieces. Each piece can be solved in a distributed and global manner. Millions of people can contribute to the problem solving through cloud sourcing. Wikipedia is one of such examples.

The transition to SRO is depicted in Fig. 7.

Softwarization in RSO is depicted in Fig. 8.

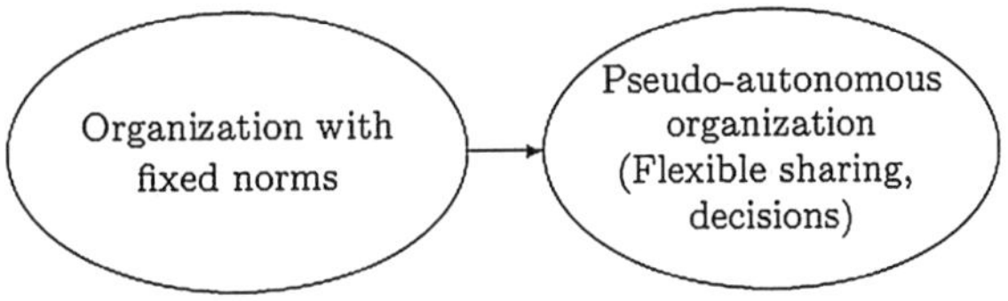

Fig. 5 Transition to flexible organization

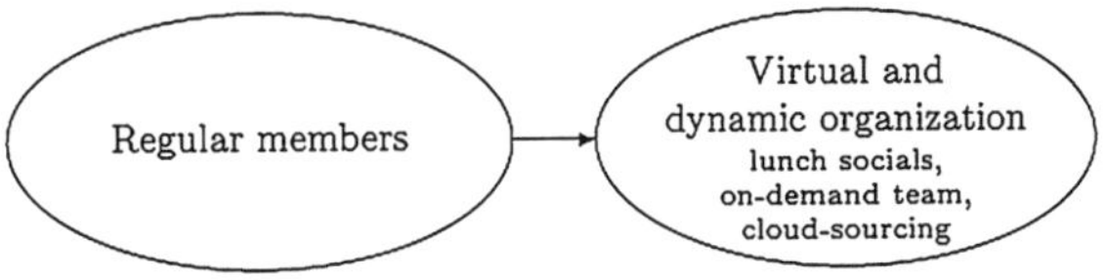

Fig. 6 Framework that facilitates RSO

Fig. 7 Transition to RSO

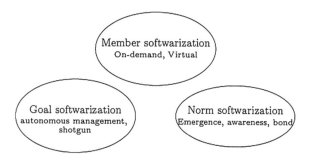

Fig. 8 Softwarization in RSO

There are three types of softwarization in RSO. One is member softwarization, where on-demand teams and virtual teams are built using an open innovation scheme. Another is goal softwarization. Relaxation of goal management is another contributing factor for RSO. The other is norm softwarization. Relaxed stability facilitates the management of flexible norms. This is a short-term instability factor. However, it contributes to the agility of the organization.

The design principles of RSO are depicted in Fig. 9.

Detailed analysis of RSO design principle is beyond the scope of this paper.

The author proposes three principles of RSO to serve as the basis for further research. One is minimum organization design where the minimum set of organizational components is established for relaxed stability. Another is tool-based dynamic team management. Relaxed stability requires constant monitoring and feedback to compensate instability of an organization. Tools are required for this purpose. The other is virtual team management, where the members of the team are flexible.

RSO-based organization management is depicted in Fig. 10.

Stability is required to maintain an organization. The main purpose of RSO is to decrease stability in order to increase organizational performance. For this

Fig. 9 Design principles of RSO

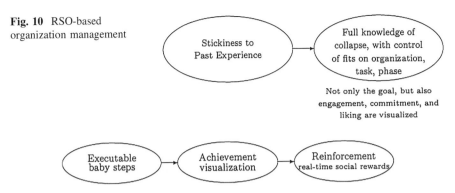

Fig. 10 RSO-based organization management

Fig. 11 Basic cycle in the micro-management

purpose, the large-scale know-how of collapses is required, just as the relaxed stability technology in military aviation required massive knowledge of aviation collapses.

Engagement, commitment, liking and culture are studied for this purpose. Dynamism of organization and organizational performance requires detailed analysis.

Gamification in enterprise environments follows the basic cycle of micro-management as depicted in Fig. 11.

For monitoring and feedback mechanisms, gamification is a promising candidate technology to be deployed in RSO. The first step is executable baby steps to facilitate achievements and changes. The second step is achievement visualization that increases engagement and emotional rewards for hard work. The third step is reinforcement using real-time social rewards. A detailed discussion of component design in this cycle remains for further research.

5 Discussion

5.1 The Advantages of the Proposed Approach

The author discusses a new umbrella concept, RSO, in analogy to RST in military aviation. There are two merits to this approach:

The factors that drive RSO are depicted in Table 3.

The concept that intentionally introduces decreased stability.

The concept that management can be maintained with decreased stability without any additional special mechanisms in office systems.

For the first point, stability is an unwritten rule of common organizations. It is difficult to manage an organization without stability. Stability serves a purpose for work metrics, workplace norms, interpersonal relationships, and so on. It is unique to place importance on the intentional decrease of stability.

Table 3 Factors that drive RSO

Factor	Summary
Open innovation	Today's organization needs to cope with the demands for open innovation, leveraging capabilities to accommodate open innovation with cross-boundary collaboration
Demands for agile operation	It is necessary to facilitate agility to cope with a changing industrial landscapes. It is also necessary to take initiative for changes
Demands for flexible management for productivity	In order to leverage workplace productivity and satisfaction, it is necessary to cope with flexible management through which detailed work contexts are considered and honored

With regard to the second point, it is analogous to the use of visualization and feedback systems in gamification. For example, the original RST utilizes real-time sensor and control systems.

Today's organization has to cope with challenges and increased demands for work efficiency, improvement of workplace satisfaction, and flexibility that deals with a changing external landscapes.

The accommodation of intentional decrease of stability (in analogy to RST in military aviation) in order to reengineer organizational management is a revolutionary idea.

5.2 Limitations

This research is a qualitative study. Quantitative measures for verifying multiple aspects of RSO discussed in this paper remain for further study.

Acceptance of RSO in a real world environment is beyond the scope of this paper. The concrete design methodology of an RSO-oriented office systems is beyond the scope of this paper.

6 Conclusion

The author discusses a new umbrella concept for corporate management, RSO, in analogy to the RST used in military aviation. In military aviation, the demands for increased flexibility in air-combat capabilities led to RST, the intentional decrease of stability.

The author proposes RSO concept based on an inspiration gained from RST in the organizational management. The emergence of gamification provides another insight into the fact that available new technologies can provide new fits in the organizational management.

Relaxed Stability Technology Approach

Acknowledgments The author expresses thanks to Toshiaki Fujii, NTT Comware, for his insightful suggestions on RST.

Reference

1. Grudin J (1994) Groupware and social dynamics: eight challenges for developers. CACM 37(1):92–105

Mapping and Optimizing 2-D Scientific Applications on a Stream Processor

Ying Zhang, Gen Li, Hongwei Zhou, Pingjing Lu, Caixia Sun and Qiang Dou

Abstract Stream processors, with the stream programming model, have demonstrated significant performance advantages in the domains signal processing, multimedia and graphics applications, and are covering scientific applications. In this paper we examine the applicability of a stream processor to 2-D stencil scientific applications, an important and widely used class of scientific applications, which compute values using neighboring array elements in a fixed stencil pattern. We first map 2-D stencil scientific applications in FORTRAN version to the stream processor in a straightforward way. In a stream processor system, the management of system resources is the programmers' responsibility. We then present several optimizations, which avail the stream program for 2-D stencil scientific applications, of various aspects of the stream processor architecture. Finally, we analyze the performance of optimized 2-D stencil scientific stream applications, with the presented optimizations. The final stream scientific programs gain from 2.56 to 7.62 times faster than the corresponding FORTRAN programs on a Xeon processor, with the optimizations playing an important role in realizing the performance improvement.

Y. Zhang (✉) · G. Li · H. Zhou · P. Lu · C. Sun · Q. Dou
School of Computer, National University of Defense Technology, Changsha 410073, China
e-mail: zhangying@nudt.edu.cn

G. Li
e-mail: genli@nudt.edu.cn

H. Zhou
e-mail: hwzh@nudt.edu.cn

P. Lu
e-mail: pjl@nudt.edu.cn

C. Sun
e-mail: cxsun@nudt.edu.cn

Q. Dou
e-mail: qd@nudt.edu.cn

J. J. (Jong Hyuk) Park et al. (eds.), *Multimedia and Ubiquitous Engineering*,
Lecture Notes in Electrical Engineering 240, DOI: 10.1007/978-94-007-6738-6_55,
© Springer Science+Business Media Dordrecht(Outside the USA) 2013

1 Introduction

Stream processors [1, 2] have demonstrated significant performance advantages in media applications [3]. Many researchers are interested in the applicability of stream processors to scientific computing applications [4].

The stream processor architecture, which has many differences from the architecture of a conventional system, is designed to implement the stream programming model [5]. Although language implementation exploit the model's features well, they do so at such a comparatively low-level; it is mainly the programmer's responsibility to manage system resources. Moreover, compared to other stream applications, such as media applications, scientific computing applications have more complex data traces and stronger data dependence. Therefore, writing a high-performance scientific stream program is rather hard and important to get right and high performance.

2-D stencil scientific applications, an important and widely used class of scientific applications, have special data access trace and compute values using neighboring array elements in a fixed stencil pattern. This stencil pattern of data accesses is then repeated for each element of the array. This paper first take 2D Jacobi iteration as an example to illuminate our mapping and optimization, and our presented methods can be used to any 2D stencil scientific application. Figure 1 presents the code for 2-D Jacobi iteration. The Jacobi iteration kernel consists of a simple 4-point stencil in two dimensions, shown in the first part of Fig. 2. On each loop iteration, four elements of the array are accessed in the 4-point diamond stencil pattern shown on the left. As the computation progresses, the stencil pattern is repeatedly applied to array elements in the column, sweeping through the array, as shown in the second part of Fig. 2.

Fig. 1 Code for 2-D Jacobi iteration

```
A(N,N), B(N,N)
do J=2,N-1, I=2,N-1
   A(I,J)=
   C*(B(I-1,J)+
   B(I+1,J)+
   B(I,J-1)+
   B(I,J+1))
```

Fig. 2 Data traces for 2D Jacobi

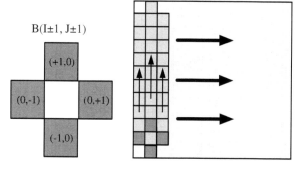

In this paper, we use the language streamC/kernelC [1] to map FORTRAN version of 2-D stencil scientific application to the stream processor. A straightforward mapping method is first given to map 2-D Jacobi iteration to the stream processor; optimizations are then proposed to improve the overall performance of the mapped stream program. Finally, the performance of the stream programs for some typical-D stencil scientific applications, and the effectiveness of our optimizations are measured through a number of experiments. Compared with FORTRAN program on a Xeon, our stream programs finally achieve from 2.56 to 7.62 times speedup.

2 Background

The stream processor architecture is developed to speed up stream applications with intensive computations. The stream programming model divides a application into a stream-level program that specifies the high-level structure of the application and one or more kernels that define each processing step [6]. Each kernel is a function that operates on streams, sequences of records (Fig. 3).

Popular languages implementing the stream programming model include StreamC/KernelC [1] for Imagine and Merrimac processor, Brook [7] for GPU and SF95 for FT64 processor. These languages can also be used to develop stream programs for Cell Processors [8]. All these stream architectures have the characteristic of SIMD stream coprocessors with a large local memory for stream buffering. Figure 4 shows a simplified diagram of such a stream processor. A stream-level program is run on the host while kernels are run on the stream processor. A single kernel that operates sequentially on records of streams is executed on clusters of ALUs, in a SIMD fashion. Only data in the local register files (LRFs), immediately adjacent to the arithmetic units, can be used by the clusters. Data passed to the LRFs is from the Stream Register File (SRF) that directly access memory. On-chip memory is used for application inputs, outputs and for intermediate streams that cannot fit in the SRF.

Fig. 3 Stream programming model

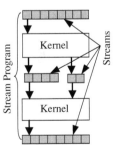

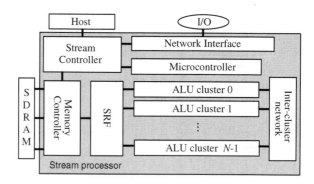

Fig. 4 A stream processor

3 Straightforward Map and Optimization

3.1 Straightforward Map

The implementation of mapping applications to the stream programming model can be thought of as a code transformation on programs that consist of a series of loops that process arrays of records. The data traces of different references in the innermost loop are extracted into different streams, and the computations performed by each loop are encapsulated inside a kernel. The remaining code composes the stream program.

When 2-D Jacobi iteration is mapped, the corresponding stream program declares five streams: four that correspond to the data accessed by four references to array B in I-loop and a fifth, a_0, that corresponds to the data accessed by the array a in I-loop. After declaring the streams, the stream program then calls a kernel that processes the streams b_0, b_1, b_2 and b_3 to produce the stream a_0. The kernel is declared, taking four input streams ("istreams"), one output stream ("ostream") and one microcontroller variables as arguments. If first reads the values of the microcontroller variables, then loops over the records in the input streams computing records in the output stream.

3.2 Exploiting the Reuse of Record

The SRF is banked into lanes such that each lane supplies data only for its connected cluster. Records of a stream are interleaved among lanes. If a cluster needs data residing in another cluster, it gets the data by inter-cluster communication.

In the FORTRAN code of 2-D Jacobi iteration, two neighboring elements, i.e. $B(I + 1, J)$ and $B(I - 1, J)$, of array B are involved in each iteration of I-loop. When mapping such loops, we have two choices as follow:

- Organize data as different streams to start and to end at different offsets into the original data arrays. The data covered by the two references, $B(I - 1, J)$ and

Mapping and Optimizing 2-D

B(I + 1, J), in I-loop are organized into the streams b_0 and b_1 in referred order. Although the data in the records of every stream are almost the same, just displaced, all the streams must be loaded from off-chip memory.

- Organizing data covered by the two references in I-loop as a single stream. In this way, the streams b_0 and b_1 in *StreamJacobi* are merged into one stream $b((j + 1) * N, (j + 2) * N - 1)$. However, during the kernel example execution, $cluster_i$ must communicate with $cluster_{i+2}$ to gather the needed record by inter-cluster communication.

We reorganize the data distribution, with adjacent records distributed on the same lane. Thus, the optimized stream program has the same number of memory transfers with the second choice, but does not require any inter-cluster communication. The steps of exploiting the reuse of records for *StreamJacobi* are given below.

***Step* 1** Organize all records covered by the two array references B(I + 1, J) and B(I − 1, J) as a new stream, with the same order as the array B.

***Step* 2** Set the stride of the new stream be $Length_{stream}/N_{cluster}$ and the record length be $Length_{stream}/N_{cluster} + 2$, where $N_{cluster}$ is the number of clusters in the stream processor and $Length_{stream}$ is the length of the new stream. This means the original records from $i \times Length_{stream}/N_{cluster}$ to $(i + 1) \times Length_{stream}/N_{cluster} + 2$ in the new stream become the ith record of the new derived stream, b_0', residing in the ith cluster. Data distribution is transposed as shown, with neighboring records distributed on the same lane, such that $cluster_i$ gets neighboring records from the lane of itself without any inter-lane communication.

***Step* 3** Divide all other stream references into $N_{cluster}$ parts by setting the stride be $Length_{stream}/N_{cluster}$ and the record length be $Length_{stream}/N_{cluster}$.

***Step* 4** Update original kernel to process the records in the corresponding new order.

After the optimization, the final stream-level program, has little inter-cluster communication and less memory transfers.

3.3 Exploiting the Reuse of Streams

The stream length, record length and stride of the streams b_2' and b_3' are changed as those of the stream b_0', with the changed streams named b_2'' and b_3''; correspondingly, the kernel is updated to process the correct records, with the changed code. The relationship among the locations, accessed by the streams b_0', b_2'' and b_3'' on neighboring iterations. The stream b_3'' on iteration i, the stream b_2'' on iteration $i + 1$ and the stream b_0' on iteration $i + 2$ access the same locations. Since the values of the basic stream b are unchanged, the stream b_2'' does not require accessing off-chip memory but accesses the SRF to get the values that are used by the stream b_3'' in a previous iteration. Similarly, b_0' can access the SRF to get the values that are used by the stream b_3'' in two previous iterations.

However, stream compilers cannot recognize and utilize the reuse supplied by the streams b_0', b_2'' and b_3''. This is because the start and end bound of these streams are variables, which means they are unknown when stream compilers allocate the SRF for streams.

We optimize it by introducing four basic streams, b_{00}, b_{01}, b_{02} and a_{00}, initializing b_{02} and b_{01} before the loop, defining b_3'' and a_0', loading the values referred to by b_3'' to b_{00}, replacing the references b_0', b_2'', b_3'' and a_0' with basic streams b_{02}, b_{01}, b_{02} and a_{00} respectively, saving the output a_{00} to the locations defined by a_0' and moving the values of b_{01} and b_{00} to b_{02} and b_{01} at the end of the loop body. The function *streamCopy(s, t)* copies records of s to t. An SRF-to-memory copy generates a save of s to memory; a memory-to-SRF copy generates a load of s to the SRF; an SRF-to-SRF copy generates a save of s to memory and a load of s to the SRF buffer that holds t. We can effect the reuse by replacing streams that have unknown starts and ends with streams that have constant starts and ends, i.e. basic streams, and explicitly transferring original reuse to the reuse among streams with constant starts and ends. The stream compiler will recognize and utilize the reuse in the transformed code. However, it does so at the expense of introducing two expensive SRF-to-SRF data moves. Since these moves implement a permutation of values in the SRF, we can eliminate the need for moves by unrolling to the cycle length of the permutation, i.e. 3 times and permuting the stream references in each unrolled loop bodies.

4 Experimental Setup and Performance Evaluation

In our experiments, we use Isim [2], a cycle-accurate stream processor simulator supplied by Stanford University, to get the performance of the stream applications with different versions. Table 1 summarizes the applications we used. The baseline configuration of the simulated stream processor and its memory system is detailed in Table 2, and is used for all experiments unless noted otherwise. For comparison, 2-D stencil scientific applications in FORTRAN is compiled by Intel's IA32 compiler (with max speed optimization option), and run on a Xeon processor, one class of the most popular machines used for scientific computing applications now. Table 3 shows the configuration of the Xeon processor.

Table 1 Application programs

	Problem size
2-D Jacobi	256*256
2-D Laplace	1 K*1 K
MG	128*128*128
QMRCGSTAB	800*800
MVM	832*832

Table 2 Baseline parameter of Isim

Parameter	Value
Cluster number	8
LRF	38.4 KB
SRF	512 KB
Off-chip DRAM	4 GB
Frequency	2 GHz

Table 3 Configuration of the Xeon processor

Parameter	Value
Core number	8
Frequency	2.7 GHz
L1 Cache	8*32 K*2
L2 Cache	8*256 K
L3 Cache	20 M

4.1 Overall Performance

The performance of 2-D stencil scientific stream applications with all optimizations is first presented to evaluate the stream processor's ability to process 2-D stencil scientific applications. Figure 5 shows the speedup yielded by final stream programs, over FORTRAN programs. Optimized stream programs yield from 2.56 to 7.62 times speedup, which indicates the stream processor can successfully process such class of scientific applications. This is because plenty of ALUs process computations in 2-D stencil scientific applications; data reuse is all exploited; memory transfers are overlapped with kernel execution perfectly.

Compared with other applications, MG has much record reuse and stream reuse, and thus gets highest speedup, 7.62. But for QMRCGSTAB, only part of its computation is accelerated, and thus only gets a speedup of 2.56.

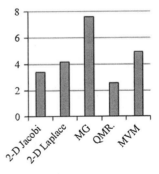

Fig. 5 Speedup yielded by Isim over Xeon

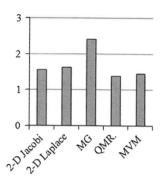

Fig. 6 Speedup from exploiting the reuse of records

4.2 Exploiting the Reuse of Records

We now demonstrate the effectiveness of exploiting the reuse of records that reorganizes streams to reduce off-chip memory transfers. Figure 6 demonstrates marginal speedup, due to the reduction of memory transfers. One of two input records of Jacobi iteration reuses previous generated data; one of two input records of Laplace reuses previous generated data; two of three input records of MG reuse previous generated data; one of two input records of QMRCGSTAB reuses previous generated data; one of two input records of MVM reuses previous generated data. Correspondingly, they yield speedup of similar trend.

4.3 Exploiting the Reuse of Streams

As a stream processor reduces memory transfers only by capturing the reuse among streams in the SRF, this optimization is important. We evaluate the impact of exploiting the reuse of streams on program performance. Our stream programs benefit greatly from this optimization. Figure 7 demonstrates the speedup attained with this optimization. One out of two input streams of Jacobi iteration reuse the

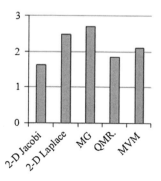

Fig. 7 Speedup from exploiting the reuse of streams

data generated on the previous iteration; two out of three input streams of Laplace reuse the data generated on the previous iteration; two out of three input streams of MG reuse the data generated on the previous iteration; two out of three input streams of QMRCGSTAB reuse the data generated on the previous iteration; two out of three input streams of MVM reuse the data generated on the previous iteration. Thus, a lot of memory transfers are reduced for each stream program. Without the optimization, the stream compiler cannot identify the reuse, all input streams must be loaded from off-chip memory and each kernel must wait until its input streams are loaded from off-chip memory. Stream reuse removes the appearance of streams with unknown starts and ends, thus making the stream compiler able to identify reuse.

Acknowledgments This work was supported by NSFC (61003075, 61103193,61103011, 61103014).

References

1. Rixner S (2001) Stream processor architecture. Kluwer Academic Publishers, Boston
2. Kapasi U, Dally W, Rixner S, Owens J, Khailany B (2002) The imagine stream processor. In: Proceedings of 2002 IEEE international conference on computer design, pp 282–288
3. Gordon M, Maze D, Amarasinghe S, Thies W, Karczmarek M, Lin J, Meli A, Lamb A, Leger C, Wong J et al (2002) A stream compiler for communication-exposed architectures. ACM SIGARCH Comput Archit News 30(5):291–303
4. Fatica M, Jameson A, Alonso J STREAMFLO: an Euler solver for streaming architectures, submitted to AIAA conference
5. Kapasi U, Rixner S, Dally W, Khailany B, Ahn J, Mattson P, Owens J (2003) Programmable stream processors. Computer 36(8):54–62
6. Das A, Dally WJ, Mattson P (2006) Compiling for stream processing. In: proceedings of the 15th international conference on parallel architectures and compilation techniques PACT '06. ACM Press, New York, pp 33–42
7. Buck I, Foley T, Horn D, Sugerman J, Fatahalian K, Houston M, Hanrahan P (2004) Brook for gpus: stream computing on graphics hardware. ACM Trans Graph 23(3):777–786
8. Kahle JA, Day MN, Hofstee HP, Johns CR, Maeurer TR, Shippy D (2005) Introduction to the cellmultiprocessor. IBM J Res Dev 49(4/5):589–604

Development of an Android Field Trip Support Application Using Augmented Reality and Google Maps

DongLim Hyun, EunGil Kim and JongHoon Kim

Abstract In this study, an application to support field learning was developed to apply to education Location Based Service (LBS) which is expanding with the spread of smartphones. To select functions in tune with the purpose, a requirements survey was conducted for current elementary school teachers, and the application was developed by reflecting the requirements based on an analysis of the survey results. Through the developed application, teachers can provide information to students and students can acquire information (location, pictures, description) required for field learning and carry out autonomous activities. The developed application was explained and demonstrated to current elementary school teachers who were given an experience of using it. A survey to verify its effectiveness was conducted and it was found that the application had a high potential for being utilized in field learning.

Keywords Field learning · Augmented reality · Android

1 Introduction

Ubiquitous Internet access is changing the users of smartphones. These social changes are leading to changes in curriculums and educational methods, and particularly, field trips are becoming more important. This study developed a field

---

D. Hyun · E. Kim · J. Kim (✉)
Department of Computer Education, Teachers College, Jeju National University,
Jeju, South Korea
e-mail: jkim0858@jejunu.ac.kr

D. Hyun
e-mail: gody5@naver.com

E. Kim
e-mail: computing@korea.kr

J. J. (Jong Hyuk) Park et al. (eds.), *Multimedia and Ubiquitous Engineering,*
Lecture Notes in Electrical Engineering 240, DOI: 10.1007/978-94-007-6738-6_56,
© Springer Science+Business Media Dordrecht(Outside the USA) 2013

trip support application that can be used in the future, following the changes in society and the educational environment, using augmented reality and Google Maps in the Android platform. Such application was developed by experienced teachers for use in schools. It will promote cooperative learning by providing information to teachers and students and enabling information sharing among students using the features of smartphones.

2 Theoretical Background

2.1 Augmented Reality

There are mainly three methods of implementing augmented reality: the layer method, the marker recognition method, and the markerless method. The layer method identifies the location and position of smartphones using the phone camera, GPS, and sensors, and adds information to the images captured by the camera. The marker recognition method positions specific markers at the locations where information must be displayed, and the markers captured by the camera are recognized and the corresponding information, displayed. The markerless method extracts the feature points from the images captured by the camera that replaced the markers. The application in this study was implemented with the augmented reality of the layer method, considering the limited computation capacity of smartphones and users' need for easy addition of information.

2.2 Related Studies

So-Hee Kim (2004) and Yoon-Kyung Min (2005) designed and developed a field learning support system [1, 2]. In these two studies, however, the actual activities of students are carried out through the Web and wireless terminals only send information to the Web.

Seung-Ah Lee (2010) designed and developed a field learning support application for performing missions of a scenario using Android-based smartphones [3]. However, it simply provides students with missions and students carry out the missions using the functions of smartphones.

Many studies presented applications for supporting field learning. However, they just use MMS service or Web pages and are not optimized to the widespread smartphones and other smart devices or they just designed and proposed the systems.

A difference between our study and previous studies is the use of LBS applications through smartphones. In this study, the information supply and sharing and missions which were implemented in previous studies were combined with the status and location of users, and they were applied to field learning in a more diverse and effective ways such as augmented reality and maps. Furthermore, user convenience was improved by providing all these functions through a single application.

3 Scope and Functions of the Application

To define the scope and functions of the field learning support application, a survey on the required functions of the field learning application was conducted with 15 current elementary school teachers who were using smartphones or had high interest in them. The key questions and results of the survey are shown in Table 1.

This survey found that elementary schools were performing field learning once or twice per semester and up to once a month depending on the circumstances of the school and for the percentage of places for field learning, the percentage of outdoor places was a little higher than that of indoor places. Furthermore, for the functions of field learning support applications, they required information sharing between teachers and students and between students, destination search through maps and augmented reality, and notification about approaching the destination.

4 Implementation of Field Learning Support Application

4.1 Implementation of Augmented Reality

To synchronize the camera image with the image of the destination, the coordinates to be indicated on the screen were calculated by combining the values of the sensors.

The azimuth of the destination was computed by substituting the GPS coordinates of the destination with those of the smartphone using the computation algorithm issued by the National Geographic Information Institute [4].

Table 1 Content and results of requirements survey

Question 1. How often do you perform field learning a year on average?			
• 3–4 times	47 %	• 5–6 times	40 %
• 7–8 times	7 %	• 9 times or more	7 %
Question 2. What are the percentages of indoor and outdoor places for field learning?			
• Indoor	40 %	• Outdoor	60 %
Question 3. Which functions of the application would assist field learning? (multiple answers are allowed)			
• Information input and sharing through maps			80 %
• Search for destinations through augmented reality			60 %
• Display of destinations and current locations on maps			53 %
• Notification of arrival at destinations and confirmation of inputted information			33 %

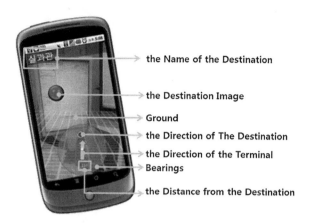

Fig. 1 Augmented reality screen

Compass and land prices were added to the developed application for user convenience. The application screen is shown in Fig. 1.

4.2 Use of Google Maps

A destination can also be added to the map by touching such destination on the map and getting the GPS values. The destination objects are managed as a list and can be easily added or removed. Users can add the name, description, and images of destinations in the Alert Dialog Window to add a destination. The name and description can be entered through the smartphone keyboard, and the image can be added by taking a picture with the smartphone camera or selecting one of the existing images. The application screen is shown in Fig. 2.

4.3 Sharing Through the Web

The application allowed the sharing through the Web of information created by teachers. Students can also share their data and use them for small group learning and cooperative learning activities. The application screen is shown in Fig. 3.

4.4 Development of the Destination Approach Notification Feature

The user must be notified when the destination set in AR Activity or Map Activity is near. This feature is provided as an Android API, so the developer only needs to set the destination coordinates, approach distance, action on notification, etc.

Development of an Android Field Trip

Fig. 2 Screens created using Google Maps

Fig. 3 Screen for sharing through the Web

In the developed application, the approach distance was set at 10 m and the observation time, at always; and the action on notification was set at viewing the destination information on a Alert Dialog Window. The application screen is shown in Fig. 4.

Fig. 4 Screen for destination approach notification

5 Expert Assessment

The feasibility and improvements of the developed field learning application were diagnosed through expert assessment. 15 current class 1 elementary school teachers were selected for the experts.

For this expert assessment, the application was explained to the experts and they practiced it before answering a questionnaire by checking one of the answers in five-step Likert scale with 1 point intervals. The expert assessment areas and results are shown in Table 2.

The expert assessment results showed generally high satisfaction levels on the application developed in this study, and the utilization potential of the application

Table 2 Expert assessment areas and results

Area 1. Satisfaction on screen layout	
• Satisfaction on the augmented reality screen layout and menu	4.5
• Satisfaction on the map screen layout and menu	4.5
• Satisfaction on the displayed information and buttons in the notification window	4.6
Area 2. Satisfaction on the implemented functions	
• Sharing and expression of information (location, pictures, description) on the destinations through a server	4.8
• Search for destinations through augmented reality	4.4
• Display of destinations and current location on the map screen	4.6
• Notification for approaching the destination	4.7
Area 3. Intuitiveness and convenience of operation	
• Intuitiveness and convenience of operation on the augmented reality screen	4.6
• Intuitiveness and convenience of operation on the map screen	4.6
• Intuitiveness and convenience of operation for switching between augmented reality screen and map screen	4.7

in field learning was found to be high. In particular, information sharing and notification of approaching a destination were highly evaluated as they could be applied to outdoor learning as well as field learning.

6 Conclusion

Society is changing fast, and schools are seeking changes in their educational methods in line with the social changes. This study paid attention to the rapid propagation of smartphones and the emphasis on field trips in the changed curriculums. It is expected that learning handsets similar to smartphones will be supplied to students and that the importance of field trips in schools will increase.

The developed application can be useful for the future educational environment. More active studies in this area are needed to prepare for the future educational environment.

References

1. Google Public Interface Map. http://developer.android.com/reference/java/util/Map.html
2. Kim SH (2003) A wireless/wired field-experience learning support system using mobile handsets. Master's Thesis, Ewha Womans University
3. Min YK (2005) Design and implementation of support system for field education learning based ubiquitous. Master's Thesis, Ewha Womans University
4. National Geographic Information Institute. http://www.ngii.go.kr/kor/board/download.do?rbsIdx=31&idx=278&fidx=1

Implementation of Automotive Media Streaming Service Adapted to Vehicular Environment

Sang Yub Lee, Sang Hyun Park and Hyo Sub Choi

Abstract Among the variety of vehicle technology trend issues, the biggest one is focus on automotive network system. Especially, optical network system is preferred. The aim in optical network system including vehicular environment information is that realization of media streaming service beyond the current car audio system. This paper is introduced the implementation of media streaming service which is consist of optical network as called MOST (Media Oriented System Transport) and realization of car audio system adapted to vehicular environment information collected from (On-Board Diagnosis) OBD via (Controller Area Network) CAN.

Keywords In-vehicle network systems · MOST · OBD · CAN · Car sound systems · Media streaming service

1 Introduction

With the passage of time, vehicle technology in modern times has been developed rapidly. Recent of today's automobile research area has been gradually changed from mechanics to electronics to the way of offering entertainment service to customs that can serve the connection to smart device easily and conveniently.

S. Y. Lee (✉) · S. H. Park · H. S. Choi
Jeonbuk Embedded System Research Centre, Korea Electronics Technology
Institute, Dunsan-ri, Bongdong-eup, Wanju-gun, Jeolabuk-do, Pyeongtaek-si 565-902,
Republic of Korea
e-mail: syublee@keti.re.kr

S. H. Park
e-mail: shpark@keti.re.kr

H. S. Choi
e-mail: hschoi@keti.re.kr

J. J. (Jong Hyuk) Park et al. (eds.), *Multimedia and Ubiquitous Engineering*, Lecture Notes in Electrical Engineering 240, DOI: 10.1007/978-94-007-6738-6_57, © Springer Science+Business Media Dordrecht(Outside the USA) 2013

In particular, most of people want to be experienced in high quality audio streaming service while they drive. According to customer's demands, MOST system is developed to provide an efficient and cost effective fabric to transmit audio data between any devices attached to the harsh environment of automobile. Compared to conventional car audio system which has been used in many of audio electrical line, bundle of lines per an audio channel, dissimilarly being with the one optical line in order to transit the audio data source, the advanced streaming service adopted optical network system makes light harness system. As the simply network construction, it can be obtained the high fuel efficiency from the usage of plastic optical fiber and for optical signal characteristics, it makes be free on electro-magnetic problems. Considered on driving environment, car audio system as the streaming service gathering automobile information via CAN enables to tune the volume automatically for their vehicular conditions. To develop the self-contained audio system, the complex information having vehicular status is needed to be accessed more comfortable.

In Sect. 2, MOST network system and MOST data processing are introduced for vehicle environment scenario. The vehicle information processing module which classified into vehicular status information to be collected from CAN network is described in Sect. 3. In Sect. 4, platform and demonstration of streaming service are shown. This paper is concluded in Sect. 5.

2 MOST Network System

MOST is the de-factor standard for efficient and cost effective networking of automotive multimedia and infotainment system [1, 2]. The current MOST standards released MOST150, 150 means that 150 Mbps network bandwidth with quality of services is available. To meet the demands from various automotive applications, MOST network system provides three different message channels: control, synchronous used in streaming service and asynchronous channel only for packet data transfer. In describe in Fig. 1, proposed network system is consist of MOST devices with CAN bus line. In conception of the network topology, MOST devices are divided into master and slave mode. Master mode device with CAN bus can be realized the optimum sound effect depending on vehicular environment information gathered by OBD interface in car.

2.1 MOST Steaming Data Processing

For the data processing, it is explained in streaming data part and network service one. Being presented the streaming data part; the bandwidth of the streaming data channel can be calculated using the following formula:

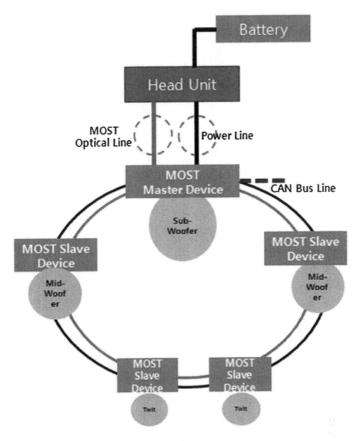

Fig. 1 MOST Network System with CAN bus line

$$\text{Bandwidth} = 93 \times 8\,\text{bit} \times 48\,\text{kHz} \qquad (1)$$

Synchronous data is used for the real-time transmission of audio data. Before data transmission, the data transmitted for the synchronous connection must be established by the connection master in network system. For this method, one socket has to be created and connected at the interfaces to the frame and to the local resource. Thus, up to 93 stereo connections can be established simultaneously on a MOST 150 frame. The content of the frame remains unmodified until the frame arrives back at the sending node. The quasi-static establishment of connections on a channel is denominated as (Time Division Multiplexing) TDM. The data are transmitted cyclically in a specified time pattern at the same frame position. There is no repetition in the case of communication errors. A valid value is then available the next cycle.

On the network service side, MOST devices have the unique interface which transfers the data between the physical layer network and processor as an external host controller [3]. Working in conjunction with the clock manager, the network

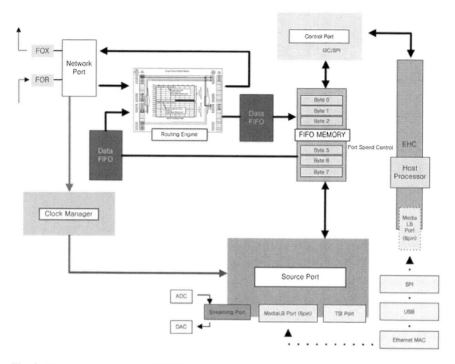

Fig. 2 The Data Process for MOST Frames

port recovers the network clock for time synchronization. And then, receiving data is decoded and delivered to the microprocessor via (Media Local Bus) MLB. Transferred data is routed into appropriate memory destination on and off in platform (Fig. 2).

3 Vehicular Status Information

3.1 Processing Module for Vehicle Information

As described in this paper, the specific module collecting automobile data is called as the vehicle information processing module. In the processing module, vehicle data transmission and channel connection is served in communication module and vehicle communicator executes the classification of information which factors on streaming service can be affected by while driving. Through the vehicle data analyzer and container, required vehicle status data is transferred to the platform using internal bus line without delay.

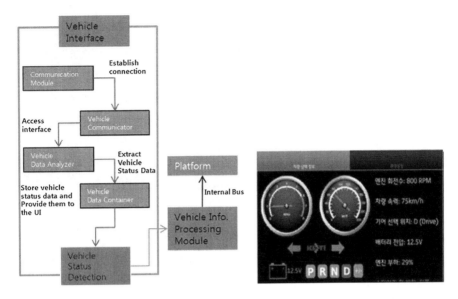

Fig. 3 The process of CAN Transceiver and Interpreter

Vehicular sensing module as depicted in Fig. 3 is coded into vehicular status information and status transit method. For applications, expressed black box on right side in Fig. 3, vehicle processing module is displayed engine rpm, vehicle speed, gear information, battery status and engine load to panel of platform.

4 Implementation of Streaming Service

4.1 System Architecture

The external host controller communicates with network interface controller via I2C bus, MLB and connected with OBD via CAN bus [4, 5]. As shown below, the streaming port included in network interface controller can be used for stereo audio exchange between the network and physical audio port. The external host controller may support for managing the streaming audio exchange remotely.

As described in Fig. 6, in order to make the data connection path, SRAM interfacing method is needed generally. Data transferred from the host bus accesses into SRAM memory space and linked in network interface controller through the MLB embedded in FPGA area. MLB is an on-PCB or inter-chip communication bus, designed to a common hardware interface. Especially, media local bus supports the MOST streaming data [6] Fig. 4.

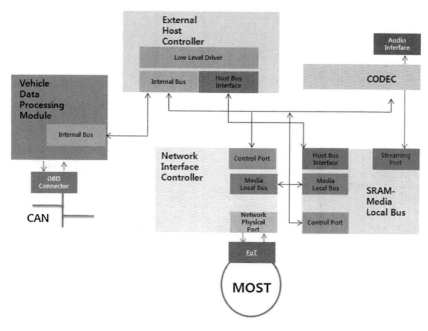

Fig. 4 System Architecture of Streaming Service

4.2 Streaming Service Platform

As mentioned that media local bus having SRAM memory interface is enable to exchange data frame. Designed logic block is mapped to Xilinx Vertex chipset and applied to streaming service platform. Table 1 provides a summary of the developed field programmable gate array utilization which can be shown that small memory usage can be expected the area effective size when this is made into system on chip level.

Table 1 The summary of device utilization

Target device		
	XC5VLX110	
Slice logic utilization		
	Number of slice registers	6,986
	Number of slice LUTs	15,347
	Number of route-thrus	897
	Number of occupied slices	6,103
	Number of LUT filp flop pairs used	17,233
	Number of BlockRAM/FIFO	18
	Number of BUFG/BUFGCTRLs	8
	Number of DCM_ADVs	2
	Total memory used	630 KB

Implementation of Automotive Media Streaming

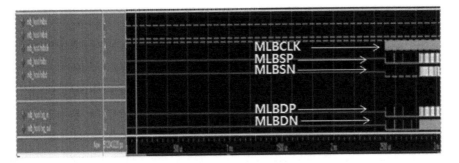

Fig. 5 The timing analysis of MLB

1: Media Local Bus

2: MOST network interface controller

3: CAN interface

4: External host controller

Fig. 6 Streaming Service Platform

All designed logic functions are showing the data and transmission commands when streaming service begins (Figs. 5, 6).

4.3 Streaming Service based on User Demanded Scenario

A proposed streaming service is reached to scenario as types of outside and inside for drivers and passengers. The environment of car audio system to be applied to proposed service are defined by below

- Outside Scenario1: Automotive audio sound can be affected by vehicle speed when moving car goes into high speed level, for being windy outside, the streaming sounds seems to be low with noise and it makes same situation when being rainy outside. Thus volume of sound system has to be up for itself.
- Outside Scenario2: Because driver has to be attention without sound disturbance, Automotive audio sound has to be down when driver want to move back, that is the gear shift paddle is located in rear mode.

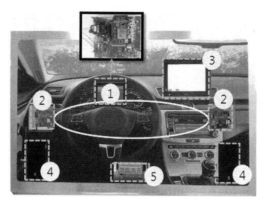

Fig. 7 System Demonstration

- Inside Scenario: Existed car audio system cannot control the woofers and speakers independently, however proposed streaming system which is assigned the network id for each device supports as a self-contained equipment. Especially, it can be the realization of optimum sounds effect according to the user's position and need in car.

As shown in Fig. 7, system demonstration is built with designed platform, audio platform connected with audio speaker and amplifier, CAN frame generator which makes the virtual environment via OBD interface applied to car. With running this set, it realizes that driver and passenger experience the car audio sound depending on vehicular status automatically without voluntary control.

5 Conclusion

For the trend of car infotainment system is moving to the high quality audio sound system, this paper is introduced the development of the audio streaming service based on optical network system. Particularly, optical network system, MOST, to be optimized sound level depending on vehicle environment is satisfied with reducing the weight and ensuring the reliability for the free of electro-magnetic problems. With vehicle status information, designed network platform realized and demonstrated that self-contained MOST speaker and woofer can be controlled and tuned their volume level automatically to be served to passengers more conveniently. Especially, outstanding point for implementation of adaptive audio streaming service is represented to hardware architecture and software frame depending on demanded vehicle environment scenario.

References

1. Grzemba A (2012) MOST Book from MOST25 to MOST150. MOST Cooperation, FRANZIS
2. Strobel O, Rejeb R, Lubkoo J (2007) Communication in automotive system principles, limits and new trends for vehicles, airplanes and vessels. In: IEEE ICTON, pp 1–6
3. Wegmuller M, von der Weid JP, Oberson P, Gisin N (2000) High resolution fiber distributed measurements with coherent OFDR. In: Proceedings of ECOC'00, paper 11.3.4, p 109
4. Lee SY (2010) A method of vehicular data collection and processing using wireless communications interface. In: KICS winter conference
5. Godavarty S, Broyles S, Parten M (2000) Interfacing to the on-board diagnostic system. In: Proceedings of IEEE vehicular technology conference 52nd-VTC, vol 4, pp 24–28
6. Lee SY, Park SH, Choi HS, Lee CD (2012) Most network system supporting full-duplexing communication. In: IEEE ICACT, Korea, pp 1272–1275

The Evaluation of the Transmission Power Consumption Laxity-Based (TPCLB) Algorithm

Tomoya Enokido, Ailixier Aikebaier and Makoto Takizawa

Abstract In order to realize energy-aware information systems, it is critical to discuss how to reduce the total electric power consumption of information systems. We consider applications of a communication type where a server transmits a large volume of data to clients. A client first selects a server in a cluster of servers and issues a file transmission request to the server. In this paper, we newly propose the transmission power consumption laxity-based (TPCLB) algorithm to select a server for a cluster of servers so that the total power consumption in a cluster of servers can be reduced. We evaluate the TPCLB algorithm in terms of the total power consumption and elapse time compared with the round-robin (RR) algorithm.

Keywords Green IT technology · Energy-aware information systems · TPC model · ETPC model · TPCLB algorithm

1 Introduction

In information systems, a client issues a request to a server and the server mainly consumes the power to perform the request as a process. We discuss software-oriented aspects of information systems to reduce the electric power consumption

T. Enokido (✉)
Rissho University, Shinagawa, Japan
e-mail: eno@ris.ac.jp

A. Aikebaier
National Institute of Information and Communications Technology (NICT),
Koganei, Japan
e-mail: alisher@nict.go.jp

M. Takizawa
Seikei University, Musashino, Japan
e-mail: makoto.takizawa@computer.org

J. J. (Jong Hyuk) Park et al. (eds.), *Multimedia and Ubiquitous Engineering*,
Lecture Notes in Electrical Engineering 240, DOI: 10.1007/978-94-007-6738-6_58,
© Springer Science+Business Media Dordrecht(Outside the USA) 2013

in a cluster of servers to realize energy-aware information systems [1–3]. In this paper, we consider communication type applications [2, 3] where a server transmits a large volume of data to clients.

The *transmission power consumption* (*TPC*) model of a server to transmit files is discussed in the paper [2]. In the TPC model, the electric power consumption of a server to transmit files to clients depends on the total transmission rate of the server. The approximated linear function to show how much a server consumes the electric power for transmission rates is derived from experimental studies of file transfer between servers and clients [2]. In the TPC model, the rotation speed of each fan is assumed to be fixed, i.e. the power consumption of each fan is constant. In current servers, the rotation speed of each fan can be changed to keep the temperature of each device. Thus, the total power consumption of a server depends on not only the power consumption of computation and communication devices but also cooling devices. The *extended TPC* (*ETPC*) model of a server is proposed to perform processes of communication type, where the power consumption of cooling devices are considered [3]. In this paper, we newly propose the *transmission power consumption laxity-based* (*TPCLB*) algorithm based on the TPC and ETPC models to select one of servers for communication type applications so that the total power consumption of servers can be reduced. We evaluate the TPCLB algorithm compared with the basic round-robin (RR) algorithm [4].

In Sect. 2, we present the file transmission model of a server. In Sect. 3, we present the power consumption models of a server. In Sect. 4, we discuss the TPCLB algorithm based on the TPC and ETPC models. In Sect. 5, we evaluate the TPCLB algorithm compared with the RR algorithm.

2 Transmission Model

Let S be a cluster of multiple data transmission servers $s_1,\ldots, s_n$ ($n \geq 1$), each of which provides clients with the same data d. Each server s_t holds a full replica of the data d. Let C be a set of clients $c_1,\ldots, c_m$ ($m \geq 1$). A client c_s issues a data transmission request to a load balancer K. The load balancer K selects one server s_t in the server cluster S and forwards the request to the server s_t. On receipt of a request, the server s_t transmits a reply file f_s to the requesting client c_s. Each request from a client c_s is performed as a process p_{ts}. Here, a notation p_{ts} shows a process p_s performed on a server s_t for a client c_s. A term *process* means an *application process* created for a request in this paper.

Let $CT_t(\tau)$ be a set of current transmission processes on a server s_t at time τ. $NT_t(\tau)$ shows the number of current processes, $NT_t(\tau) = |CT_t(\tau)|$. Suppose a server s_t concurrently transmits files $f_1,\ldots, f_m$ to a set C_t ($\subseteq C$) of clients $c_1,\ldots, c_m$ at rates $tr_{t1}(\tau),\ldots, tr_{tm}(\tau)$ ($m \geq 1$), respectively, at time τ. Let b_{ts} be the maximum network bandwidth [bps] between a server s_t and a client c_s. Let $Maxtr_t$ be the maximum transmission rate [bps] of the server s_t ($\leq b_{ts}$). The total transmission rate $tr_t(\tau)$ of the server s_t at time τ is given as $tr_t(\tau) = tr_{t1}(\tau) + \cdots + tr_{tm}(\tau)$. Here,

The Evaluation of the Transmission Power Consumption

$0 \leq tr_t(\tau) \leq Maxtr_t$ for each server s_t. Each client c_s receives a file f_s at receipt rate $rr_s(\tau)$ at time τ. Let $Maxrr_s$ indicate the maximum receipt rate of the client c_s. $rr_s(\tau) \leq Maxrr_s$.

Let TR_{ts} be the total transmission time [s] of a file f_s from a server s_t to a client c_s. Let $minTR_{ts}$ show the minimum transmission time $|f_s|/min(Maxrr_s, Maxtr_t)$ [s] of a file f_s from a server s_t to a client c_s where $|f_s|$ indicates the size [bit] of the file f_s. $TR_{ts} \geq minTR_{ts}$. Let $tr_{ts}(\tau)$ be the transmission rate [bps] of a file f_s from a server s_t to a client c_s at time τ. Suppose a server s_t starts and ends transmitting a file f_s to a client c_s at time st_{ts} and et_{ts}, respectively. Here, $\int_{st_{ts}}^{et_{ts}} tr_{ts}(\tau)d\tau = |f_s|$ and the transmission time TR_{ts} of the server s_t to the client c_s is $(et_{ts}-st_{ts})$. The transmission laxity $lt_{ts}(\tau)$ [bit] of transmission time is $lt_{ts}(\tau) = |f_s| - \int_{\tau}^{et_{ts}} tr_{ts}(\tau)d\tau$ at time τ $(st_{ts} \leq \tau \leq et_{ts})$, i.e. how many bits of the file f_s the server s_t still has to transmit to the client c_s at time τ.

First, we consider a model where a server s_t satisfies the following properties: **[Server-bound model]** If $Maxrr_1 + \cdots + Maxrr_m \geq Maxtr_t$, $\sum_{c_s \in CT_t(\tau)} tr_{ts}(\tau) = \sigma_t(\tau) \cdot Maxtr_t$ at every time τ.

Here, $\sigma_t(\tau)$ (≤ 1) is the transmission degradation ratio of a server s_t. In this paper, we assume $\sigma_t(\tau) = \gamma_t^{NT_t(\tau) - 1}$ $(0 < \gamma_t \leq 1)$ at time τ. Here, the *effective* transmission rate $maxtr_t(\tau)$ of the server s_t is $\sigma_t(\tau) \cdot Maxtr_t$ at time τ.

Suppose a client c_s cannot receive a file f_s from a server s_t at the maximum transmission rate $Maxtr_t$, i.e. $Maxrr_s < Maxtr_t$. Here, $tr_{ts}(\tau) = Maxrr_s$.

[Client-bound model] If $Maxrr_1 + \cdots + Maxrr_m \leq Maxtr_t$, $\sum_{c_s \in CT_t(\tau)} tr_{ts}(\tau) = Maxtr_t \cdot (Maxrr_1 + \cdots + Maxrr_m)/Maxtr_t$ at time τ.

Even if every client c_s receives a file f_s at the maximum rate $Maxrr_s$, the effective transmission rate is not degraded, i.e. $\sigma_t(\tau) = 1$.

In a fair allocation algorithm, the transmission rate $tr_{ts}(\tau)$ for each transmission process p_{ts} in the set $CT_t(\tau)$ is the same, i.e. $tr_{ts}(\tau) = maxtr_t(\tau)/NT_t(\tau)$. However, the maximum receipt rate $Maxrr_s$ of the client c_s might be smaller than $maxtr_t(\tau)/NT_t(\tau)$. Here, the rate $(maxtr_t(\tau)/NT_t(\tau) - Maxrr_s)$ is not used. In order to more efficiently use the total transmission rate $maxtr_t(\tau)$, at the higher receipt rate a client c_s would like to receive, at the higher transmission rate a server s_t allocates to the client c_s. In this paper, the transmission rate $tr_{ts}(\tau)$ for each client c_s at time τ is allocated by the following algorithm:

1. $V = 0; R = 0; TS = maxtr_t/NT_t(\tau);$
2. For each client c_s, $tr_t(\tau) = TS$ and $R = R + (TS - Maxrr_s)$ if $Maxrr_s \leq TS$. Otherwise, $tr_t(\tau) = Maxrr_s$ and $V = V + (Maxrr_s - TS)$.
3. For each client c_s, $tr_t(\tau) = tr_t(\tau) + V \cdot (Maxrr_s - tr_t(\tau))/R$ if $tr_t(\tau) < Maxrr_s$.

3 Power Consumption Model

We would like to discuss how much electric power a server s_t consumes to transmit files to clients. $maxE_t$ is the maximum electric power consumption rate [W] of the server s_t to transmit files. $minE_t$ is the minimum electric power consumption rate [W], i.e. the server s_t is in idle state. $minETR_t(\tau)$ is the minimum electric power consumption rate [W] of the server s_t to transmit a file. If $tr_t(\tau) > 0$, $minETR_t$ is constant. Otherwise, $minETR_t = 0$. $E_t(\tau)$ is the electric power consumption rate [W] of the server s_t at time τ where $minE_t \leq E_t(\tau) \leq maxE_t$.

In our previous studies [2], the *transmission power consumption* (*TPC*) model for a server s_t is proposed. Let $PC_t(tr_t(\tau))$ show the electric power consumption rate [W] of a server s_t at time τ where the server s_t transmits files to clients at the total transmission rate $tr_t(\tau)$. Here, the TPC model for a server s_t is given as follows:

[Transmission power consumption (TPC) model]

$$E_t(\tau) = PC_t(tr_t(\tau)) = \beta_t(m) \cdot \delta_t \cdot tr_t(\tau) + (minETR_t + minE_t). \tag{1}$$

Here, δ_t is the power consumption rate of a server s_t to transmit one Mbits [W/Mb]. m is the number $NT_t(\tau)$ of transmission processes on the server s_t. $\beta_t(m)$ shows how much power consumption rate is increased for the number m of transmission processes, $\beta_t(m) \geq 1$ and $\beta_t(m) > \beta_t(m-1)$. There is a fixed point $maxm_t$ such that $\beta_t(maxm_t - 1) \leq \beta_t(maxm_t) = \beta_t(maxm_t + h)$ for $h > 0$. The number $maxm_t$ shows the maximum number of processes which can be performed on a server s_t. $maxPC_t = \beta_t(maxm_t) \cdot \delta_t \cdot Maxtr_t + (minETR_t + minE_t)$ gives the maximum power consumption rate $maxE_t$ of the server s_t to transmit files.

In the TPC model, the rotation speed of each fan is assumed to be fixed. The *extended transmission power consumption* (*ETPC*) model for a server s_t where the rotation speed of each fan can be changed is given as follows:

[Extended transmission power consumption (ETPC) model]

$$E_t(\tau) = \begin{cases} maxE_t & \text{if } tr_t(\tau) > 0. \\ minE_t & \text{otherwise.} \end{cases} \tag{2}$$

The rotation speed of each fan is the maximum to transmit a file f_s if at least one process p_{ts} is performed on a server s_t. The amount of the power consumption rate to rev up to the maximum rotation speed of fans for transmitting files is so large that the power consumption rate to transmit files can be neglected. Hence, the server s_t consumes the power at the maximum power consumption rate $maxE_t$ if at least one process is performed on the server s_t.

4 TPCLB Algorithm

We discuss how to estimate the total power consumption laxity $lpc_t(\tau)$ [Ws] of a server s_t at time τ. Suppose a new transmission process p_{ts} is started on a server s_t at time τ. Here, the transmission laxity $lt_{ts}(\tau)$ of each current process p_{ts} in the set $CT_t(\tau)$ is decremented by the transmission rate $tr_{ts}(\tau)$ at time τ. If the transmission laxity $lt_{ts}(\tau)$ gets 0 at time τ, the process p_{ts} terminates.

The Evaluation of the Transmission Power Consumption 481

Given a process set $CT_t(\tau)$ at time τ, we can estimate time when each process in $CT_t(\tau)$ to terminate and the power consumption laxity $lpc_t(\tau)$ of a server s_t. The power consumption laxity $lpc_t(\tau)$ shows how much power a server s_t consumes to perform every current processes in $C_t(\tau)$ at time τ, which is given by the following procedure **CommLaxity**(s_t, τ):

```
CommLaxity (s_t, τ) {
    if CT_t (τ) = φ, return (0);
    tr_t (τ) = Σ_{c_s∈CT_t(τ)} tr_ts(τ); /* total transmission rate.*/
    lpc_t = E_t (τ); /* formulas (1) or (2). */
    for each current process p_ts in CT_t (τ), {
        lt_ts (τ + 1) = lt_ts (τ) - tr_ts (τ);
        if lt_ts (τ + 1) = 0, CT_t (τ + 1) = CT_t (τ) - {p_ts};
    }
    return (lpc_t + CommLaxity (s_t, τ) + 1 [unit time]);
}
```

In the *transmission power consumption laxity-based (TPCLB)* algorithm, a load balancer K selects a server s_t where **CommLaxity**(s_t, τ) is the minimum in the server cluster S at time τ. The load balancer K issues the request to the server s_t.

5 Evaluation

We evaluate the TPCLB algorithm in terms of the total power consumption and elapse time compared with the round-robin (RR) algorithm [4]. There are eight servers $s_1,\ldots s_8$ each of which holds full replicas of files f_1, f_2, f_3, and f_4 of 1024, 512, 256, and 103 [MB], respectively, as shown in Table 1, $S = \{s_1,\ldots, s_8\}$. Parameters of each server are defined based on our previous experimentations. The ETPC model holds for the servers s_3 and s_7. On the other hand, the TPC model holds for the other servers.

A number m of clients randomly download files from one server s_t in the server cluster S. The maximum receipt rate $Maxrr_s$ of each client c_s is randomly selected between 10 and 100 [Mbps]. Each client issues a transfer request of a file f_h ($h \in \{1, 2, 3, 4\}$) to a load balancer K at time st_s. The file f_h is randomly selected in $f_1,\ldots, f_4$. The starting time st_s of each client is also randomly selected between 1 and 1000 [s] at the simulation time. Each client c_s issues one file transfer request at time st_s in the simulation. In the evaluation, the TPCLB and RR algorithms are performed on the same traffic pattern.

Figure 1a shows the total net power consumption [KWs] which is obtained by subtracting the power consumption $minE_t$ of each server s_t in the idle state from the total power consumption during the simulation, i.e. the power consumption for

Table 1 Servers

Servers	s_1	s_2	s_3	s_4	s_5	s_6	s_7	s_8
Model	TPC	TPC	ETPC	TPC	TPC	TPC	ETPC	TPC
$Maxtr_t$	160	447	778	802	160	447	778	802
δ_t	0.11	0.02	–	0.012	0.11	0.02	–	0.012
γ_t	1	1	1	1	1	1	1	1
$minE_t$	105	149	97	96	105	149	97	96
$maxE_t$	282	273	141	230	282	273	141	230
$minETR_t$	4	3	–	2	4	3	–	2
$maxm_t$	10	10	–	10	10	10	–	10
$\beta_t(m)$	1.09	1.5	–	1.42	1.09	1.5	–	1.42

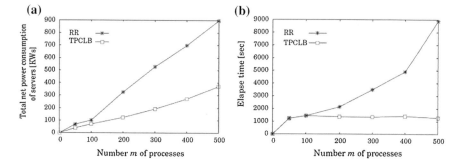

Fig. 1 Total power consumption and elapse time. **a** Total power consumption [KWs]; **b** Elapse time [s]

number m of processes. In the TPCLB algorithm, the total net power consumption can be more reduced than the RR algorithm.

Figure 1b shows the elapse time [s] of the TPCLB and RR algorithms for number m of processes. For $m \leq 100$, the elapse time of TPCLB and RR algorithms are almost the same since servers hold enough transmission rates. For $m > 100$, the number of transmission overloaded servers in the RR algorithm increases. However, the elapse time of TPCLB algorithm does not increase. This means the transmission rate of servers is efficiently used in the TPCLB algorithm than the RR algorithm.

From the evaluation results, the TPCLB algorithm is more useful than the RR algorithm.

6 Concluding Remarks

In this paper, we newly proposed the transmission power consumption laxity-based (TPCLB) algorithm for a cluster of heterogeneous servers which follow the TPC and ETPC models to reduce the total power consumption of a cluster of servers for communication type applications. In the TPCLB algorithm, a server s_t whose power consumption laxity $lpc_t(\tau)$ is the minimum in a cluster of servers is selected for a new request from a client c_s at time τ. We evaluated the TPCLB algorithm in terms of the total power consumption of a cluster of servers and the elapse time compared with the RR algorithm. From the evaluation, the TPCLB algorithm is more useful than the RR algorithm.

References

1. Enokido T, Aikebaier A, Takizawa M (2011) Process allocation algorithms for saving power consumption in peer-to-peer systems. IEEE Trans Ind Electron 58(6):2097–2105
2. Enokido T, Aikebaier A, Takizawa M (2011) An extended power consumption-based algorithm for communication-based applications. J Ambient Intell Humanized Comput 2(4):263–270
3. Enokido T, Takizawa M (2012) The extended transmission power consumption model for communication-based applications. In: Proceedings of the 15th international conference on network-based information systems (NBiS-2012), pp 112–119
4. Job scheduling algorithms in linux virtual server (2010). http://www.linuxvirtualserver.org/docs/scheduling.html

The Methodology for Hardening SCADA Security Using Countermeasure Ordering

Sung-Hwan Kim, Min-Woo Park, Jung-Ho Eom and Tai-Myoung Chung

Abstract In this paper, we considered that SCADA system has few authorized users and access control is one of the most important values for cyber security. We propose the method which reducing the success probability of attacker's penetration using ordered countermeasures. We assume that any system has two or more safety countermeasures for authentication. It follows that setting multiple countermeasures in chain and making a causal relationship before and after action. And then, we making an access procedure matrix for it and sharing them among authorized users. As doing so, we can prevent attacker's penetration and reduce risk level by hacking.

Keywords Security hardening · Penetration success probability · Ordered countermeasure

---

Jung-Ho Eom is co-author of this paper.

S.-H. Kim · M.-W. Park · T.-M. Chung
Department of Computer Engineering, School of Information and Communication
Engineering, Sungkyunkwan University, Suwon-si, Republic of Korea
e-mail: shkim47@imtl.skku.ac.kr

M.-W. Park
e-mail: mwpark@imtl.skku.ac.kr

T.-M. Chung
e-mail: tmchung@ece.skku.ac.kr

J.-H. Eom (✉)
Military Studies, Daejeon University, 62 Daehakro, Dong-Gu, Daejeon, Republic of Korea
e-mail: eomhun@gmail.com

J. J. (Jong Hyuk) Park et al. (eds.), *Multimedia and Ubiquitous Engineering*,
Lecture Notes in Electrical Engineering 240, DOI: 10.1007/978-94-007-6738-6_59,
© Springer Science+Business Media Dordrecht(Outside the USA) 2013

1 Introduction

In recent years, there has been an increasing interest in cyber attacks on SCADA such as 'Stuxnet' in 2010, 'Duqu' in 2011 and 'Flame' in 2012.

The latest cyber-attacks have a variety of target, Attacker's penetration is becoming more precise. According to this trend, there is a great deal of research on the cyber defense method for SCADA.

In case of cyber attacks taking control of the critical infrastructure, the damage caused by the malicious use will be very critical. Because of this risk, The United States government has been paying attention to cyber security since the early 2000s.

In this paper, we look at the major features and security-related issues of SCADA system. And we propose a method to reduce the attacker's penetration success probability.

This paper is organized as follows. In Sect. 2, we present the related works on SCADA security. In Sect. 3, we deal with security issues on SCADA and major features of SCADA related to cyber defense. In Sect. 4, we explain the method to reduce attacker's penetration success probability. In Sect. 5, we demonstrate a more detailed process with a case study. In Sect. 6, we summarize this paper and provide a conclusion and suggestions for future work.

2 Related Work

One of the most significant current discussions in cyber security is SCADA cyber defense. Many kinds of studies including key management, multiple password, and attack/defense tree have been conducted to strengthen SCADA security.

Beaver et al. [1], Dawson et al. [2], Pietre-Cambacedes and Sitbon [3] conducted a study on key management for SCADA network and Ni et al. [4], Adar and Wuchner [5], Taylor et al. [6] and Haimes and Chittester [7] conducted a study on cryptographic algorithms from the point of risk management. Chiasson et al. [8], Topkara et al. [9] studied multiple password and proposed the graphical passwords and passwords separation. We present a new approach to SCADA security to build up this idea.

3 SCADA System Security and Access Control

3.1 Security Issues in SCADA System

SCADA is a supervisory and control system for large-scale facilities such as power plants and steel mills. The initial SCADA system has no concept on network, but nowadays most SCADA systems are operating in a network environment [10].

As already mentioned, Cyber attacks on SCADA systems continue to increase due to the generalization of network and evolution of cyber attacks. Types of cyber attacks on SCADA systems are as follows.

First, look at the proportion of internal and external attacks, prior to 2000 insider attack by a disgruntled insider occupied 70 % of total. But outsider attacks through the network have occupied 70 % of the total since 2000 [11]. In recent years, the SCADA system can be accessed through the Internet. For this reason, it is becoming increasingly important issue that network protection for SCADA system [12].

Second, the major form of SCADA attack aims to intercept the system administrative authority. Malicious codes such as 'Stuxnet' attacks on the control authority for PLC (Programmable Logic Controllers) or RTU (Remote Terminal Units). PLC and RTU control the field devices directly by converting electrical signals into physical signals [13]. The final goal of the SCADA attack would be a great confusion rather than financial gain. So, it can be seen that countermeasures of access control are very important for both attack and defense.

3.2 Internet SCADA Architecture

Undoubtedly, some security consultants didn't know the characteristics of the system and could not provide effective advice about the countermeasures for hacking attacks. In the view of cyber defense, the features of SCADA system are as follows.

First, the scale of SCADA system users is smaller than other common systems. SCADA systems communicated not with users but with devices mainly. So, there are few users in the SCADA system. Thus, key management is simple and access load is low.

4 SCADA System Security Hardening Methodology

4.1 Background

Let us consider the following. The act of opening the door is access and the locking mechanism is the control method for access. Only authorized users can open the lock in normal conditions. There exists a possibility that abnormal access using an illegal key or breaking the lock can open. The locking device may be only one. Also depending on the level of security, it would be set two or more.

The following figures depict the door lock that has two block devices. One is a number device another is key.

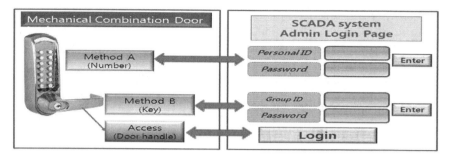

Fig. 1 Countermeasure combination

The open door (access) process in Fig. 1 can be represented by the following formula.

- Common access procedure

①$Open(KeyLock)$ → ②$Open(NumberLock)$ → ③$Pull(handle)$ or
①$Open(NumberLock)$ → ②$Open(KeyLock)$ → ③$Pull(handle)$

Let ith Countermeasure in system X = $CM(x_i)$,
Attacker's Penetration Success Probability on $CM(x_i) = PSP(CM(x_i))$,
Authorized user's unlocking success probability is 1,
n = total number of countermeasures $(n \geq 2)$
Unauthorized user penetration success probability. Then

$$PSP(SystemX) = PSP(CM(x_1)) \times \cdots \times PSP(CM(x_n))$$
$$= \prod_{i=1}^{n} PSP(CM(x_n))(n \geq 2) \quad (1)$$

4.2 Ordered Countermeasures

Service this section will explain the security hardening method step by step based on the previous section. The core of the proposed method is setting a causal and procedural relationship between the countermeasures. Generating the dependency condition between the countermeasures is the first step. In case that 'a' is a prerequisite of 'b', it is expressed as follows.

$$a \Rightarrow b \neq b \Rightarrow a$$

Any system X has two countermeasure; countermeasure 'a' and 'b'. The countermeasure 'a' must be unlocked before countermeasure 'b'. That case can be presented as follow formula:

$$CM(x_a) \Rightarrow CM(x_b)$$

The Methodology for Hardening SCADA

It means that CM(a) should be open (unlock) before the execution of CM(b). Let's suppose some system has $n(n \geq 2)$ countermeasures. And the all countermeasure of that system have to execute only once. That system can be created procedure number of n!.

$$nPr = \frac{n!}{(n-r)!}, \quad nPr = n!(n = r)$$

Also, let X' is the security hardening system on X, attacker's penetration success probability is as follows:

$$PSP(X') = \frac{\prod_{i=1}^{n} PSP(CM(x_n))}{n!} \tag{2}$$

Since n is 2 or more, the hardening method could reduce the attacker's success probability below 50 %.

4.3 Access Procedure Matrix and Main Process

The second is the process of creating a matrix of all possible procedures as shown below table. After creating the procedure matrix, it is being distributed to all authorized users. As we have mentioned previously, SCADA systems have a small and limited user group. Thus, the system shares the access matrix easier than other common system (Table 1).

4.4 Access Procedure Using Ordered Countermeasures

Where AC = Authorized Client, SCS = Control Server of SCADA
  AP(number) = Access Procedure number
  $CM(SCS_a)$ = Countermeasure A of SCADA control server system
  $AP(1) = CM(SCS_a) \Rightarrow CM(SCS_b)$
  $P(CM(SCS_a))$ = Password of Countermeasure A in system SCS
  The main procedure described so far can be expressed as follows.

(1) Authorized user request access procedure matrix number for SCS

  AC → SCS : AC request number of AP for SCS

(2) SCS provide the procedure number (Let #1 : $CM(SCS_a) \Rightarrow CM(SCS_b)$

  SCS → AC : AP(1)

Table 1 Access procedure matrix

Procedure number	First	Second	...	n th
1	CM 1	CM 2	...	CM n
2	CM 2	CM 1	...	...
...	...	...	...	...
$n!$	CM n	...	...	CM 1

(3) Authorized user access countermeasure A with password of A

$AC \rightarrow CM(SCS_a)$: AC unlock $CM(SCS_a)$ by $P(CM(SCS_a))$

(4) Authorized user access countermeasure B with password of B

$AC \rightarrow CM(SCS_b)$: AC unlock $CM(SCS_b)$ by $P(CM(SCS_b))$.

(5) AC access the SCS : Finish.

5 Demonstration

In this chapter, we explain the security hardening method using the example case. We assume a virtual power generation SCADA system as below. And we compare the penetration success probability before and after using the proposed hardening method (Table 2).

Firstly, we apply the example table data to the formula 1.

$$PSP(PP) = \prod_{i=1}^{n} PSP(CM(PP_n))$$
$$= PSP(CM(PP_1)) \times PSP(CM(PP_2)) \times PSP(CM(PP_3)) = 0.3 \times 0.2 \times 0.1 = 0.006$$

For comparison, we substitute the formula 2.

$$PSP \text{ (Hardening PP)} = \frac{\prod_{i=1}^{n} PSP(CM(PP_n))}{n!} = 0.3 \times 0.2 \times 0.1/6 = 0.001$$
$$= 0.1(\%)$$

Table 2 Power plant system specification

Item	Description
Purpose of SCADA	Power plant (PP) management
Total Countermeasure number	$3[(CM(PP_1), CM(PP_2), CM(PP_3)]$
Total number of procedure	$3! = 6$
Penetration Success Probability	$CM(PP_1) = 0.3$, $CM(PP_2) = 0.2$, $CM(PP_3) = 0.1$

As a result, we can see that the penetration success probability was reduced by a factor of n!.

6 Conclusion and Future Work

In this paper, we explain the security hardening method by setting the causal relationship between the access control countermeasures. The primary target of SCADA cyber attack is obtaining an access control. For this reason, we focused on the reducing the attacker's access control penetration probability. It is difficult to say that the proposed method is absolute method. But it is clear that proposed method is a useful methodology for SCADA system security. In the future, we will refine algorithm for ordered countermeasures and evaluating the validity. We will also conduct a further study to apply this in risk management.

Acknowledgments This work was supported by the IT R&D program of MKE/KEIT. [10041244, Smart TV 2.0 Software Platform].

References

1. Beaver C, Gallup D, Neumann W et al (2002) Key management for SCADA. Cryptog information systems security dept, Sandia Nat. Labs, Technical Report SAND 2001–3252
2. Dawson R, Boyd C, Dawson E et al (2006) SKMA: a key management architecture for SCADA systems. In: Proceedings of the 2006 Australasian workshops on grid computing and e-research ACSW Frontiers '06, vol 54, pp 183–192
3. Pietre-Cambacedes L, Sitbon P (2008) Cryptographic key management for SCADA systems-issues and perspectives. International conference on information security and assurance ISA 2008. pp 156–161
4. Ni M, McCalley JD, Vittal V et al (2003) Online risk-based security assessment. IEEE Trans Power Syst 18:258–265
5. Adar E, Wuchner A (2005) Risk management for critical infrastructure protection (CIP) challenges, best practices and tools. First IEEE international workshop on critical infrastructure protection
6. Taylor C, Krings A, Alves-Foss J (2002) Risk analysis and probabilistic survivability assessment (RAPSA) an assessment approach for power substation hardening
7. Haimes YY, Chittester CG (2005) A Roadmap for quantifying the efficacy of risk management of information security and interdependent SCADA systems. J Homel Secur Emerg Manage 2:1–21
8. Chiasson S, Forget A, Stobert E et al (2009) Multiple password interference in text passwords and click-based graphical passwords. In: Proceedings of the 16th ACM conference on computer and communications security CCS '09. pp 500–511
9. Topkara U, Atallah MJ, Topkara M (2006) Passwords decay, words endure: secure and re-usable multiple password mnemonics. In: Proceedings of the 2007 ACM symposium on applied computing SAC '07. pp 292–299

10. Cai N, Wang J, Yu X (2008) SCADA System security: complexity, history and new developments, industrial informatics. INDIN 2008. 6th IEEE international conference on 2008. pp 569–574
11. Igure VM, Laughter SA, Williams RD (2006) Security issues in SCADA networks. Computer and security 2006. pp 498–506
12. Qiu B, Gooi HB (2000) Web-based SCADA display systems (WSDS) for access via internet. IEEE transactions on power systems, vol 15. pp 681–686
13. Chunlei W, Lan F, Yiqi D (2010) A simulation environment for SCADA security analysis and assessment. International conference on measuring technology and mechatronics automation (ICMTMA) 2010, vol 1. pp 342–347

Development and Application of STEAM Based Education Program Using Scratch: Focus on 6th Graders' Science in Elementary School

JungCheol Oh, JiHwon Lee and JongHoon Kim

Abstract For this study, we reviewed theoretical background of STEAM education and domestic and international case studies in STEAM education. By doing so, we developed and applied the STEAM Education Program through the use of Scratch. This program is designed for the 3rd ("Energy and Tools") and 4th ("Combustion and Extinguishing") lessons of 6th graders' science in elementary school. As a result, the creativity index and positive attitude about science of the students who went through the researched program increased with meaningful difference compared to that of the sample population. The result of this study shows that 'The STEAM Education Program,' using Scratch, can improve creativity. And it is sure that it brings positive changes for the Science Related Affective Domains.

Keywords STEAM · Scratch · Creativity · Scientific attitude · Fluency

1 Introduction

In the present century, science and technology has combined with human life in a more humane and artistic way than ever before. The iPhone, introduced by Steve Jobs, an artistic engineer, is a good example of how science and technology combine with human life in a human-friendly and artistic way. This societal

---

J. Oh · J. Lee · J. Kim (✉)
Department of Computer Education, Teachers College,
Jeju National University, Jeju, Korea
e-mail: jkim0858@jejunu.ac.kr

J. Oh
e-mail: lov0502@naver.com

J. Lee
e-mail: torchere@naver.com

J. J. (Jong Hyuk) Park et al. (eds.), *Multimedia and Ubiquitous Engineering*,
Lecture Notes in Electrical Engineering 240, DOI: 10.1007/978-94-007-6738-6_60,
© Springer Science+Business Media Dordrecht(Outside the USA) 2013

demand has been reflected in the educational community, and the Ministry of Education and Science of South Korea has established various educational strategies for educating students to create fused talents [1].

In 2009, Partnership for Twenty first-Century Skills, an organization in the U.S., suggested essential skills that students need to learn and master in order to succeed in the twenty first century [2]. The organization suggested that students should learn such skills as creativity, critical thinking, problem solving, communication, and collaboration through art, mathematics, science, economics, and history. That is, learners need to develop the ability to fuse diverse skills holistically based on creativity to be able to adapt to the rapidly changing society of the twenty first century and to get ahead of the times.

The government stressed the importance of STEAM education to train people with such holistic talents and is now preparing various strategies for educating them. This study aimed to develop a STEAM education program that can be applied in the field, using technology and engineering Scratch, which can be easily accessed by learners in the aforementioned context.

2 Theoretical Background

2.1 Need for and Definition of STEAM

Smart STEAM stands for Science, Technology, Engineering, Arts, Mathematics and means learning the fused knowledge of various fields. The effort to find the cause of the economic crisis in the U.S. led to the identification of the decrease in the academic performance of mathematics and science learners as the cause [3, 4]. To address this problem, STEAM aims to promote the learners' motivation for learning and to educate people so as to help them become capable of solving multidisciplinary problems.

Regarding STEAM fusion education, Yakman (2008) presented a pyramid model consisting of several levels, from continuing education to the classification of the detailed study contents and stated that the interdisciplinary integrative level was appropriate for elementary school education [5].

Although in most of the current STEAM education programs computers serve only as auxiliary tools, in this study, it was used as a main activity in the fusion education for applying and utilizing the science lessons. That is, the students can clearly understand the lesson contents and can see the process and result of making programs real-time. Thus, the possible errors in the real-life experiment can be reduced, and the students can have opportunities to apply and express the scientific principle in various ways using divergent and creative methods, without temporal and spatial restrictions.

Development and Application of STEAM 495

3 Design and Making of the Steam Education Program

3.1 Design of the Stages of Steam Teaching and Learning

Study stages were established as follows to develop the STEAM education program using Scratch, and to apply the developed program. The study was conducted for 6 months, according to the study stages. The teaching and learning stages for science and technology presented by Miaoulis (2009) and the creative comprehensive design stages for people with inter-disciplinary talents presented by Korea Foundation for the Advancement of Science and Creativity (2011) were reviewed to design the stages for STEAM teaching and learning that will be applied to the education field.

Table 1 shows the stages for creative design presented by Korea Foundation for Advancement of Science and Creativity when it introduced the concept of creative comprehensive design education. The foundation explained that the stages are the characteristics of the new STEAM education.

Based on this, six stages of STEAM teaching and learning Table 2 were determined, and the program was conducted. In particular, during the "making or synthesizing" and "testing" stages, creative work was accomplished by frequently making and modifying the activities. Thus, continuous testing and feedback are required.

Table 1 Stages for creative design presented by Korea foundation for advancement of science and creativity

1. Setting of objective	2. Planning and designing	3. Analysis of design	4. Making	5. Test	6. Evaluation

Table 2 Stages of STEAM teaching and learning

1.	Experiencing priming water for an idea	Define the issues and experience priming water for an idea related with the issue
2.	Coming up with an idea	Create various ideas and share them with collaborators
3.	Planning and design fusion	Establish a plan for materializing the idea, and make a design by fusing related studies
4.	Making or synthesizing	Make or synthesize works based on the idea using scientific, technological, engineering, and artistic methods
5.	Testing	Test, and Feedback or modifying
6.	Evaluation	Inter-collaborators' evaluation, and refinement of the idea through evaluation

3.2 Formulation of the Steam Education Program

The units were constructed, using the stages of the STEAM teaching and learning activities and the categorized science experiment themes. Activities for making games using Scratch were included considering that the learners are elementary school students, who are highly interested in games, and that various artistic activities, such as plotting a story, selecting background music, and drawing characters, are complexly performed for game creation.

At the "experiencing priming water for an idea" stage of each unit, the teachers pre-sent Scratch games to help the learners come up with an idea. At the "coming up with an idea" stage, the learners are allowed to present and discuss the games that they want to make, and by doing so, to share the ideas with one another.

At the "planning and integrative design" stage, the learners are allowed to make a storyboard based on the various science experiments presented in the textbook. Many storyboards can be made, according to the experiment, or only one storyboard can be made in detail. The learners are allowed to make storyboards freely, without restrictions in theme and expression. At the "making and synthesizing" stage, the learners are allowed to formulate "instructions on how to play the game made" based on the storyboard made, and to make games using Scratch.

At the "testing" stage, the learners can modify the game by playing it with their collaborators, and by reviewing it. There was no boundary between the "making or synthesizing" and "testing" stages, allowing the learners to make, synthesize, and review activities whenever they needed to do so, and by doing so, to complete creative works like those shown in Fig. 1.

In addition, the learners can write what they felt through these two stages in "Instructions on how to play the game made." The learners who made a maze game through several stages added mazes to the game by reviewing the game together with their collaborators (Fig. 2) and by completing the final game.

At the "evaluation" stage, the learners can display the games that they made, and can play these, so as to evaluate one another's games. The learners can also refine their ideas by sharing their completed ideas with others.

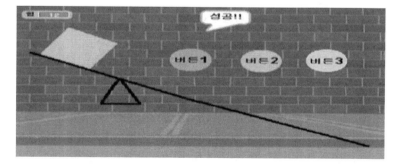

Fig. 1 The game completed after the review with the collaborators, and after modification

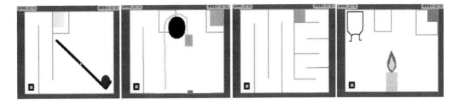

Fig. 2 Games made by other learners

4 Application of the Program and Analysis of the Results

4.1 Study Design and Control of Variables

Two classes in the 6th grade of Elementary School in Jeju City, Jeju Island, South Korea were included in the study (Table 3). The classes were assigned as the experimental and control groups, respectively. For 10 weeks, from the fourth week of September 2011 to the first week of December 2011, the Scratch-based STAEM education program was conducted in the experimental group, on the two units of the science textbook.

As shown in Table 3, a test of baseline creativity and an evaluation of the affective characteristics associated with science were performed in the two groups, and the homogeneity of the two groups was validated. For the control group, a normal science class was conducted as planned.

4.2 Test Tools

The area that this study intended to identify through objective validation after the application of the Scratch-based STEAM education program was the creativity and affective characteristics associated with science. To determine if the creativity had been enhanced, Torrance's TTCT (diagram) Creativity Test Type A was performed before and after the experiment. In addition, "Evaluation System for Affective Characteristics (Attitude) Related with National Science" developed by Hyonam et al. [6] at the Science Education Institute of Korea National University of Education, based on the theory on science-related attitude presented by Klopfer

Table 3 Experimental group

Division	Number of students		
	Male	Female	Total
The experimental group	12	13	25
Comparison group	12	13	25
Total	24	26	50

and the criteria for evaluation items presented by Edward, was used for the test of affective characteristics related with science. Type A consisted of items evaluating the awareness of and interest in science, and type B consisted of items evaluating scientific attitude. For the reliability of the items, the Chronbach alpha coefficients were 0.83 and 0.86 for types A and B, respectively. As both exceeded 0.8, they were deemed reliable.

4.3 Results of the Creativity Test and Interpretation of the Results

To verify the homogeneity of the factors of creativity between the experimental and control groups, the mean of each area of creativity in each group was tested through a t test, using SPSS 12.0 for Windows, before the experimental treatment ($p = 0.05$). As shown in Table 4, the significance probability for the creativity index was $p = 0.929$; thus, there was no significant difference in creativity between the experimental and control groups. The other areas of creativity, such as fluency, creativity, abstracting ability, and delicacy, were not significantly different between the two groups as the significance probability was higher than 0.05.

Ten weeks later, the creativity test was performed again on the experimental and control groups. As shown in Table 5, the creativity and originality were significantly different between the groups, with $p = 0.036$ and $p = 0.039$, respectively. In particular, the post-experiment creativity increased by 10.17, and the significance level was remarkably different, with $p = 0.000$ ($p < 0.05$) in the experimental group.

Then, the difference in creativity before and after the experiment was compared within the experimental group and was analyzed. As shown in Table 6, in the

Table 4 Test of baseline creativity

Domain	Class	N	Average	The standard deviation	T	Note that the probability
Fluency	Pre	25	115.32	24.81	−1.619	N.S.
	Post	25	122.72	18.57		0.118
Originality	Pre	25	104.80	25.51	−0.202	N.S.
	Post	25	106.04	23.09		0.842
Abstractness of title	Pre	25	95.44	29.87	0.507	N.S.
	Post	25	90.36	37.30		0.617
Elaboration	Pre	25	94.20	18.45	0.676	N.S.
	Post	25	91.36	14.17		0.505
Resistance to premature closure	Pre	25	99.60	17.60	−0.663	N.S.
	Post	25	101.56	11.88		0.513
Creativity index	Pre	25	105.99	19.25	−0.090	N.S.
	Post	25	106.37	16.15		0.929

[*] $p < 0.05$, *N.S.* not significant, *N* number of cases

Development and Application of STEAM

Table 5 Post results comparing

Domain	Class	N	Average	The standard deviation	T	Note that the probability
Fluency	Pre	25	125.52	19.12	−2.225	0.039*
	Post	25	132.96	13.89		
Originality	Pre	25	117.00	25.82	−2.187	0.039*
	Post	25	129.24	22.81		
Abstractness of title	Pre	25	100.64	22.93	0.724	N.S.
	Post	25	95.28	29.65		0.476
Elaboration	Pre	25	91.64	18.45	−0.593	N.S.
	Post	25	93.96	14.88		0.559
Resistance to premature closure	Pre	25	104.28	15.62	1.212	N.S.
	Post	25	99.92	15.19		0.237
Creativity Index	Pre	25	114.29	14.67	−4.104	0.000*
	Post	25	116.54	16.29		

*$p < 0.05$, N.S. not significant, N number of cases

comparison of the fluency, originality, and creativity index before and after the experiment within the experimental group, the significance probability was less than 0.05; thus, the difference was significant. It was found that Scratch-based STEAM education had a positive effect on the improvement of the fluency, originality, and creativity index.

4.4 Results of the Test of Affective Characteristics Related with Science

The evaluation of affective characteristics developed by Hyonam et al. [6] consists of three categories (awareness, interest, and scientific attitude), and each item is graded based on a 5-point Likert scale. The results of the pre-experiment test showed that compared with the control group, the awareness of science was low (0.548 points), the interest in science was high (0.008), and the scientific attitude was high (0.284) in the experimental group (Table 7).

The post-experiment test showed that the mean of awareness (C) of science in-creased by 0.168, the interest in science (I) by 0.231, and the scientific attitude (A) by 0.281. In particular, compared with the control group, the awareness and interest of the experimental group considerably increased. The results showed that the Scratch-based STEAM education program had a positive effect on the affective characteristics related with science in the experimental group.

Table 6 Results of the test of creativity by time point

Domain	Class	N	Average	The standard deviation	T	Note that the probability
Fluency	Pre	25	122.72	18.57	−3.556	0.002*
	Post	25	132.96	13.89		
Originality	Pre	25	106.04	23.09	−5.705	0.000*
	Post	25	129.24	22.81		
Abstractness of title	Pre	25	90.36	37.30	−0.571	N.S.
	Post	25	95.28	29.65		0.573
Elaboration	Pre	25	91.36	14.17	−0.732	N.S.
	Post	25	93.96	14.88		0.472
Resistance to premature closure	Pre	25	101.56	11.88	−0.543	N.S.
	Post	25	99.92	15.19		0.592
Creativity index	Pre	25	106.37	16.15	−3.323	0.003*
	Post	25	116.54	16.29		

*$p < 0.05$, N.S. not significant, N number of cases

Table 7 Results of the test of affective characteristics related with science

Domain	Class		N	Average	The standard deviation	T	Note that the probability
Cognition	Comparison	Pre	25	3.599	0.419	+0.014	
		Post	25	3.613	0.351		
	Experiment	Pre	25	3.051	0.366	+0.168	
		Post	25	3.219	0.440		
Interest	Comparison	Pre	25	3.293	0.585	+0.003	
		Post	25	3.296	0.586		
	Experiment	Pre	25	3.301	0.579	+0.231	
		Post	25	3.532	0.615		
Attitude	Comparison	Pre	25	3.081	0.424	+0.231	
		Post	25	3.312	0.460		
	Experiment	Pre	25	3.365	0.443	+0.281	
		Post	25	3.646	0.661		

5 Conclusions

This study aimed to develop and apply a STEAM education program that can be applied to elementary school education, and to show the effect of the program according to the education for people with integrated talents stressed by the Ministry of Education and Science. Towards these ends, foreign and local studies on STEAM were reviewed, and appropriate subjects and contents that could be applied to the education were selected and combined with Scratch, an educational programming language. In addition, to increase the appropriateness of the program, discussion with and inter-views of experts in schools and colleges were performed during the selection of education contents and the designing of the stages of teaching and learning. Then, the developed STEAM education program was applied to the students for 10 weeks, under the condition where all the possible variables were controlled. The fluency, originality, and creativity index significantly increased in the experimental group, which used the STEAM education program, as opposed to the control group, and the positive answers in the area of awareness of and interest in the affective area considerably increased in the experimental group. These results indicate that the Scratch-based STEAM education program has a positive effect on the creativity and affective characteristics related with science. In addition, the STEAM education program developed in this study has the two following meanings: (1) STEAM education was utilized without interrupting the flow of the units of the science textbook, to allow the smooth progression of the current curriculum; and (2) using the educational programming language, methods of fusing science, technology, engineering, arts, and mathematics to attain what STEAM intends to achieve were presented.

References

1. Ministry of Education Science and Technology (2010) The future Republic of Korea to open using creative talent and advanced science technology
2. P21 Framework Definitions (2009) The partnership for 21st century skills
3. Tarnoff J (2010) STEM to STEAM—recognizing the value of creative skills in the competitiveness debate
4. Puffenberger A (2010) The STEAM movement: it's about more than hot air
5. Yakman G (2008) STEAM education: an overview of creating a model of integrative education. In: Proceeding of PATT on 19th ITEEA conference, pp 335–358
6. Hyonam K, WanHo C, JinWoo J (1998) National assessment system development of science-related affective domain. Korean J Sci Edu 18(3)357–369

Printed by Publishers' Graphics LLC